Recent Titles in This Series

(*Continued in the back of this publication*)

Mathematical Aspects of Conformal and Topological Field Theories and Quantum Groups

CONTEMPORARY
MATHEMATICS

175

Mathematical Aspects of Conformal and Topological Field Theories and Quantum Groups

AMS-IMS-SIAM Summer Research Conference
on Conformal Field Theory, Topological Field Theory
and Quantum Groups
June 13–19, 1992
Mount Holyoke College

Paul J. Sally, Jr.

Moshe Flato
James Lepowsky
Nicolai Reshetikhin
Gregg J. Zuckerman
Editors

American Mathematical Society
Providence, Rhode Island

The AMS-IMS-SIAM Joint Summer Research Conference in the Mathematical Sciences on Conformal Field Theory, Topological Field Theory and Quantum Groups was held at Mount Holyoke College, South Hadley, Massachusetts, June 13–19, 1992, with support from the National Science Foundation, Grant DMS-8918200 02.

1991 *Mathematics Subject Classification*. Primary 08–XX, 16–XX, 17–XX, 20–XX, 33–XX, 35–XX, 58–XX, 81–XX.

Library of Congress Cataloging-in-Publication Data

AMS-IMS-SIAM Summer Research Conference on Conformal Field Theory, Topological Field Theory, and Quantum Groups (1992: Mount Holyoke College)

Mathematical aspects of conformal and topological field theories and quantum groups / AMS-IMS-SIAM Summer Research Conference on Conformal Field Theory, Topological Field Theory, and Quantum Groups, June 13–19, 1992, Mount Holyoke College : Paul J. Sally, Jr. ... [et al.], editors.

p. cm. — (Contemporary mathematics, ISSN 0271-4132; v. 175)

Includes bibliographical references.

ISBN 0-8218-5186-1 (acid-free)

1. Conformal invariants—Congresses. 2. Quantum field theory— Congresses. 3. Quantum groups—Congresses. 4. Mathematical physics—Congresses. I. Sally, Paul. II. Title. III. Series: Contemporary mathematics (American Mathematical Society); v. 175.

QC174.52.C66A47 1992

512′.55—dc20 94-27537

CIP

♾ The paper used in this book is acid-free and falls within the guidelines
established to ensure permanence and durability.

♻ Printed on recycled paper.

All articles in this volume were printed from copy prepared by the authors.
Some articles were typeset using $\mathcal{A}\mathcal{M}\mathcal{S}$-TEX or $\mathcal{A}\mathcal{M}\mathcal{S}$-LATEX,
the American Mathematical Society's TEX macro systems.

10 9 8 7 6 5 4 3 2 1 99 98 97 96 95 94

Contents

PREFACE

In July 1982, the 14th AMS-SIAM Summer Seminar on *Applications of Group Theory in Physics and Mathematical Physics* was held in Chicago. A volume of papers submitted by speakers at the Seminar was published by the AMS in 1985 under the same title as Volume 21 of the Series *Lectures in Applied Mathematics.* This volume was edited by three of the editors of the present volume (M. Flato, P. Sally and G. Zuckerman). The aim of the Chicago meeting was "to bring together a broad spectrum of scientists from theoretical physics, mathematical physics and various branches of pure and applied mathematics in order to promote interaction and an exchange of ideas and results of common interest".

At that time, the idea may have appeared to be somewhat revolutionary. Eight years later, it was obvious to everyone that this attitude had become part of mainstream mathematics, to such an extent that three of the four 1990 Fields medalists were inspired by such an interaction.

As a result, the editors of the Seminar volume decided that it was time for an update on the 1982 conference. Thus was organized (with J. Lepowsky and N. Reshetikhin added to the Organizing Committee) the June 13-19, 1992 AMS-IMS-SIAM Joint Summer Research Conference at Mount Holyoke College, entitled "Conformal Field Theory, Topological Field Theory and Quantum Groups". The present volume contains papers submitted by speakers at the Conference. The papers do not always coincide 100% with the actual lectures, but in most cases are close to the lectures, and they always deal with matter that was the subject of the Conference. Some of the speakers have not yet submitted papers (for a variety of reasons) but we feel that this volume should be published without further delay.

The Conference turned out to be much more mathematically oriented than the previous Seminar, reflecting the fact that many of the areas of theoretical physics which were discussed in the Seminar became leading areas in mathematics in the ten years between the meetings. (One of the subjects treated at the Conference, Quantum Groups, was barely born at the time of the Seminar). We therefore added the words "Mathematical Aspects" to the title to reflect that situation. Simultaneously, the partition of the contributions into well-defined subjects became almost impossible because several mathematical methods are common to two or more physical applications. We shall thus present the contributions as a continuous list.

The first group of papers deals with one of the aspects of Conformal Field Theory (CFT), the so-called Vertex Operator Algebras (VOA) or superalgebras (SVOA) and their representations:

1. Y.-Z. Huang and J. Lepowsky reformulate the notion of VOA, interpreted geometrically by one of the authors (YZH) using certain moduli spaces of spheres with punctures and local coordinates, in terms of structures called "operads," which first arose in the homotopy-theoretic characterization of iterated loop spaces. Doing so, they show that the rich geometric structure of CFT and the rich algebraic structure of VOA share a precise common foundation in basic operations associated with a certain two-dimensional object. This paper presents a step in an approach which was just beginning around the time of the Seminar and has developed into a full-fledged theory with numerous ramifications.

2. C. Dong deals with the representations of the moonshine module VOA, whose automorphism group is the Monster finite simple group. He proves the significant result that the adjoint representation is the unique irreducible one (part of a conjecture of Frenkel-Lepowsky-Meurman) and that every representation is a multiple of it.

3. C. Dong and G. Mason present a new (and rigorous) construction of the moonshine module VOA as a $\mathbb{Z}_3$-orbifold structure in the sense of CFT, and more generally, they discuss its construction as a $\mathbb{Z}_p$-orbifold structure. (The construction of Frenkel-Lepowsky-Meurman was as a $\mathbb{Z}_2$-orbifold structure.) The main theorem provides a nontrivial instance of "mirror symmetry". This work includes a characterization of this VOA as an irreducible module for the affinization of the Griess algebra.

4. V. Kac and W. Wang. Recently, Zhu has constructed an associative algebra $A(V)$ corresponding to a VOA V such that up to equivalence, the irreducible representations of V and $A(V)$ are in one-to-one correspondence. In addition, to any V-module M, Frenkel and Zhu have associated an $A(V)$-module $A(M)$ in terms of which the fusion rules can be determined. An advantage of these constructions is that $A(V)$ and $A(M)$ can usually be computed explicitly. This was used by several authors (e.g., Wang, and Dong, Mason and Zhu) to prove rationality and calculate fusion rules for series of representations of the Virasoro algebra. In this paper, the authors extend these results to superalgebras (SVOA's) and concentrate on those of Neveu-Schwarz type ($N = 1$). In particular they prove rationality and study fusion rules for SVOA's corresponding to an affine Kac-Moody superalgebra introduced by Kac and Todorov in relation to superconformal current algebras. They also show rationality for the SVOA's generated by charged and neutral free fermionic fields. This paper represents an extension of the VOA theory to superalgebras, needed in several physical applications.

A central object in CFT is the Khnizhnik-Zamolodchikov (KZ) equation, which has been extensively studied since it was introduced in 1984 in relation to current algebra Wess-Zumino-Novikov-Witten models. Here:

5. Y. Stanev and I.T. Todorov review the classification (that they obtained in collaboration with L. Michel) of all polynomial solutions of the KZ equation for the $SU(1,1)$-invariant 4-point amplitude of a primary field in $\mathfrak{su}(2)$ conformal current algebra.

The theory of knots in 3-manifolds is related to all the subjects dealt with at the Conference, including CFT, the Topological Field Theory (TFT) of Witten and

Quantum Groups at roots of unity. Here:

6. T. Kohno gives a lower bound for a certain topological invariant of a knot, the tunnel number, in terms of the Jones-Witten invariants for a knot in a closed oriented 3-manifold. In particular for the manifold S^3 and Lie algebra $\mathfrak{sl}(2,\mathbb{C})$ he shows that the lower bound is described by special values of the Jones polynomial.

TFT arose from the Chern-Simons gauge theory, the classical phase space of which can be viewed as the moduli space $\bar{S}_g$ of flat connections on the trivial principal G-bundle (G a compact simple Lie group) on a compact oriented 2-manifold of genus g. $\bar{S}_g$ contains an open dense subset S_g that is a symplectic manifold and is conjectured to be related to invariants of 3-manifolds. Here:

7. L. Jeffrey and J. Weitsman summarize some of their recent work on the symplectic geometry of $\bar{S}_g$, for $G = SU(2)$. They use the existence of Hamiltonian torus actions in these $\bar{S}_g$ and the images of the corresponding moment maps to find a simple description of $\bar{S}_g$, and they use this to compute such quantities as symplectic volumes.

A concept of tensor product of representations is suggested by fusion rules in CFT, and tensor products appear in many ways (arising from coproducts) in Quantum Groups. Computing and understanding suitably-defined tensor product multiplicities is therefore an important theme in both subjects. Here:

8. G. Georgiev and O. Mathieu compute some modified tensor product multiplicities for indecomposable tilting modules of Chevalley groups (over a field of characteristic p large enough), showing that they are the same as those obtained (from the recently-studied conformal-field-theoretic tensor product) for an associated complex affine Kac-Moody algebra. They conjecture that this coincidence can be explained through a lifting to quantum groups. This tensor category also turns out to be the same as one constructed recently by S. Gelfand and D. Kazhdan that was discussed by Kazhdan at the Conference. (The talks of Huang and Lepowsky at the Conference also presented the construction and properties of a tensor category of representations of a VOA.)

Another paper touching both CFT and Quantum Groups is the following:

9. K. Aomoto and Y. Kato deal with some mathematical problems around the KZ equation in 2-dimensional CFT and Quantum Groups. Considering a holonomic system of q-difference equations, they construct special q-cycles ($q = e^{2\pi i\tau}, \mathrm{Im}(\tau) > 0$) for n-dimensional Jackson integrals on which the canonical action of the permutation group gives a linear representation satisfying the Yang-Baxter equation.

The remaining four papers deal mostly with the "quantum groups" part of the Conference:

10. M. Flato and D. Sternheimer review the deformation-quantization approach both in quantum mechanics and quantum field theory, i.e., the star-product theory that can now be viewed as a predecessor of quantum groups. They also review the related notion of star-representation of Lie groups and the case of closed star-products, classified by cyclic cohomology and related to noncommutative geometry. Finally they explain the relations with quantum groups, showing in what sense quantum groups are special cases of star-product algebras.

11. M.A. Semenov-Tian-Shansky describes various notions of double for a quantum group in the FRT (Faddeev-Reshetikhin-Takhtajan) approach (with R-matrices supposed known in advance) and concentrates on what is called the Heisenberg double, the quantum analogue of the Hopf algebra of functions on T^*G (the cotangent bundle on a Poisson Lie group G), and its twisted versions (nontrivial deformations associated with outer automorphisms of the underlying algebras). The quantum duality principle (a pairing between the quantum Hopf algebras of functions on G and G^*) then permits him to define a quantum Fourier transform.

12. C. Frønsdal and A. Galindo base their study on an abstraction (definition in the structure and not in a representation) of the T-matrices of integrable models, that they call "the universal T-matrix" and is the Hopf algebra dual form expressed in terms of generators. They start with the Woronowicz quantum group $\mathcal{U}_{q,q'}(\mathfrak{gl}(2))$ to make their formalism more transparent, and get its (solvable) bialgebra dual. For $\mathfrak{gl}(3)$ they then discover a new (esoteric) quantum deformation, not of the coboundary type (and with nonsolvable dual) but for which a Yang-Baxter R-matrix still exists. They then extend these results to $\mathfrak{gl}(n)$, obtain also the corresponding doubles and end with a few remarks on multiparameter deformations.

13. V. Ginzburg, N. Reshetikhin and E. Vasserot obtain an analogue of the Weyl correspondence (between irreducible representations of the symmetric groups and certain finite-dimensional representations of the general linear groups, in the complex case) in the quantum affine setup, and provide its geometric interpretation. The q-analogue of the symmetric group algebra is a Hecke algebra, and the q-Weyl correspondence was described already by Jimbo and interpreted geometrically by Lusztig and coworkers. Here the authors extend the formalism to the affine case using a notion of polynomial tensor representation which they interpret geometrically using affine flag varieties.

Finally, we want to express our thanks to the American Mathematical Society for its excellent logistical support, and especially to Carol Kohanski, who helped in the organization at Mount Holyoke, and to Donna Harmon, who helped in collecting the contributions. Last but not least we thank all the participants, speakers with written contributions, speakers without written contributions, and all those who attended, for the very fruitful atmosphere that was created at Mount Holyoke College during the Conference.

The Editors

Paul Sally

Moshé Flato, James Lepowsky, Nicolai Reshetikhin and Gregg Zuckerman

Contemporary Mathematics
Volume 175, 1994

Connection coefficients for A-type Jackson integral and Yang-Baxter equation

Abstract. We construct special q-cycles called "charateristic cycles" for Jackson integrals and show that the canonical action of the permutation group $\mathfrak{S}_m$ there give a linear representation satisfying the Yang -Baxter equation.

Kazuhiko AOMOTO* and Yoshifumi KATO**

1. A-type Jackson integrals

Let $q = e^{2\pi i \tau}$, Im $\tau > 0$, be the elliptic modulus. We consider an n-dimensional Jackson integral

$$(1.1) \qquad J = \int_{\mathfrak{T}} \Phi(t|\alpha,\gamma)\, \tilde{\omega}\ , \quad \text{for} \quad \tilde{\omega} = \frac{d_q t_1}{t_1} \wedge \cdots \wedge \frac{d_q t_n}{t_n}$$

over a suitable cycle $\mathfrak{T}$ for a q-multiplicative function

---

1991 Mathematics Subject Classification, Primary 33D60.
** supported by the Ishida Foundation No. 91545.
This paper is in final form and no version of it will be submitted elsewhere.

$$(1.2)\quad \Phi(t|\alpha,\gamma) = \prod_{j=1}^{n} \{ t_j^{\alpha_j} \prod_{k=1}^{m} \frac{(t_j/x_k)_\infty}{(t_j q^{\beta_k}/x_k)_\infty} \}\cdot$$

$$\prod_{1\le i<j\le n} \frac{(q^{\gamma'_{i,j}} t_j/t_i)_\infty}{(q^{\gamma_{i,j}} t_j/t_i)_\infty}$$

defined on the n-dimensional algebraic torus $\overline{X} = (C^*)^n$. Here α_j , x_k , q^{β_k} , $\gamma_{i,j}$, $\gamma'_{i,j}$ denote complex parameters, and $(u)_\infty$ denotes the infinite product $\prod_{m=0}^{\infty} (1- uq^m)$.

We denote by X the n-dimensional lattice subgroup of $\overline{X}$ generated by $q^{\chi_j} = (1,\ldots,\underset{(j\ th)}{q},\ldots,1)$ consisting of the elements $q^\chi = (q^{\nu_1},\ldots,q^{\nu_n})$ for $\chi = \sum_{j=1}^{n} \nu_j\chi_j$, $\nu_j \in Z$. $Q^\chi = Q_1^{\nu_1}\cdots Q_n^{\nu_n}$ denotes the shift operator defined by $Q^\chi f(t) = f(q^\chi\cdot t)$. Then Φ satisfies the following rationally holonomic system of q-difference equations :

$$(1.3)\qquad Q^\chi\Phi(t|\alpha,\gamma) = b_\chi(t)\cdot \Phi(t|\alpha,\gamma),$$

where $\{ b_\chi(t)\}_{\chi\in X}$ denote the set of rational functions on $\overline{X}$ such that they define a 1-cocycle :

$$(1.4)\qquad b_{\chi+\chi'}(t) = b_\chi(t)\cdot Q^\chi b_{\chi'}(t), \qquad \text{for } \chi,\ \chi' \in X ,$$

$$(1.5) \qquad b_{\chi_j}(t) = q^{\alpha_j} \prod_{k=1}^{m} \frac{(1-q^{\beta_k} t_j/x_k)}{(1-\ t_j/x_k)}.$$

$$\prod_{1\le h<j} \frac{(1-\ q^{\gamma_{h,j}} t_j/t_h)}{(1-\ q^{\gamma'_{i,j}} t_j/t_h)} \prod_{j<h\le n} \frac{(1-\ q^{\gamma'_{j,h}-1} t_h/t_j)}{(1-\ q^{\gamma_{j,h}-1} t_h/t_j)} .$$

Hereafter we put $\gamma_{j,i} = 1 - \gamma'_{i,j}$ and $\gamma'_{j,i} = 1 - \gamma_{i,j}$ for $i < j$ so that $\gamma_{r,s} + \gamma'_{s,r} = 1$ for $r \neq s$. A permutation $\sigma \in \mathfrak{S}_n$ of the figures $\{1,2,\dots,n\}$ induces the operation for a function $\Phi(t|\alpha,\gamma)$ as follows : $\sigma\Phi(t|\alpha,\gamma) = \Phi(\sigma^{-1}(t|\alpha,\gamma))$, where $(\tilde t|\tilde\alpha,\tilde\gamma) = \sigma(t|\alpha,\gamma)$ is given by $\tilde t_j = t_{\sigma(j)}$, $\tilde\alpha_j = \alpha_{\sigma(j)} + \sum_{\substack{k<j \\ \sigma(j)<\sigma(k)}} (\gamma'_{\sigma(j),\sigma(k)} - \gamma_{\sigma(j),\sigma(k)}) - \sum_{\substack{k>j \\ \sigma(j)>\sigma(k)}} (\gamma'_{\sigma(k),\sigma(j)} - \gamma_{\sigma(k),\sigma(j)})$, $\tilde\gamma_{i,j} = \gamma_{\sigma(i),\sigma(j)}$ respectively. We also denote the function $\Phi(t|\sigma^{-1}(\alpha,\gamma))$ by $\hat\sigma\Phi(t|\alpha,\gamma)$. Then we have the equality

$$(1.6) \qquad \sigma\Phi(t|\alpha,\gamma) = U_\sigma(t)\ \Phi(t|\alpha,\gamma)$$

where $U_\sigma(t)$ denotes the pseudo constant in t, i.e., $Q^{\chi}U_\sigma(t) = U_\sigma(t)$ for any $\chi \in X$, defined by

$$(1.7) \quad U_\sigma(t) = \prod_{\substack{1\le i<j\le n \\ \sigma(i)>\sigma(j)}} \left(\frac{t_j}{t_i}\right)^{\gamma_{i,j}-\gamma'_{i,j}} \frac{\theta(q^{\gamma_{i,j}} t_j/t_i)}{\theta(q^{\gamma'_{i,j}} t_j/t_i)} .$$

We denote here by $\theta(u)$ the Jacobi elliptic theta function $(u)_\infty (q/u)_\infty\ (q)_\infty$.

Let V be the minimal subspace of the space of rational functions $R(\overline{X})\otimes \mathbb{C}(q^{\alpha}, x, q^{\beta}, q^{\gamma}, q^{\gamma'})$ in q^{α_j}, x_k, q^{β_k}, $q^{\gamma_{i,j}}$, $q^{\gamma'_{i,j}}$ such that the function space $\Phi\cdot V$ admits all the q-difference operations $\tilde{Q}^{\pm 1}_{j}$, $\tilde{Q}^{\pm 1}_{x_k}$, $\tilde{Q}^{\pm 1}_{\beta_k}$, $\tilde{Q}^{\pm 1}_{\gamma_{i,j}}$ and $\tilde{Q}^{\pm 1}_{\gamma'_{i,j}}$ for $\Phi\ \varphi$, $\varphi \in V$, induced by : $\alpha_j \to \alpha_j \pm 1$, $x_k \to x_k q^{\pm 1}$, $\beta_k \to \beta_k \pm 1$, $\gamma_{i,j} \to \gamma_{i,j} \pm 1$ $(i < j)$, and $\gamma'_{i,j} \to \gamma'_{i,j} \pm 1$ $(i < j)$ respectively. The r dimensional de Rham cohomology $H^r(\overline{X}, \Phi.\nabla_q)$ associated with Jackson integrals is defined by using the covariant q-difference operator :

$$(1.8) \qquad \nabla_q \psi = \sum_{j=1}^{n} \sum_{i_1 < \cdots < i_r} \nabla_j \psi_{i_1,\ldots,i_r}(t) \cdot \frac{d_q t_j}{t_j} \wedge \frac{d_q t_{i_1}}{t_{i_1}} \wedge \cdots \wedge \frac{d_q t_{i_r}}{t_{i_r}}$$

for an arbitrary q-difference form of r-th degree

$$(1.9) \qquad \psi = \sum_{1 \le i_1 < \cdots < i_r \le n} \psi_{i_1,\ldots,i_r}(t) \cdot \frac{d_q t_{i_1}}{t_{i_1}} \wedge \cdots\cdots \wedge \frac{d_q t_{i_r}}{t_{i_r}},$$

where ∇_j is defined as $\nabla_j f(t) = \dfrac{f(t) - b_{\chi_j}(t)\, Q_j f(t)}{1 - q}$.
We shall not here discuss general properties of these cohomologies (see [A1], [A2] and [S] for related subjects.) In particular the n-dimensional de Rham cohomology $H^n(\overline{X},\Phi,\nabla_q)$ is isomorphic to $V/\sum_{j=1}^{n} \nabla_j V$. On the other hand there exists the n-dimensinal homology $H_n(\overline{X},\Phi,\partial_q)$ with a boundary ∂_q such that

$$(1.10)\qquad \begin{array}{ccccc} H_n(\overline{X},\Phi,\partial_q) & \times & H^n(\overline{X},\Phi,\nabla_q) & \longrightarrow & \mathbb{C} \\ \Psi & & \Psi & & \Psi \\ \mathfrak{T} & \times & \varphi & \longrightarrow & \displaystyle\int_{\mathfrak{T}} \Phi\, \varphi \end{array}$$

for $\varphi = \varphi_{1,2,\ldots,n}\, \tilde{\omega}$, $\varphi_{1,\ldots,n} \in V$ so that $H_n(\overline{X},\Phi,\partial_q)$ can be regarded as the dual of $H^n(\overline{X},\Phi,\nabla_q)$. In fact, an n dimensional cycle $\mathfrak{T}$ satisfying $\partial_q \mathfrak{T} = 0$ modulo coboundaries is identified with a linear functional on $H^n(\overline{X},\Phi,\nabla_q)$. In particular the countable subset $[0, \xi\infty]_q = \{ q^{\chi}\xi \mid \chi \in X \}$ for each $\xi \in \overline{X}$ is a cycle provided that the corresponding Jackson integral (1.1) over $\mathfrak{T} = [0,\xi\infty]_q$ exists.

Since $U_\sigma(t)$ is a pseudo constant, we have from (1.6), the following

Lemma 1.

$$(1.11)\qquad \int_{[0,\xi\infty]_q} \sigma\Phi(t|\alpha,\gamma)\cdot\tilde{\omega} = U_\sigma(\xi)\int_{[0,\xi\infty]_q} \Phi(t|\alpha,\gamma)\cdot\tilde{\omega},$$

provided both hand sides are well defined.

The first basic fact can be stated as follows.

Proposition 1. Under the genericity condition for the parametres α_j , β_k, x_k and $\gamma_{i,j}$, $\dim H^n(\bar{X},\Phi,\nabla_q) = \dim H_n(\bar{X},\Phi,\partial_q) = m(m+n)^{n-1}$.

For the proof see [A1] and [A2]. In [A1] we have only proved this Proposition in case m = 1. However a similar proof is valid in the present situation.

2. α-stable cycles and α-unstable cycles.

We consider a forest F in graph theoretical sense with the following properties.

i) The set of vertices V(F) of F consists of the n figures $\{1,2,\ldots,n\}$.

ii) The set of roots R(F) of F is contained in the set of m letters $\{x_1,\ldots,x_m\}$.

iii) Each connected component T of F, which is a tree, contains only one root.

We call this forest F an admissible forest. We denote by $\mathcal{F}$ the set of all admissible forests. The total number of admissible forests are equal to $\kappa = m(m+n)^{n-1}$ due to Cayley-Moon Theorem (see [M3]). This is also equal to the dimension of $H^n(X,\Phi,\nabla_q)$. For each admissible forest F we want to construct two kinds of countable subsets $Y_F^{(\pm)}(\Phi)$

(abreviated by $Y_F^{(\pm)}$) of $\overline{X}$ as follows.

Let T be a tree, a connected component of an admissible forest F. For each vertex j of V(T), we denote by p(j) the unique vertex in V(T) which precedes j by one going from the root of T to j.

<u>Definition</u> 1. We denote by $Y_F^{(+)}$ (or $Y_F^{(-)}$) the countable set consisting of the points t of $\overline{X}$ having the following properties :

i) $t_j = qx_k$, q^2x_k , ... (or $q^{-\beta_k}x_k$, $q^{-\beta_k-1}x_k$, ...) if p(j) is a root. In this case we put p(j) = 0 by convention.

ii) $t_j/t_{p(j)} = q^{-\gamma'_{p(j),j}+1}$, $q^{-\gamma'_{p(j),j}+2}$, ... (or $q^{-\gamma_{p(j),j}}$, $q^{-\gamma_{p(j),j}-1}$, ...) otherwise.

Similarly we denote by $\xi_F = (\xi_1,\dots,\xi_n)$ (or $\eta_F = (\eta_1,\dots,\eta_n)$) the point of $\overline{X}$ defined by

i) $\xi_j = qx_k$ (or $\eta_j = q^{-\beta_k}x_k$) if p(j) is a root.

ii) $\xi_j/\xi_{p(j)} = q^{-\gamma'_{p(j),j}+1}$ (or $q^{-\gamma_{p(j),j}}$) otherwise.

Then $Y_F^{(+)}$ (or $Y_F^{(-)}$) is a subset of $[0,\xi_F\infty]_q$ (or $[0,\eta_F\infty]_q$). We call the point ξ_F (or η_F) the <u>summit of</u> $Y_F^{(+)}$(or $Y_F^{(-)}$). If the inequalties p(j)< j hold for all j $\in$ V(F), we call F <u>standard</u>.

We see that $|t_j|$ (or $|t_j|^{-1}$) are all bounded on $Y_F^{(+)}$ (or on $Y_F^{(-)}$). The Jackson integral (1.1) over $Y_F^{(+)}$ is well defined over a standard $Y_F^{(+)}$ provided $|q^{\alpha_j}| < 1$. If it is not standard, there exists a permutation $\rho_F \in \mathfrak{S}_n$

such that $Y_F^{(+)}$ is standard for $\rho_F\Phi(t)$, i.e., $\rho_F(p(j)) < \rho_F(j)$ for all $j \in V(F)$. Then using (1.6), a <u>regularized Jackson integral over</u> $Y_F^{(+)}$ is defined as :

$$\text{(2.1)} \qquad \mathrm{reg}\int_{Y_F^{(+)}} \Phi\,\tilde{\omega} = \int_{Y_F^{(+)}} \rho_F\Phi(t|\alpha,\gamma)\,\tilde{\omega}$$

$$= \lim_{\xi \longrightarrow \xi_F} U_{\rho_F}(\xi)^{-1} \int_{[0,\xi\infty]_q} \Phi(t|\alpha,\gamma)\,\tilde{\omega}$$

This is irrelevant to the choice of ρ_F except for a pseudo constant factor. We also denote the left hand side by $\int_{\mathrm{reg}Y_F^{(+)}} \Phi\,\tilde{\omega}$.

As for the integral over $Y_F^{(-)}$ is not generally well defined, since Φ has poles along $Y_F^{(-)}$. However we can also modify it as follows. For each F there exists a permutation $\rho_F^* \in \mathfrak{S}_n$ of the figures $\{1,2,\ldots,n\}$ such that $\rho_F^*(p(j)) > \rho_F^*(j)$ for every $j \in V(F)$. Then one can replace the Jackson integral over $Y_F^{(-)}$ by its regularization as follows :

$$\text{(2.2)} \quad \mathrm{reg}\int_{Y_F^{(-)}} \Phi\,\tilde{\omega} = \sum_{\nu_j \geq 0} \int_{Y_F^{(-)}} \prod_k \Big(\operatorname*{Res}_{t_j = q^{-\beta_k - \nu_j} x_k}\Big)[\rho_F^*\Phi\,\tilde{\omega}],$$

where Res means taking residues with respect to t_j at $q^{-\beta_k - \nu_j} x_k$, $\nu_j = 0,1,2,\ldots$ and j ranges over all vertices j such that p(j) is a root. (2.2) is also denoted by $\int_{\mathrm{reg}Y_F^{(-)}} \Phi\,\tilde{\omega}$. If $Y_F^{(-)}$ is standard, we can choose as ρ_F^* the transposition $\rho_0 : i \longrightarrow n-i+1$ $(1 \le i \le n)$ which does not depend on F. Then (2.2) coincides, except for a constant factor, with taking residues relative to the coordinates $\tau_j = t_j/t_{p(j)}$ at the points $t \in \overline{X}$ such that $\tau_j = q^{-\gamma_{p(j),j}}, q^{-\gamma_{p(j),j}-1}, \ldots$ and sum them up. If $Y_F^{(-)}$ is not standard, then it is standard for $\rho_0\rho_F^*\Phi$.

The meaning of reg $Y_F^{(\pm)}$ are given by the following asymptotic formulae.

Lemma 2. *For* $\alpha_j \longrightarrow +\infty$ *for all* j, *we have*

$$(2.3) \qquad \int_{\mathrm{reg}Y_F^{(+)}} \Phi\,\tilde{\omega} \sim \rho_F(\xi_1^{\alpha_1} \cdots \xi_n^{\alpha_n})\cdot C_F^{(+)} (\ne 0).$$

Likewise for all $\alpha_j \longrightarrow -\infty$, *we have*

$$(2.4) \qquad \int_{\mathrm{reg}Y_F^{(-)}} \Phi\,\tilde{\omega} \sim \rho_F^*(\eta_1^{\alpha_1} \cdots \eta_n^{\alpha_n}) \cdot C_F^{(-)} (\ne 0),$$

where $C_F^{(\pm)}$ *denote constants independent of* α.

These two kinds of asymptotics are both linearly independent. We call $\{\mathrm{reg}Y_F^{(+)}\}_{F\in\mathcal{F}}$ and $\{\mathrm{reg}Y_F^{(-)}\}_{F\in\mathcal{F}}$ α-*stable* and α-*unstable cycles* respectively. As a consequence, we get

Corollary. *The set of* α-*stable cycles* $\{\mathrm{reg}Y_F^{(+)}\}_{F\in\mathcal{F}}$

and the one of α-unstable cycles $\{\mathrm{reg}Y_F^{(-)}\}_{F\in\mathcal{F}}$ form a basis of $H_n(\overline{X},\Phi,\partial_q)$ over the field of pseudo-constants $\mathfrak{K}$ in the parametres u_j, q^{β_k}, x_k, $q^{\gamma_{i,j}}$.

For a proof see [A1] and [A2].

In particular

Proposition 2. For an arbitrary $\xi \in \overline{X}$ such that $\int_{[0,\xi\infty]_q} \Phi\,\tilde{\omega}$ is convergent, we have the formula

$$(2.5)\quad \int_{[0,\xi\infty]_q} \Phi\,\tilde{\omega} = \sum_{F\in\mathcal{F}} ([0,\xi\infty]_q:\mathrm{reg}Y_F^{(+)})_\Phi \cdot \int_{\mathrm{reg}Y_F^{(+)}} \Phi\,\tilde{\omega},$$

$$= \sum_{F\in\mathcal{F}} ([0,\xi\infty]_q : \mathrm{reg}Y_F^{(-)})_\Phi \cdot \int_{\mathrm{reg}Y_F^{(-)}} \Phi\,\tilde{\omega}$$

where the connection coefficients $([0,\xi\infty]_q:\mathrm{reg}Y_F^{(-)})_\Phi$ are expressed, in terms of pseudo constants which depend not only on u_j, q^{β_k}, x_k, $q^{\gamma_{i,j}}$ but also ξ, as follows (see [A3]):

$$(2.6)\qquad ([0,\xi\infty]_q:\ \mathrm{reg}Y_F^{(-)})_\Phi =$$

$$U_{\rho_0\rho_F^*}(\eta_F)^{-1}\sum_{\rho\in\mathfrak{S}_n} \rho\psi_n(\xi,\eta_F|\alpha)\ U_\rho(\xi)^{-1}U_\rho(\eta_F),$$

$$(2.7)\qquad \psi_n(\xi,\eta|\alpha) =$$

$$(1-q)^n \prod_{j=1}^{n} \left(\frac{\xi_j}{\eta_j}\right)^{\alpha_j} \frac{(q)_\infty^3\, \theta(q^{\alpha_j+\cdots+\alpha_n+1}\xi_j\eta_{j-1}/(\eta_j\xi_{j-1}))}{\theta(q^{\alpha_j+\cdots+\alpha_n+1})\theta(q\xi_j\eta_{j-1}/(\eta_j\xi_{j-1}))}$$

with $\xi_0 = \eta_0 = 1$.

In the right hand side of (2.6), $U_\rho(\eta_F)$ may have poles at $\eta = \eta_F$ but $U_{\rho_F}(\eta)^{-1}$ cancel them out so that every term $U_{\rho_F}(\eta)^{-1}U_\rho(\eta)$ has definite value. As a consequence, every term in the right hand side of (2.6) vanishes if there exists some j such that $\rho(j) < \rho(p(j))$.

3. Charateristic cycles.

We are now interested in the asymptotics for $x_1,\ldots,x_m$ at the infinity such that $|x_1| >> |x_2| >> \cdots >> |x_m|$, i.e., $x_{j+1}/x_j \longrightarrow 0$ for all j. For that purpose we define special cycles $W_F(\Phi)$ in $H_n(\overline{X},\Phi,\partial_q)$ over which the corresponding Jackson integrals (1.1) have monomial asymptotic expansions (see (3.1)).

Let an admissible forest F have a partition $F = T_1 + \cdots + T_m$, where T_j denotes a tree which is a connected component of F, rooted in x_j. T_j may be empty. $V(T_1),\ldots,V(T_m)$ give a partition of the whole vertices $V(F) = \{1,2,\ldots,n\}$. We can introduce the partial order $<<$ into the set $\mathcal{F}$ as follows. Let F' have a tree partition $F' = T_1' + \cdots + T_m'$. We say that F is smaller than F' if $V(T_1) = V(T_1'), \cdots, V(T_{p-1}) = V(T_{p-1}')$ and $V(T_p) \supsetneqq V(T_p')$ for some p.

On the other hand, if $V(T_i) < V(T_j)$ for $i < j$, i.e., the inequality $r < s$ holds for any pair $r \in V(T_i)$ and $s \in V(T_j)$. Then we say F is <u>in normal</u> <u>ordering</u>.

Suppose first F is in normal ordering. Then by using Proposition 3 below, we can define the cycle $W_F(\Phi) \in H_n(X,\Phi,\partial_q)$ associated with F giving the asymptotic form :

$$(3.1) \qquad \int_{W_F(\Phi)} \Phi\, \tilde{\omega} \sim \text{Const}\cdot x_1^{\alpha_{T_1}} \cdots x_m^{\alpha_{T_m}} .$$

$$\prod_{j=1}^{m-1} \prod_{k=j+1}^{m} \left(\frac{x_j}{x_k}\right)^{\beta_k |V(T_j)|},$$

where we put $\alpha_{T_j} = \sum_{j\in V(T_j)} \alpha_j$, and $|V(F)|$ denotes the number of elements of $V(F)$. We may remark that two different characteristic cycles, say $W_F(\Phi)$ and $W_{F'}(\Phi)$, may give the same asymptotics (3.1). Owing to Corollary of **Lemma 2**, $W_F(\Phi)$ is homologous to a linear combination of $\{\text{reg}Y_{F'}^{(\pm)}\}_{F'\in \mathcal{F}}$:

$$(3.2) \qquad W_F(\Phi) = \sum_{F'\in \mathcal{F}} (W_F(\Phi) : \text{reg}Y_{F'}^{(+)})_\Phi \ \text{reg}Y_{F'}^{(+)}$$

$$(3.3) \qquad = \sum_{F'\in \mathcal{F}} (W_F(\Phi) : \text{reg}Y_{F'}^{(-)})_\Phi \ \text{reg}\, Y_{F'}^{(-)} ,$$

where $\omega(F,F') = (W_F(\Phi) : \text{reg}Y_{F'}^{(\pm)})_\Phi$ denote pseudo-constants. We can normalize $W_F(\Phi)$ uniquely by the conditions (3.1) and that $\omega(F,F') = 1$ for $F' = F$, and $= 0$ otherwise unless $F' \ll F$.

Let F_0 be a forest such that $V(F_0) \subset \{1,2,\ldots,n\}$ and $R(F_0) \subset \{x_r,\ldots,x_s\}$ for some r, s such that $1 \le r \le s \le m$. We define the function $\Phi_{r,\ldots,s}(t|\alpha,\gamma)$ in the variables t_j for $j \in V(F_0)$ as

$$(3.4) \qquad \Phi_{r,\ldots,s}(t|\alpha,\gamma) = \prod_{j\in V(F_0)} t_j^{\alpha_j + \sum_{k=s+1}^{m}\beta_k} .$$

$$\prod_{\substack{r\le k\le s\\ j\in V(F_0)}} \frac{(t_j/x_k)_\infty}{(t_j q^{\beta_k}/x_k)_\infty} \cdot \prod_{\substack{i<j\\ i,j\,\in\, V(F_0)}} \frac{(q^{\gamma'_{i,j}} t_j/t_i)_\infty}{(q^{\gamma_{i,j}} t_j/t_i)_\infty} .$$

We now fix $1 \le p \le m$. then holds the following partition formula for $W_F(\Phi)$ into products of lower dimensional characteristic cycles.

Proposition 3. <u>Let</u> F <u>be an admissible forest in normal ordering</u> <u>and let</u> $F = F_1 + F_2$ <u>be a partition of F such that</u> F_1 <u>and</u> F_2 <u>are rooted in</u> $\{x_1,\ldots,x_p\}$ <u>and</u> $\{x_{p+1},\ldots,x_m\}$ <u>respectively</u>, i.e. $R(F_1) \subset \{x_1,\ldots,x_p\}$ <u>and</u> $R(F_2) \subset \{x_{p+1},\ldots,x_m\}$ <u>respectively. Then</u>

$$(3.5) \quad W_F(\Phi) = \sum_{F'_1,F'_2} (W_{F_1}(\Phi_{1,\ldots,p}) : \mathrm{reg}Y^{(-)}_{F'_1})_{\Phi_{1,\ldots,p}} \cdot C_p(\eta_{F'_1})^{-1}$$

$\cdot (W_{F_2'}(\Phi_{p+1,\dots,m}) : \mathrm{reg}Y_{F_2'}^{(+)})_{\Phi_{p+1,\dots,m}} \cdot \mathrm{reg}Y_{F_1'}^{(-)} \otimes \mathrm{reg}Y_{F_2'}^{(+)}$,

where F_1' and F_2' range over the sets of admissible forests for $\Phi_{1,\dots,p}$ and $\Phi_{p+1,\dots,m}$ respectively. $C_p(\eta_{F_1'})$ denotes the pseudo constant $\prod_{\substack{p+1\le k\le m \\ j\in V(F_1')}} \left(\frac{x_k}{\eta_j}\right)^{\beta_k} \frac{\theta(\eta_j/x_k)}{\theta(\eta_j q^{\beta_k}/x_k)}$ for the Jacobi elliptic theta function $\theta(u)$. $\eta_{F_1'}$ is a point of $(\mathbf{C}^*)^{|V(F_1)|}$ defined for F_1' as before in place of F (see Defintion 1).

This follows from the following lemma. It is valid even when F_1 or F_2 is empty. Under the same assumption as above for F , we may suppose $V(F_1) = \{t_1,\dots,t_r\}$ and $V(F_2) = \{t_{r+1},\dots,t_n\}$. Then

Lemma 3. *We take arbitrary α-unstable and α-stable cycles $\mathrm{reg}Y_{F_1}^{(-)}$ and $\mathrm{reg}Y_{F_2}^{(+)}$ with respect to $t_1,\dots,t_r$ and $t_{r+1},\dots,t_n$ respectively. When x is $|x_1| >> |x_2| >> \cdots >> |x_m|$ at the infinity, the integral (1.1) over the product cycle $\mathrm{reg}Y_{F_1}^{(-)} \otimes \mathrm{reg}Y_{F_2}^{(+)}$ is asymptotically decomposed into a product form :*

$$(3.6) \qquad \int_{\mathrm{reg}Y_{F_1}^{(-)} \otimes \mathrm{reg}Y_{F_2}^{(+)}} \Phi\, \tilde{\omega} \sim$$

$$C_p(\eta_{F_1}) \prod_{k=p+1}^{m} x_k^{-\beta_k |V(F_1)|} \cdot \int_{\mathrm{reg}Y_{F_1}^{(-)}} \Phi_{1,2,\dots,p}(t_1,\dots,t_r|\alpha,\gamma)\tilde{\omega}_1$$

$$\cdot \int_{\mathrm{reg}Y^{(+)}_{F_2}} \Phi_{p+1,\ldots,n}(t_{r+1},\ldots,t_n|\alpha,\gamma)\tilde{\omega}_2,$$

where $\tilde{\omega}_1 = \frac{d_q t_1}{t_1} \wedge\cdots\wedge \frac{d_q t_r}{t_r}$ and $\tilde{\omega}_2 = \frac{d_q t_{r+1}}{t_{r+1}} \wedge\cdots\wedge \frac{d_q t_n}{t_n}$.

This can be derived from the asymptotic properties $(q^{\gamma'_{j,k}} t_k/t_j)_\infty \sim 1 (j \le r,$ and $k \ge r+1)$, $(t_j/x_k)_\infty/(t_j q^{\beta_k}/x_k)_\infty \sim \theta(t_j/x_k)/\theta(t_j q^{\beta_k}/x_k)$ $(j \le r$, and $k \ge p+1)$, ~ 1 $(j \ge r+1$, and $k \le p)$ hold uniformly for $t \in Y^{(-)}_{F_1} \otimes Y^{(+)}_{F_2}$.

Suppose now that F is <u>not in normal ordering</u>. Then there exists a permutation $\sigma_F \in \mathfrak{S}_n$ such that F is in normal ordering for $\sigma_F\Phi(t|\alpha,\gamma)$. We can define the associated <u>characteristic cycle</u> $W_F(\Phi)$ as

$$(3.7) \qquad \int_{W_F(\Phi)} \Phi(t|\alpha,\gamma)\,\tilde{\omega} = \int_{W_F(\sigma_F\Phi)} \sigma_F\Phi(t|\alpha,\gamma)\,\tilde{\omega},$$

which has the asymptotic form according to (3.1) and (3.2):

$$(3.8) \qquad \int_{W_F(\Phi)} \Phi(t|\alpha,\gamma)\,\tilde{\omega} \sim \mathrm{Const}\cdot\sigma_F(x_1^{\alpha_{T_1}}\cdots x_m^{\alpha_{T_m}})$$

$$\cdot \prod_{j=1}^{m-1} \prod_{k=j+1}^{m} \left(\frac{x_j}{x_k}\right)^{\beta_k|V(T_j)|}.$$

(3.5) and (3.7) show the following equality

$$(3.9)\quad W_F(\Phi) = \sum_{F_1',F_2'} (W_{F_1}(\sigma_F\Phi_{1,\dots,p}) : \mathrm{reg}Y_{F_1'}^{(-)})_{\sigma_F\Phi_{1,\dots,p}}$$

$$\cdot\, C_p(\eta_{F_1'})^{-1}\, (W_{F_2}(\sigma_F\Phi_{p+1,\dots,m}) : \mathrm{reg}Y_{F_2'}^{(+)})_{\sigma_F\Phi_{p+1,\dots,m}}$$

$$\cdot\, U_{\sigma_F}((\eta_{F_1'}\times\eta_{F_2'}))\cdot\, (\mathrm{reg}Y_{F_1'}^{(-)} \otimes \mathrm{reg}Y_{F_2'}^{(+)}).$$

From Corollary to Lemma 2, for an arbitrary admissible forest F, $W_F(\Phi)$ can be expressed as

$$(3.10)\qquad W_F(\Phi) = \sum_{F\in\mathcal{F}} (W_F(\Phi) : \mathrm{reg}Y_{F'}^{(+)})_\Phi\, \mathrm{reg}Y_{F'}^{(+)}$$

$$(3.11)\qquad = \sum_{F\in\mathcal{F}} (W_F(\Phi) : \mathrm{reg}Y_{F'}^{(-)})_\Phi\, \mathrm{reg}Y_{F'}^{(-)}.$$

We denote by $\Omega = ((\omega(F,F'))_{F,F'\in\mathcal{F}}$ the connection matrix of degree κ with elements $\omega(F,F') = (W_F(\Phi):\mathrm{reg}Y_{F'}^{(-)})_\Phi$. Let $V_1 + \cdots + V_m$ be a partition of the figures $\{1,2,\dots,n\}$. An admissible forest F is said to be <u>in the class</u> $\langle V_1,\dots,V_m\rangle$ if F has a tree partition $F = T_1 + \cdots + T_m$ such that $V(T_j) = V_j$ and T_j has the root x_j for every j. We denote by $[V_1,\dots,V_m]$ the linear space spanned by all F in the class $\langle V_1,\dots,V_m\rangle$. It is isomorphic to the m tensor product $[V_1] \otimes \cdots \otimes [V_m]$. We denote by Ω_0 the submatrix of Ω restricted on the space $[V_1,\ldots,V_m]$ and by A_j the matrix $(\mathrm{reg}Y^{(+)}_{F_j} : \mathrm{reg}Y^{(-)}_{F'_j})_{\sigma_F\Phi_j}$ for admissible forests F_j and $F'_j \in [V_j]$. It is an endmorphism on $[V_j]$. The multiplication operator by $C_p(\eta_{F'})$: $Y^{(-)}_{F'} \longrightarrow C_p(\eta_{F'})\, Y^{(-)}_{F'}$ for $R(F') \subset \{x_1,\ldots,x_p\}$ induces a diagonal endomorphism expressed as $C^{(1)}_p \otimes \cdots \otimes C^{(p)}_p$ on $[V_1] \otimes \cdots \otimes [V_p]$, where $C^{(r)}_p \in \mathrm{End}\,[V_r]$. By successive applications of the formula (3.9), we have

Lemma 4.

$$(3.12) \qquad \Omega_0 = B_1\tilde{A}_1 \otimes \cdots \otimes B_m\tilde{A}_m ,$$

<u>where</u> $\tilde{A}_j$ <u>is given by</u> $A_j \cdot (C^{(j)}_j)^{-1} \cdots (C^{(j)}_{m-1})^{-1}$. B_j <u>denotes the matrix</u> $B_j = ((B_j(T_j,T'_j))$ <u>with entries</u> $B_j(T_j,T'_j) = (W_{T_j}(\sigma_F\Phi_j) : \mathrm{reg}Y^{(-)}_{T'_j})_{\sigma_F\Phi_j}$, <u>which is also an endomorphism on</u> $[V_j]$.

In fact, in the formula (3.9) $(\mathrm{reg}Y^{(-)}_{F'_1} \otimes \mathrm{reg}Y^{(-)}_{F'_2} : \mathrm{reg}Y^{(-)}_{F''})_{\sigma_F\Phi}$ equals 0 unless $F''_1 \supset F'_1$ and $V(F'') = V(F'_1) + V(F'_2)$ for a partition $F'' = F''_1 + F''_2$ such that $R(F''_1) \subset \{x_1,\ldots,x_p\}$ and $R(F''_2) \subset \{x_{p+1},\ldots,x_m\}$. Furhtermore if F'' is in the same class as $F'_1 + F'_2$, i.e., if $F''_1 = F'_1$, then $(\mathrm{reg}Y^{(-)}_{F'_1} \otimes \mathrm{reg}Y^{(+)}_{F'_2} : \mathrm{reg}Y^{(-)}_{F''})_{\sigma_F\Phi}$ reduces to $(\mathrm{reg}Y^{(+)}_{F'_2} : \mathrm{reg}Y^{(-)}_{F''_2})_{\sigma_F\Phi_{p+1,\ldots,m}}$.

By fixing $W_{T_j}(\sigma_F\Phi_j)$ in a suitable way, i.e., by taking $B_j = \tilde{A}_j^{-1}$, we get Ω_0 to be the identity. Hence we can normalize $W_F(\Phi)$ uniquely by the following expression :

$$(3.13)\quad W_F(\Phi) = \sum_{F'<<F} \omega(F,F')\ \mathrm{reg}Y_{F'}^{(-)} + U_{\sigma_F}(\eta_F)\ \mathrm{reg}Y_F^{(-)}.$$

In particular $\omega(F,F') = 0$, if $F \neq F'$ and if $V(T_j') = V(T_j)$ for all j concerning the tree partitions $F = T_1 + \cdots + T_m$ and $F' = T_1' + \cdots + T_m'$.

$W_F(\Phi)$ which has thus been obtained does not depend on any manner of construction using partition of F.

4. Yang Baxter equation.

For a permutation $\tau \in \mathfrak{S}_m$ of the m letters $\{x_1,\dots,x_m\}$ we can define an $\mathfrak{R}$-endomorphism on the space $H_n(\overline{X},\Phi,\partial_q)$:

$$(4.1)\qquad \tau\, W_F(\Phi) = \sum_{F'\in \mathcal{F}} P_\tau^{-1}(F,F')\ W_{F'}(\Phi)$$

where $P_\tau(F,F')$ denotes a $\kappa \times \kappa$ matrix with elements in $\mathfrak{R}$. This gives the connection formula among the Jackson integrals $\int_{W_F(\Phi)} \Phi\ \tilde{\omega}$. These matrices can be expressed in the following manner.

Let $V_1 + V_2 + V_3$ a partition of the set of vertices $\{1,2,\dots,n\}$. Consider the linear space (denoted by $[V_1,V_2,V_3]$) spanned by admissible forests F such that F is decomposed into $F_1 + F_2 + F_3$ with $V(F_1) = V_1$, $V(F_2) = V_2$, and $V(F_3) = V_3$ and that $R(F_1) = \{x_1,\dots,x_{r-1}\}$, $R(F_2)$

$= \{x_r, x_{r+1}\}$ and $R(F_3) = \{x_{r+2},\ldots,x_m\}$ respectively. We can take a permutation $\sigma = \sigma_{V_1,V_2,V_3} \in \mathfrak{S}_n$ such that $\sigma\Phi$ is in normal ordering for all $F \in [V_1,V_2,V_3]$, i.e., $\sigma_F = \sigma$. It is ovbious that $[V_1,V_2,V_3] \simeq [V_1] \otimes [V_2] \otimes [V_3]$.

Theorem. i) We have

$$(4.2) \qquad P_\tau = \Omega \cdot S_\tau \cdot (\tau\Omega)^{-1},$$

where S_τ denotes the linear representation of dimension κ on $H_n(\overline{X},\Phi,\partial_q)$ as to the basis $\{\mathrm{reg}Y_F^{(-)}\}_{F\in\mathfrak{F}}$: $S_{\tau\tau'} = S_\tau \cdot S_{\tau'}$ for τ, $\tau' \in \mathfrak{S}_m$.

ii) For the transposition τ_r between the letters x_r and x_{r+1}, P_{τ_r} can be represented as

$$(4.3) \qquad P_{\tau_r} = \sum_{V_1,V_2,V_3} 1^{[V_1]} \otimes P_{\tau_r}^{[V_2]} \otimes 1^{[V_3]}$$

where $\langle V_1, V_2, V_3\rangle$ ranges over all partitions of $\{1,2,\ldots,n\}$ such that V_1, V_2 and V_3 are rooted in $\{x_1,\ldots,x_{r-1}\}$, $\{x_r,x_{r+1}\}$ and $\{x_{r+1},\ldots,x_m\}$ respectively. $1^{[V_1]}$, $1^{[V_3]}$ denote the identity operators and $P_{\tau_r}^{[V_2]}$ denotes the endomorphism induced by τ_r, acting on $[V_1]$, $[V_3]$ and $[V_2]$ respectively.

In fact by using (3.9) successively we get

(4.4) $$W_F(\Phi) =$$

$$\sum_{F'_{1,2}, F'_3} (W_{F_1+F_2}(\sigma\Phi_{1,2,\dots,r+1}) : \mathrm{reg}Y^{(-)}_{F'_{1,2}})_{\sigma\Phi_{1,2,\dots,r+1}} \cdot$$

$$(W_{F_3}(\sigma\Phi_{r+2,\dots,m}) : \mathrm{reg}Y^{(+)}_{F'_3})_{\sigma\Phi_{r+2,\dots,m}} C_{r+1}(\eta_{F'_{1,2}})^{-1} \cdot$$

$$(\mathrm{reg}Y^{(-)}_{F'_{1,2}} \otimes \mathrm{reg}Y^{(+)}_{F'_3})$$

$$= \sum_{F'_{1,2}, F'_3} \{ \sum_{F''_1, F''_2} (W_{F_1}(\sigma\Phi_{1,\dots,r-1}) : \mathrm{reg}Y^{(-)}_{F''_1})_{\sigma\Phi_{1,\dots,r-1}}$$

$$\cdot (W_{F_2}(\sigma\Phi_{r,r+1}) : \mathrm{reg}Y^{(+)}_{F''_2})_{\sigma\Phi_{r,r+1}}$$

$$\cdot W_{F_3}(\sigma\Phi_{r+2,\dots,m}) : \mathrm{reg}Y^{(+)}_{F'_3})_{\sigma\Phi_{r+2,\dots,m}}$$

$$\cdot (\mathrm{reg}Y^{(-)}_{F''_1} \otimes \mathrm{reg}Y^{(+)}_{F''_2} : \mathrm{reg}Y^{(-)}_{F'_{1,2}})_{\sigma\Phi_{1,\dots,r+1}}$$

$$\cdot C_{r-1}(\eta_{F''_1})^{-1} C_{r+1}(\eta_{F'_{1,2}})^{-1} \} (\mathrm{reg}Y^{(-)}_{F'_{1,2}} \otimes \mathrm{reg}Y^{(+)}_{F'_3}),$$

where $F'_{1,2}$, F'_3, F''_1 and F''_2 range over the sets of admissible forests for $\sigma\Phi_{1,\dots,r+1}$, $\sigma\Phi_{r+2,\dots,m}$, $\sigma\Phi_{1,\dots,r-1}$ and $\sigma\Phi_{r,r+1}$ such that $V(F'_{1,2}) = V_1$, $V(F'_3) = V_3$, $V(F''_1) = V_1$ and $V(F''_2) = V_2$ respectively.

Since $\sigma\Phi_{1,\dots,r-1}$, $\sigma\Phi_{1,\dots,r+1}$, $\sigma\Phi_{r,r+1}$, $\sigma\Phi_{r+2,\dots,m}$, F''_1, F'_3, $C_{r-1}(\eta_{F''_1})$ and $C_{r+1}(\eta_{F'_{1,2}})$ leave fixed by the action of τ_r, we have from (4.4)

(4.5) $\tau_r W_F(\Phi) =$

$$\sum_{F'_{1,2},F'_3} \{ \sum_{F''_1,F''_2} (W_{F_1}(\sigma\Phi_{1,\dots,r-1}): \mathrm{reg}Y^{(-)}_{F''_1})_{\sigma\Phi_{1,\dots,r-1}}$$

$$\cdot(\tau_r W_{F_2}(\sigma\Phi_{r,r+1}):\mathrm{reg}Y^{(+)}_{F''_2})_{\sigma\Phi_{r,r+1}}$$

$$\cdot(W_{F_3}(\sigma\Phi_{r+2,\dots,m}):\mathrm{reg}Y^{(+)}_{F'_3})_{\sigma\Phi_{r+2,\dots,m}}$$

$$\cdot(\mathrm{reg}Y^{(-)}_{F''_1} \otimes \mathrm{reg}Y^{(+)}_{F''_2}:\mathrm{reg}Y^{(-)}_{F'_{1,2}})_{\sigma\Phi_{1,\dots,r+1}}$$

$$\cdot C_{r-1}(\eta_{F''_1})^{-1} C_{r+1}(\eta_{F'_{1,2}})^{-1}\} \, (\mathrm{reg}Y^{(-)}_{F'_{1,2}} \otimes \mathrm{reg}Y^{(+)}_{F'_3}),$$

where $F'_{1,2}$, F'_3 , F''_1 and F''_2 range over the sets of admissible forests for $\sigma\Phi_{1,\dots,r+1}$, $\sigma\Phi_{r+2,\dots,m}$, $\sigma\Phi_{1,\dots,r-1}$ and $\sigma\Phi_{r,r+1}$ respectively. Hence the matrix $P_{\tau_r}(F,F')$ is expressed as follows, only by using the block matrices of degree $2(2+|V_2|)^{|V_2|-1}$

$$(4.6) \quad \tau_r W_F(\Phi) = \sum_{F'_2\in[V_2]} (P^{[V_2]}_{\tau_r})^{-1}(F_2,F'_2) \; W_{(F_1+F'_2+F_3)}(\Phi)$$

$$(4.7) (P^{[V_2]}_{\tau_r})^{-1}(F_2,F'_2)) = (\tau_r W_{F_2}(\sigma\Phi_{r,r+1}):W_{F'_2}(\sigma\Phi_{r,r+1}))_{\sigma\Phi_{r,r+1}}$$

such that $V(F'_2) = V_2$ and $R(F'_2) = \{x_r,x_{r+1}\}$.

Hence the evaluation of P_{τ_r} reduces to the case of m = 2 for the function $\sigma\Phi_{r,r+1}$. It depends only on x_{r+1}/x_r, as a function of x. As a result, in view of the cocycle condition for $\{P_\tau\}_{\tau\in\mathfrak{S}_m}$ which follows from (4.2),

$$(4.8) \qquad P_{\tau\tau'} = P_\tau \cdot \tau P_{\tau'} \qquad , \text{for } \tau,\ \tau' \in \mathfrak{S}_m ,$$

we see that <u>the</u> m-1 <u>matrices</u> $P_r(\frac{x_{r+1}}{x_r}) = P_{\tau_r}(x)$, $1\leq r\leq m-1$, <u>satisfy</u> <u>the Yang-Baxter equation</u>

$$(4.9) \qquad P_r(u)\ P_{r+1}(uv)\ P_r(v) = P_{r+1}(v)\ P_r(uv)\ P_{r+1}(u).$$

5. Explicite formula

(i) Case of m = 2.

First remark that an admissible forest F has only 2 components T_1 and T_2 , both trees rooted in x_1 and x_2 respectively. Suppose that F is in normal ordering for $\sigma\Phi$ for some $\sigma \in \mathfrak{S}_n$. Then the formula (3.9) gives

$$(5.1) \qquad W_F(\Phi) = \sum_{T_1',T_2'} C_1(n_{T_1'})^{-1} \ (W_{T_1}(\sigma\Phi_1):\mathrm{reg}Y_{T_1'}^{(-)})_{\sigma\Phi_1} \cdot$$

$$(W_{T_2}(\sigma\Phi_2):\mathrm{reg}Y_{T_2'}^{(+)})_{\sigma\Phi_2} \cdot (\mathrm{reg}\ Y_{T_1'}^{(-)} \otimes \mathrm{reg}\ Y_{T_2'}^{(+)}).$$

In this case we can take $W_{T_1}(\sigma\Phi_1) = \sum_{T_1'} B_1(T_1,T_1')\ \mathrm{reg}Y^{(-)}_{T_1'}$, $W_{T_2}(\sigma\Phi_2) = \sum_{T_2'} B_2(T_2,T_2')\ \mathrm{reg}Y^{(+)}_{T_2'}$ such that $B_1(T_1,T_1') = C_1(\eta_{T_1'})\delta_{T_1,T_1'}$ and that $\sum_{T_2'} B_2(T_2,T_2')\cdot(\mathrm{reg}Y^{(+)}_{T_2'} : \mathrm{reg}Y^{(-)}_{T_2''})_{\Phi_2} = \delta_{T_2,T_2''}$, where δ denotes the Kronecker delta function with respect to the above bases in $[V_1]$ and $[V_2]$ respectively.

(ii) <u>Case of</u> $n = 1$.

$V(F)$ (or $V(F')$) consists of only one vertex and one root say x_j (or x_k). Then (2.6) reduces to the following formula for $\eta_1 = q^{-\beta_k}x_k$ the summit of $Y^{(-)}_{F'}$:

$$(5.2)\quad ([0,\xi\infty]_q : \mathrm{reg}Y^{(-)}_{F'})_\Phi = (1-q)\left(\frac{\xi_1}{\eta_1}\right)^{\alpha_1} \frac{\theta(q^{\alpha_1+1}\xi_1/\eta_1)(q)^3_\infty}{\theta(q^{\alpha_1+1})\theta(q\xi_1/\eta_1)}.$$

Hence (4.4) shows that $(W_F(\Phi) : \mathrm{reg}Y^{(-)}_{F'})_\Phi$ equals

$$(5.3)\qquad \psi_1(\xi_1,\eta_1|\alpha) = C_j(\eta_1)^{-1}B_j(Y^{(+)}_F)^{-1}.$$

$$(1-q)\left(\frac{x_j}{x_k}q^{\beta_k}\right)^{\alpha_1} \frac{\theta(q^{\alpha_1+\beta_k+1}x_j/x_k)\ (q)^3_\infty}{\theta(q^{\alpha_1+1})\ \theta(q^{\beta_k+1}x_j/x_k)},$$

if $j \geq k$ and equals 0 otherwise, where $B_j(Y^{(+)}_F)$ is given by

$$(5.4)\qquad \int_{[0,1]_q} t_1^{\alpha_j+\beta_{j+1}+\cdots+\beta_m}\frac{(t_1)_\infty}{(t_1q^{\beta_j})_\infty}\frac{d_qt_1}{t_1}$$

$$= q^{\alpha_j+\beta_{j+1}+\cdots+\beta_m}\frac{\Gamma_q(\alpha_j+\beta_{j+1}+\cdots+\beta_m)\Gamma_q(\beta_j+1)}{\Gamma_q(\alpha_j+\beta_j+\cdots+\beta_m+1)}$$

$P_{\tau_1}^{(V_2)}$ is a 2 by 2 matrix given explicitely in [M2],[F] and [A4].

So far we have not discussed the symmetric or anti-symmetric case where $\gamma_{i,j}$ are all equal and $\gamma'_{i,j}$ are all equal. In this case the regularization of Jackson integrals seems more complicated. It is not certain if Jackson integrals discussed here satisfy quantum KZ-equations using R-matrices formulated in [F], although remarkable results have been obtained by N.Yu.Reshetikhin and A.Matsuo (see [R] and [M1]).

References.

[A1] K.Aomoto, Finiteness of a cohomology associated with certain Jackson integrals, Tôhoku Math. Jour., 43(1991), 75-101.

[A2] K.Aomoto and Y.Kato, q-analog of de Rham cohomology associated with Jackson integrals, in Special Functions, ICM-90 Satellite Conference Proc., M.Kashiwara and T.Miwa eds., Springer, 1991, 30-62.

[A3] K. Aomoto, On connection coefficients for q-difference system of A-type Jackson integral, preprint,

1992.

[A4] K.Aomoto, Y.Kato and K.Mimachi, A solution of the Yang-Baxter equation as connection coefficients of a holonomic q-difference system, International Math. Research Notices, No.1, Duke math. Journ., 7-15, 1992.

[A5] R.Askey, Beta integrals in Ramanujan's papers, his unpublished work and further examples, Ramanujan Revisited, Academic Press, 1988.

[C] I.Cherednik, Quantum Knizhnik-Zamolodchikov equations and affine root systems, preprint, 1992.

[F] I.B.Frenkel and N.Yu.Reshetikhin, Quantum affine alegebras and holonomic q-difference equations, In Differential Geometric Methods in Theoretical Physics, (S.Catto and A.Rocha eds.), World Sci. 1992.

[M1] A.Matsuo, Quantum algebra structure of certain Jackson integrals, preprint, 1992.

[M2] K.Mimachi, Holonomic q-difference system of 1st order associated with Jackson integrals of Selberg type, preprint, 1992.

[M3] J.W.Moon, Various proofs of Cayley's formula for counting trees, in Seminar on Graph Theory (F.Harary Ed.) Holt, Rinehart and Winton, Texas, 1967, 70-78.

[R] N.Yu.Reshetikhin, Jackson type integrals, Bethe vectors, and solutions to a difference analog of the Kniznik-Zamolodchikov system, to appear in this volume.

[S] C.Sabbah, Systèmes holonomes d'équations aux q-différences, to appear in Proc. of the Conference on $\mathcal{D}$-modules (Lisbon, 1990).

* Dept. of Math., Nagoya Univ., Nagoya, Japan

** Dept. of Math., Meijo Univ., Nagoya, Japan

Current Adress : the same as above

Contemporary Mathematics
Volume **175**, 1994

Representations of the moonshine module vertex operator algebra

Chongying Dong

Abstract. We consider the moonshine module vertex operator algebra constructed by Frenkel-Lepowsky-Meurman. This algebra has a unique irreducible module, the adjoint module, and any module is completely reducible. This proves the first part of the FLM's conjecture on the moonshine module.

1 Introduction

The moonshine module $V^\natural$ constructed by Frenkel-Lepowsky-Meurman is a **Z**-graded representation of the monster simple group with the modular function J as generating function of the dimensions of its homogeneous spaces. This construction led Borcherds and FLM to the notion of a vertex operator algebra. In particular, $V^\natural$ is a vertex operator algebra whose automorphism group is precisely the monster simple group (see [1], [11], [12]). Roughly speaking, $V^\natural$ is the subspace of fixed points of an involution (which is based on an order 2 isometry of the Leech lattice without nonzero fixed points) of a given space. This space is a direct sum of the untwisted and twisted spaces constructed from the Leech lattice. Recently, another moonshine module, which turns out to be isomorphic to $V^\natural$, has been constructed based on an order 3 isometry of the Leech lattice without nonzero fixed points [6].

1991 Mathematics Subject Classification. Primary 17B65, 17B68.

Supported by Regents of the University of California and by NSA grant MDA904-92-H-3099.

The detailed version of this paper will be submitted for publication elsewhere.

In fact, the FLM moonshine module structure proved one form of a conjecture of McKay and Thompson [16]. At the same time, based on the uniqueness of the Leech lattice, they put forward a uniqueness conjecture on the moonshine module: (a) $V^\natural$ is holomorphic in the sense that the only irreducible module is itself. (b) Any vertex operator algebra V satisfying the following conditions is isomorphic to $V^\natural$: (i) V is holomorphic. (ii) The rank of V is 24. (iii) There are no weight 1 elements. We announce a proof of part (a) of their conjecture in this paper.

Part (a) of the FLM conjecture is a special case of a conjecture in the "orbifold" conformal field theory. Roughly speaking, an orbifold theory is a conformal field theory which is obtained from a given conformal field theory modulo the action of a finite symmetry group (see [8]). The main new feature of the orbifold theory is the introduction of twisted modules. Let $(V, Y, \mathbf{1}, \omega)$ be a vertex operator algebra and G be a finite subgroup of the automorphisms of V. Denote by V^G the subspace of V consisting of the fixed points under the action of G. Then V^G is a vertex operator subalgebra of V. One of the most important problems in orbifold theory is to determine the irreducible representations of V^G. Now we assume that V is a holomorphic and that $G = \{1, g, g^2, ..., g^{n-1}\}$ is a subgroup of $\mathrm{Aut}\,(V)$ generated by g of order n. The following conjecture concerning the x-twisted V-modules for $x \in G$ and the representations of V^G is well known: (1) For each $x \in G$, there is a unique irreducible x-twisted module $V(x)$ such that $V(1) = V$, (2) Each $V(x)$ decomposes into

$$V(x) = V(x)^0 \oplus \cdots \oplus V(x)^{n-1}$$

where $V(x)^k = \{v \in V(x) | g \cdot v = e^{2k\pi i/n} v\}$. Then the $V(x)^k$ are irreducible V^0-modules and are the only irreducible V^0-modules. (3) The space $V_g = V(1)^0 \oplus V(g)^0 \oplus \cdots \oplus V(g^{n-1})^0$ has the structure of a vertex operator algebra and this vertex operator algebra is holomorphic.

It is known from [2] that the vertex operator algebra V_L associated with an even self dual lattice L is holomorphic. In particular, the vertex operator algebra V_Λ associated with the Leech lattice Λ is holomorphic. In [3] part (1) of the second conjecture for the holomorphic vertex operator algebra V_L with g constructed from the involution -1 of the L is proved. Take $V = V_\Lambda$ and $g = \theta$ where θ is an automorphism of V_Λ based the involution -1 of Λ. Recall from [12] that the moonshine module $V^\natural$ is defined to be $V_\Lambda^0 \oplus V_\Lambda(\theta)^0$ where $V_\Lambda(\theta)$ is the unique irreducible θ-twisted V_Λ-module (cf [3]). In [7] we found 48 mutually commutative Virasoro algebras with central charge $1/2$ inside the moonshine module such that the original Virasoro algebra is a sum of these 48 Virasoro algebras. This discovery allows us to study $V^\natural$ in terms of the representations of the

Virasoro algebra with central charge 1/2. In fact, any $V^\natural$-module is a direct sum of finitely many irreducible modules for the tensor product of these 48 Virasoro algebras. Using the results in [2] and [3] on the representations (including the twisted representation) of V_Λ, we show that the untwisted part V_Λ^0 of $V^\natural$ has exactly four irreducible representations, namely V_Λ^0, V_Λ^1, $V_\Lambda(\theta)^0$ and $V_\Lambda(\theta)^1$. Moreover, any V_Λ^0-module is completely reducible. This result, together with the fusion rules for the representations of the Virasoro algebra with central charge 1/2 [7], imply immediately part (a) of the FLM conjecture. Combining this with the result that $V^\natural$ is a vertex operator algebra proves the second conjecture for the holomorphic vertex operator algebra $V = V_\Lambda$ with $g = \theta$.

We assume that the reader has some familiarity with the theory of vertex operator algebras and their representations (including twisted representations, intertwining operators and fusion rules) as presented in [3], [9], [10] and [12].

2 Virasoro algebra with central charge 1/2

Recall that the Virasoro algebra Vir is an infinite-dimensional Lie algebra with a basis $\{L_n | n \in \mathbf{Z}\} \cup \{\mathbf{c}\}$ and commutation relations

$$[L_m, L_n] = (m-n)L_{m+n} + \delta_{m+n,0}\frac{m^3-m}{12}\mathbf{c},$$

where $\mathbf{c}$ is a central element. Given two complex numbers c and h, denote by $L(c,h)$ the unique irreducible Vir-module with $\mathbf{c}$ acting as multiplication by c (called the central charge) and with highest weight h. It is proved in [13] that $L(c,0)$ with $c \neq 0$ is a vertex operator algebra. Then Vir has just three irreducible unitary representations $L(\frac{1}{2}, h)$ with highest weight $h = 0, \frac{1}{2}, \frac{1}{16}$ and with central charge 1/2 (see [14] and [15]). The following theorem can be found in [7]. (The definitions of intertwining operators and fusion rules are given in [10] in the mathematics literature.)

Theorem 1 *(1) The vertex operator algebra $L(\frac{1}{2},0)$ has exactly three irreducible representations $L(\frac{1}{2},h)$ for $h = 0, \frac{1}{2}, \frac{1}{16}$, and any module is completely reducible.*

(2) The fusion rules among these modules are given by: $L(\frac{1}{2},\frac{1}{2}) \times L(\frac{1}{2},\frac{1}{2}) = L(\frac{1}{2},0)$, $L(\frac{1}{2},\frac{1}{2}) \times L(\frac{1}{2},\frac{1}{16}) = L(\frac{1}{2},\frac{1}{16})$ and $L(\frac{1}{2},\frac{1}{16}) \times L(\frac{1}{2},\frac{1}{16}) = L(\frac{1}{2},0) + L(\frac{1}{2},\frac{1}{2})$.

(3) The tensor product $L = L(\frac{1}{2},0)^{\otimes n}$ (n a positive integer) again is a vertex operator algebra (cf [10]) and any module for this tensor product algebra is a direct sum of irreducible modules $L(h_1,...,h_n) = L(\frac{1}{2},h_1) \otimes \cdots \otimes L(\frac{1}{2},h_n)$ for $h_i \in \{0, \frac{1}{2}, \frac{1}{16}\}$.

Fix a positive integer $q \in 8\mathbf{Z}$. Let S_u be the set of of irreducible L-modules $L(h_1, ..., h_q)$ with property that (h_{2i-1}, h_{2i}) is one of the following pairs $(0,0), (0, \frac{1}{2}), (\frac{1}{2}, 0), (\frac{1}{2}, \frac{1}{2}), (\frac{1}{16}, \frac{1}{16})$ for $i = 1, ..., \frac{q}{2}$. Let S_t be the set of irreducible modules $L(h_1, ..., h_q)$ with the property that $(h_{2i-1}, h_{2i}) = (0, \frac{1}{16}), (\frac{1}{16}, 0), (\frac{1}{2}, \frac{1}{16}), (\frac{1}{16}, \frac{1}{2})$. Set $W_1 = L(\frac{1}{2}, \frac{1}{2}, 0, \cdots, 0)$ and let $W_2 = L(x_1, \cdots, x_q) \in S_u$ such that $\sum_{i=1}^{q} x_i \in \frac{1}{2}\mathbf{Z}$. Let $M_i = L(h_1^i, \cdots, h_q^i)$ for $i \in \{1,2,3,4,5\}$ such that either the all $M_i \in S_u$ or the all $M_i \in S_t$, and there exist nonzero intertwining operators of types $\left[\begin{matrix} M_2 \\ W_1\, M_1 \end{matrix}\right]$, $\left[\begin{matrix} M_3 \\ W_2\, M_2 \end{matrix}\right]$, $\left[\begin{matrix} M_4 \\ W_2\, M_1 \end{matrix}\right]$, $\left[\begin{matrix} M_3 \\ W_1\, M_4 \end{matrix}\right]$, and $\left[\begin{matrix} M_5 \\ W_1\, W_2 \end{matrix}\right]$.

Theorem 2 *Let $Y_2(\cdot, z)$, $Y_3(\cdot, z)$, $Y_4(\cdot, z)$ and $Y_5(\cdot, z)$ be nonzero intertwining operators of types* $\left[\begin{matrix} M_2 \\ W_1\, M_1 \end{matrix}\right]$, $\left[\begin{matrix} M_3 \\ W_1\, M_4 \end{matrix}\right]$, $\left[\begin{matrix} M_4 \\ W_2\, M_1 \end{matrix}\right]$, *and* $\left[\begin{matrix} M_5 \\ W_1\, W_2 \end{matrix}\right]$, *respectively. Then there exist unique intertwining operators $Y_1(\cdot, z)$ and $Y_6(\cdot, z)$ of types* $\left[\begin{matrix} M_3 \\ W_2\, M_2 \end{matrix}\right]$ *and* $\left[\begin{matrix} M_3 \\ M_5\, M_1 \end{matrix}\right]$ *such that if $M_1 \in S_u$, the following Jacobi identity holds for $w_1 \in W_1$, $w_2 \in W_2$ acting on M_1*

$$z_0^{-1}\delta\left(\frac{z_1 - z_2}{z_0}\right) Y_3(w_1, z_1)Y_4(w_2, z_2) - z_0^{-1}\delta\left(\frac{z_2 - z_1}{-z_0}\right) Y_1(w_2, z_2)Y_2(w_1, z_1)$$
$$= z_2^{-1}\delta\left(\frac{z_1 - z_0}{z_2}\right) Y_6(Y_5(w_1, z_0)w_2, z_2)$$

and that if $M_1 \in S_t$, the following twisted Jacobi identity holds

$$z_0^{-1}\delta\left(\frac{z_1 - z_2}{z_0}\right) Y_3(w_1, z_1)Y_4(w_2, z_2) - z_0^{-1}\delta\left(\frac{z_2 - z_1}{-z_0}\right) Y_1(w_2, z_2)Y_2(w_1, z_1)$$
$$= z_2^{-1}\left(\frac{z_1 - z_0}{z_2}\right)^{1/2} \delta\left(\frac{z_1 - z_0}{z_2}\right) Y_6(Y_5(w_1, z_0)w_2, z_2),$$

where $\delta(z) = \sum_{n\in\mathbf{Z}} z^n$ and where $(z_1 - z_2)^r$ $(r \in \mathbf{Q})$ is to be expanded as a formal power series in the second variable, z_2, and analogously for the other expressions.

In [7] we introduced and studied a special class $\mathcal{C}$ of vertex operator algebras $(V, Y, \mathbf{1}, \omega)$ (see [12] for the notation). The vertex operator algebra $V = \coprod_{n\in\mathbf{N}} V_n$ is in class $\mathcal{C}$ if: (a) There exist $\omega_i \in V$ for $i = i, ..., 2r$ such that $\omega = \omega_1 + \cdots + \omega_{2r}$ where $r = \operatorname{rank} V$; each vertex operator $Y(\omega_i, z)$ generates a copy of the Virasoro algebra with central charge $\frac{1}{2}$ for $i = 1, ..., 2r$, and these Virasoro algebras are mutually commutative; (b)

The module for the $2r$ Virasoro algebras generated by $\mathbf{1}$ is isomorphic to the tensor product vertex operator algebra $L = L(\frac{1}{2}, 0)^{\otimes 2r}$. Let $V \in \mathcal{C}$ and W be a V-module. Since W is a finite sum of irreducible modules for the tensor product vertex operator algebra $L \subset V$ (cf. [7]), we have:

Proposition 3 *The module W has only finitely many submodules and there exists a finite filtration: $0 = W_0 \subset W_1 \subset W_2 \subset \cdots \subset W_n = W$ for some n such that each factor W_{i+1}/W_i is irreducible. Moreover, if we also assume that V is holomorphic with one-dimensional weight 0 subspace $V_0 \subset V$, then W is a direct sum of several copies of V and is completely reducible.*

3 The moonshine module

Let Λ be the Leech lattice. Recall from [12] that the untwisted space $V_\Lambda = (V_\Lambda, Y, \mathbf{1}, \omega)$ associated with Λ is a vertex operator algebra and that there is an automorphism θ of V_Λ of order 2. Also recall from [12] the θ-twisted module $V_\Lambda^T = (V_\Lambda^T, Y)$ on which θ acts. Then we have decomposition $V = V^+ \oplus V^-$ where $V^\pm = \{v \in V | \theta v = \pm v\}$ for $V = V_\Lambda$ or V_Λ^T. The moonshine module vertex operator algebra is defined to be $V^\natural = V_\Lambda^+ \oplus (V_\Lambda^T)^+$. It was proved in [12] that $V^\natural$ is a vertex operator algebra with the modular function J as generating function of the dimensions of its homogeneous spaces. Moreover, the group of grading preserving automorphisms of $V^\natural$ is the Monster simple group. In [2] and [3] we classified the irreducible modules and irreducible $\mathbf{Z}_2$-twisted modules for the vertex operator algebra associated with an even lattice. Applying these results to the vertex operator algebra V_Λ and using Proposition 3, we have:

Theorem 4 *(1) The vertex operator algebra V_Λ is holomorphic and any module is a finite direct sum of V_Λ.*
(2) The space V_Λ^T is the unique irreducible θ-twisted V_Λ-module and any θ-twisted module is a finite direct sum of V_Λ^T.
(3) The space V_Λ^+ is a vertex operator subalgebra of both V_Λ and $V^\natural$ and the spaces $V_\Lambda^\pm$, $(V_\Lambda^T)^\pm$ are irreducible modules for V_Λ^+.

In [7], we have found 48 Virasoro elements $\omega_i \in V_\Lambda^+$ for $i = 1, ..., 48$ such that the vertex operators $Y(\omega_i, z) = Y(\omega_i, z) \oplus Y_\theta(\omega_i, z)$ on $V = V_\Lambda \oplus V_\Lambda^T$ are generating functions of the Virasoro algebra with central charge $\frac{1}{2}$ where the operators $Y_\theta(v, z)$ for $v \in V_\Lambda$ define the θ-twisted module structure on V_Λ^T. These Virasoro algebras are mutually commutative and $\omega = \omega_1 + \cdots + \omega_{48}$. The following information on decomposition of V_Λ and V_Λ^T into the direct sum of irreducible modules for $L(\frac{1}{2}, 0)^{\otimes 48}$ is crucial in studying the representations of V_Λ^+ and $V^\natural$.

Proposition 5 *Let $W = V_\Lambda$ or V_Λ^T. Then as a $L = L(\frac{1}{2},0)^{\otimes 48}$-module, W is completely reducible:*

$$W = \coprod_{h_i \in \{0,\frac{1}{2},\frac{1}{16}\}} c_{h_1 \cdots h_{48}} L(h_1, ..., h_{48}),$$

where the nonnegative integer $c_{h_1 \cdots h_{48}}$ is the multiplicity of $L(h_1, ..., h_{48})$ in W. If $W = V_\Lambda$ then the all irreducible components $L(h_1, ..., h_{48}) \in S_u$; if $W = V_\Lambda^T$ then the all $L(h_1, ..., h_{48}) \in S_t$. Moreover, for any $1 \le i \le 24$, the multiplicity of $L(x_1^i, \cdots x_{48}^i) \subset V_\Lambda^-$ is 1 where $x_{2j-1}^i = x_{2j}^i = 0$ if $j \ne i$ and $x_{2i-1}^i = x_{2i}^i = \frac{1}{2}$.

4 Representations of V_Λ^+

Let $W = (W, Y_W)$ be an irreducible V_Λ^+-module and we have $W = \coprod_{i \in I} W_i$ where $W_i \simeq L(h_1^i, \cdots, h_{48}^i)$ for some $h_j^i \in \{0, \frac{1}{2}, \frac{1}{16}\}$. Using Proposition 5, we have:

Lemma 6 *(1) If there is $i \in I$ and $j \in \{1, ..., 24\}$ such that $(h_{2j-1}^i, h_{2j}^i) = (\frac{1}{16}, \frac{1}{16})$, then $W_{i'} \in S_u$ for any $i' \in I$.*
(2) If there is $i \in I$ and $j \in \{1, ..., 24\}$ such that one of h_{2j-1}^i or h_{2j}^i is $\frac{1}{16}$ and the other is not, then $W_{i'} \in S_t$ for any $i' \in I$.

We shall prove that if case (1) in Lemma 6 occurs, W is isomorphic to either V_Λ^+ or V_Λ^-, and if case (2) occurs, W is isomorphic to either $(V_\Lambda^T)^+$ or $(V_\Lambda^T)^-$. We achieve this by introducing a space $\bar{W}$ and by showing that $W \oplus \bar{W}$ on which V_Λ acts is either isomorphic to V_Λ or V_Λ^T.

Recall $W_1 = L(\frac{1}{2}, \frac{1}{2}, 0, \cdots, 0)$. Set $\bar{W} = \coprod_{i \in I} \bar{W}_i$ where the space $\bar{W}_i$ is defined to be $L(\bar{h}_1^i, \bar{h}_2^i, h_3^i, \cdots, h_{48}^i)$ such that there exists a nonzero intertwining operator of type $\begin{bmatrix} \bar{W}_i \\ W_1 \, W_i \end{bmatrix}$.

Lemma 7 *There exist unique nonzero intertwining operators Y_1 and Y_2 of types $\begin{bmatrix} \bar{W} \\ W_1 \, W \end{bmatrix}$ and $\begin{bmatrix} W \\ W_1 \, \bar{W} \end{bmatrix}$ such that $Y_1(w, z) W_i \subset \bar{W}_i[[z^{1/2}, z^{-1/2}]]$ and $Y_2(w, z) \bar{W}_i \subset W_i[[z^{1/2}, z^{-1/2}]]$ ($i \in I$) for $w \in W_1$ and for $u, v \in W_1$, $Y_1(u, z_1)$ and $Y(v, z_2)$ satisfy the Jacobi identity*

$$z_0^{-1}\delta\left(\frac{z_1 - z_2}{z_0}\right) Y_2(u, , z_1) Y_1(v, z_2) - z_0^{-1}\delta\left(\frac{z_2 - z_1}{-z_0}\right) Y_2(v, z_2) Y_1(u, z_1)$$

$$= z_2^{-1}\delta\left(\frac{z_1 - z_0}{z_2}\right) Y_W(Y(u, z_0)v, z_2)$$

in case (1) of Lemma 6 and the twisted Jacobi identity

$$z_0^{-1}\delta\left(\frac{z_1-z_2}{z_0}\right)Y_2(u,z_1)Y_1(v,z_2)-z_0^{-1}\delta\left(\frac{z_2-z_1}{-z_0}\right)Y_2(v,z_2)Y_1(u,z_1)$$

$$=z_2^{-1}\left(\frac{z_1-z_0}{z_2}\right)^{1/2}\delta\left(\frac{z_1-z_0}{z_2}\right)Y_W(Y(u,z_0)v,z_2)$$

in case (2) of Lemma 6, acting on W*, where we consider* W_1 *as a subspace of* V_Λ *and* $Y(u,z_0)$ *is the operator on* V_Λ.

Let $U=L(u_1,\cdots,u_{48})\subset V_\Lambda^+$ and $u\in U$. Then the restriction $Y(u,z)|_{W_i}$ for $i\in I$ is an intertwining operators of type $\begin{bmatrix} W \\ U\,W_i\end{bmatrix}$. By Theorem 2, there exist intertwining operators $Y_{\bar{W}_i}$ and $Y_{W_i,\bar{W}}$ of types $\begin{bmatrix} \bar{W} \\ U\,\bar{W}_i\end{bmatrix}$ and $\begin{bmatrix} \bar{W} \\ \bar{U}\,W_i\end{bmatrix}$ (where $\bar{U}\subset V_\Lambda^-$ is defined analogously) such that $Y_1(w,z_1)$, $Y_W(u,z_2)$, $Y_{\bar{W}_i}(u,z_2)$ and $Y_{W_i,\bar{W}}(Y(w,z_0)u,z_2)$ satisfy either the Jacobi identity or the twisted Jacobi identity for $u\in U$ and $w\in W_1$, and that $Y_{W_i,\bar{W}}(w,z)=Y_1(w,z)$. Thus we have linear maps $Y_{\bar{W}}(\cdot,z)\colon V_\Lambda^+\to \operatorname{End}\bar{W}[[z,z^{-1}]]$ and $Y_{W,\bar{W}}(\cdot,z)\colon V_\Lambda^-\to(\operatorname{Hom}(W,\bar{W}))[[z^{1/2},z^{-1/2}]]$ so that $Y_{\bar{W}}|_{\bar{W}_i}=Y_{\bar{W}_i}$ and $Y_{W,\bar{W}}|_{W_i}=Y_{W_i,\bar{W}}$. Since $Y_{\bar{W}}(u,z)$ $(u\in V_\Lambda^+)$ acting on $\bar{W}$ is an intertwining operator of type $\begin{bmatrix} \bar{W} \\ V_\Lambda^+\,\bar{W}\end{bmatrix}$, the action of V_Λ^+ on $\bar{W}$ satisfies all the axioms of a module for the vertex operator algebra V_Λ^+ except the Jacobi identity which is equivalent to commutativity and associativity.

Let $V=(V,Y,\mathbf{1},\omega)$ be an arbitrary vertex operator algebra and $M=(M,Y)$ be a V-module. Let $u_1,u_2\in V$. Then the Jacobi identity for $Y(u_1,z_1)$ and $Y(u_2,z_2)$ on W implies the commutativity relation

$$Y(u_1,z_1)Y(u_2,z_2)(z_1-z_2)^t=Y(u_2,z_2)Y(u_1,z_1)(z_2-z_1)^t$$

for some integer t (cf. [4] and [5]). Applying this to V_Λ^+ and its module W, we associate an integer t with any $u_1,u_2\in V_\Lambda^+$. By Theorem 2, there exist integers t_1,t_2 such that

$$Y_{\bar{W}}(u_1,z_1)Y(v,z_3)(z_1-z_3)^{t_1}=Y(v,z_3)Y(u_1,z_1)(z_3-z_1)^{t_1}$$

and

$$Y_{\bar{W}}(u_2,z_2)Y(v,z_3)(z_2-z_3)^{t_2}=Y(v,z_3)Y(u_2,z_2)(z_3-z_1)^{t_2}$$

as operators from W to $\bar{W}$. Thus we have

$$Y_{\bar{W}}(u_1,z_1)Y_{\bar{W}}(u_2,z_2)Y(v,z_3)(z_1-z_2)^t(z_1-z_3)^{t_1}(z_2-z_3)^{t_2}$$

$$= Y_{\bar{W}}(u_2, z_2)Y_{\bar{W}}(u_1, z_1)Y(v, z_3)(z_2 - z_1)^t(z_1 - z_3)^{t_1}(z_2 - z_3)^{t_2}$$

as operators from W to $\bar{W}$. The last identity gives the rationality of products, and commutativity for $Y_{\bar{W}}(u_1, z_1)$ and $Y_{\bar{W}}(u_2, z_2)$ on $\bar{W}$. Similarly, one can show the rationality of iterates, and associativity.

Lemma 8 *The space $\bar{W} = (\bar{W}, Y_{\bar{W}})$ is an irreducible module for V_Λ^+.*

Set $M = W \oplus \bar{W}$. Then M is a module for V_Λ^+. We next define an action of V_Λ^- on M which interchanges W and $\bar{W}$ and show that M is either isomorphic to V_Λ or V_Λ^T. The action of V_Λ^- on W has been defined already so we need only define the action on $\bar{W}$. Since V_Λ^- is an irreducible V_Λ^+-module, the action of V_Λ^- on M is determined uniquely by the actions of V_Λ^+ and $W_1 \subset V_\Lambda^-$. We have:

Lemma 9 *There exists a unique intertwining operator $Y_{\bar{W},W}$ of type $\begin{bmatrix} W \\ V_\Lambda^- \, \bar{W} \end{bmatrix}$ such that the operators $Y_W(u, z_1)$, $Y_{\bar{W}}(u, z_1)$, $Y_2(w, z)$ and $Y_{\bar{W},W}(Y(u, z_0)w, z_2)$ satisfy the Jacobi identity acting on $\bar{W}$ for $u \in V_\Lambda^+$ and $w \in W_1$, and that $Y_{\bar{W},W}(w, z) = Y_2(w, z)$.*

We now have a linear map from V_Λ to $(\operatorname{End} M)[[z^{1/2}, z^{-1/2}]]$ given by $Y(u, z)|_W = Y_W(u, z)$, $Y(u, z)|_{\bar{W}} = Y_{\bar{W}}(u, z)$, $Y(v, z)|_W = Y_{W,\bar{W}}(v, z)$ and $Y(v, z)|_{\bar{W}} = Y_{\bar{W},W}(v, z)$ for $u \in V_\Lambda^+$ and $v \in V_\Lambda^-$. Then for $v \in V_\Lambda^-$, $Y(v, z) = \sum_{n\in\mathbf{Z}} v_n z^{-n-1}$ if case (1) of Lemma 6 occurs, and $Y(v, z) = \sum_{n\in\frac{1}{2}+\mathbf{Z}} v_n z^{-n-1}$ if case (2) occurs. Again the action of V_Λ on M satisfies all the axioms of a module (or twisted module) for V_Λ except the Jacobi identity (or twisted Jacobi identity).

Using Theorem 2, Lemmas 7, 8 and 9 we have

Lemma 10 *If case (1) occurs, (Y, M) is an irreducible V_Λ−module. If the case (2) occurs, (Y, M) is an irreducible θ-twisted V_Λ-module.*

The following theorem is an immediate corollary of Theorem 4 and Lemma 10.

Theorem 11 *If case (1) occurs, W is isomorphic to V_Λ^+ if $L(0, \cdots, 0) \subset W$, and isomorphic to V_Λ^- if $L(0, \cdots, 0) \not\subset W$. If case (2) occurs, W is isomorphic to $(V_\Lambda^T)^+$ if W is $\mathbf{Z}$-graded, and isomorphic to $(V_\Lambda^T)^-$ if W is not $\mathbf{Z}$-graded.*

We have classified all irreducible modules for the vertex operator algebra V_Λ^+. Next we discuss the complete reducibility of an arbitrary module W for V_Λ^+. Let $W = \coprod_{i\in I} W_i$ be an arbitrary V_Λ^+-module. Set $I_2 = \{i \in$

$I|W_i \subset V_\Lambda^T\}$ and $I_1 = I \backslash I_2$. Also set $W^j = \coprod_{i \in I_j} W_i$ for $j = 1, 2$. Then each W_j is a module for V_Λ^+ and $W = W^1 \oplus W^2$. Note that the above procedure for constructing $\bar{W}$ and M from W works for W^j also. Thus we get V_Λ^+-modules $\bar{W}^j$ and $M^j = W^j \oplus \bar{W}^j$. Then M^1 is a V_Λ-module and M^2 is a θ-twisted V_Λ-module. By Theorem 4, M^1 is a finite direct sum of V_Λ, and M^2 is a finite direct sum of V_Λ^T. Therefore we have:

Theorem 12 *The space W^1 is a finite direct sum of V_Λ^+ and V_Λ^- and W^2 is a finite direct sum of $(V_\Lambda^T)^+$ and $(V_\Lambda^T)^-$. In particular, W is completely reducible.*

5 Representations of $V^\natural$

The main theorem of this announcement is the following:

Theorem 13 *Any irreducible module for $V^\natural$ is isomorphic to the adjoint module $V^\natural$. Moreover, any module is a finite direct sum of $V^\natural$ and is completely reducible.*

The second assertion follows from the first assertion and Proposition 2. We sketch the proof of the first assertion. Let W be an irreducible $V^\natural$-module. By Theorems 11 and 12, $W = c_1 V_\Lambda^+ \oplus c_2 V_\Lambda^- \oplus c_3 (V_\Lambda^T)^+ \oplus c_4 (V_\Lambda^T)^-$ where the nonnegative integers c_j are the multiplicities. If $c_4 \neq 0$ then there exist nonzero vectors $u \in (V_\Lambda^T)^+ \subset V^\natural$ and $v \in (V_\Lambda^T)^-$ such that $Y(u, z)v \subset (c_1 V_\Lambda^+ + c_2 V_\Lambda^-)[[z^{1/2}, z^{-1/2}]]$. However, V_Λ is **Z**-graded and $(V_\Lambda^T)^-$ is not **Z**-graded. This contradicts to the irreducibility of W. Similarly, $c_2 = 0$. Clearly $c_1 \neq 0$. In particular, $L(0, \cdots, 0) \subset W$ and therefore $V^\natural \subset W$. Since W is irreducible, $W = V^\natural$.

I would like to thank A. Feingold, I. Frenkel, J. Lepowsky, G. Mason, R. Wilson and Y. Zhu for useful discussions.

REFERENCES

1. R. E. Borcherds, Vertex algebras, Kac-Moody algebras, and the Monster, *Proc. Natl. Acad. Sci. USA* **83** (1986), 3068-3071.
2. C. Dong, Vertex algebras associated with even lattices, *J. Algebra* **160** (1993), 245-265.
3. C. Dong, Twisted modules for vertex algebras associated with even lattices, *J. Algebra* **165** (1994), 90-112.
4. C. Dong and J. Lepowsky, Abelian intertwining algebras – a generalization of vertex operator algebras, *Proc. Symp. Pure. Math., American Math. Soc.* **56** II (1994), 261-294.

5. C. Dong and J. Lepowsky, Generalized Vertex Algebras and Relative Vertex Operators, *Progress in Math.* Vol. 112, Birkhäuser, Boston, 1993.

6. C. Dong and G. Mason, On the construction of the moonshine module as a $\mathbf{Z}_p$-orbifold, in this volume, *Contemporary Math.*

7. C. Dong, G. Mason and Y. Zhu, Discrete series of the Virasoro algebra and the moonshine module, *Proc. Symp. Pure. Math., American Math. Soc.* **56** II (1994), 295-316.

8. R. Dijkgraaf, C. Vafa, E. Verlinde and H. Verlinde, The operator algebra of orbifold models, *Comm. Math. Phys.* **123** (1989), 485-526.

9. A. J. Feingold, I. B. Frenkel and J. F. X. Ries, Spinor construction of vertex operator algebras, triality and $E_8^{(1)}$, *Contemporary Math.* **121,** 1991.

10. I. B. Frenkel, Y.-Z. Huang and J. Lepowsky, On axiomatic approaches to vertex operator algebras and modules, *Memoirs American Math. Soc.* **494,** 1993.

11. I. B. Frenkel, J. Lepowsky and A. Meurman, A natural representation of the Fischer-Griess Monster with modular function J as character, *Proc. Natl. Acad. Sci. USA* **81** (1984), 3256-3260.

12. I. B. Frenkel, J. Lepowsky and A. Meurman, Vertex Operator Algebras and the Monster, *Pure and Applied Math.,* Vol. **134**, Academic Press, 1988.

13. I. B. Frenkel and Y. Zhu, Vertex operator algebras associated to representations of affine and Virasoro algebras, *Duke Math. J.* **66** (1992), 123-168.

14. D. Friedan, Z. Qiu and S. Shenker, Details of the non-unitarity proof for highest weight representations of Virasoro algebra, *Comm. Math. Phys.* **107** (1986), 535-542.

15. P. Goddard, A. Kent and D. Olive, Unitary representations of the Virasoro and Super-Virasoro algebras, *Comm. Math. Phys.* **103** (1986), 105-119.

16. J. G. Thompson, Some numerology between the Fischer-Griess Monster and elliptic modular functions, *Bull. London Math. Soc.* **11** (1979), 352-353.

Department of Mathematics, University of California, Santa Cruz, CA 95064

E-mail address: dong@cats.ucsc.edu

Contemporary Mathematics
Volume 175, 1994

The construction of the Moonshine Module as a Z_p-orbifold

Chongying Dong[1] Geoffrey Mason[2]

Abstract. We discuss our recent work (currently being prepared for publication) on the Monstrous Moonshine Module first constructed by Frenkel-Lepowsky-Meurman. As is well-known, the work of FLM can be loosely described as the construction of the conformal field theory $V_{\Lambda/\mathbb{Z}_2}$ based on a $\mathbb{Z}_2$-orbifold $T^\Lambda/\mathbb{Z}_2$ where Λ is the Leech lattice and $T^\Lambda = \mathbb{R}^{24}/\Lambda$. We consider

(i) Conformal field theory $V_{\Lambda/\mathbb{Z}_p}$ based on certain $\mathbb{Z}_p$-orbifolds of T^Λ, including a completely rigorous construction for $p = 3$.

(ii) The identification $V_{\Lambda/\mathbb{Z}_2} \simeq V_{\Lambda/\mathbb{Z}_3}$ and characterization of the Moonshine Module as an irreducible module for the affine Griess algebra.

1 Introduction

In April of 1992 we circulated an announcement [11] of results concerning conformal field theory on the orbifolds $T^\Lambda/\mathbb{Z}_p$ for various symmetry groups $\mathbb{Z}_p$ of Λ of prime order p. Because the complete proofs of our results are so long and technically complex, it may be a while before they are fit for publication. We therefore thought it a good idea to publish a discussion of our work which is somewhat fuller than [11]. This will also

1991 Mathematics Subject Classification. Primary 17B65, 17B68, 20C34
[1]Supported by Regents of the University of California and by NSA grant MDA904-92-H-3099.
[2]Supported by NSF grant DMS-9122030.
The detailed version of this paper will be submitted for publication elsewhere.

allow us to include some refinements obtained since [11] was circulated, and which pertain to item (ii) in the abstract.

Researchers from various disciplines and with somewhat diverse mathematical backgrounds are concerned with the kind of phenomenon dealt with here – we have in mind group-theorists, Lie algebraists and physicists – and accordingly we have tried to make our exposition understandable to a wide audience. At bottom, however, our work is but an example – albeit a particularly intricate one – of something common to all three disciplines, namely the search for symmetry.

We have not mentioned the Monster group M, but it plays a critical rôle throughout our work. In fact our interests reside as much in understanding M as they do in the conformal field theory; indeed, the two are inseparable.

For the rest of this introduction we will anticipate some definitions and language to be reviewed later in order to explain our main results. Thus we use the notion of *vertex (operator) algebra* (VOA) as introduced and used in [1] and [13] as a rigorous algebraic formulation of conformal field theory. In particular, if Λ is the Leech lattice, propagation of a bosonic string on the Leech torus $\mathbb{R}^{24}/\Lambda$ corresponds to a VOA V_Λ. Furthermore V_Λ is holomorphic in the sense that V_Λ has a unique irreducible module (namely itself) (see [7]).

In general, given a VOA V and a (finite) group G of automorphisms of V, the theory of *orbifolds* concerns itself with the VOA V^G of G-invariants of V, together with a description of the irreducible modules for V^G. These are closely related to the *g-twisted modules* for V and $g \in G$ (see [8]). The theory of orbifolds was initiated in [6] and continued in [5]. But from a mathematical point-of-view these papers are heuristic, and the general theory remains a web of beautiful conjectures of which very little has been untangled.

Suppose, for example, that V is a holomorphic theory and that $G \simeq \mathbb{Z}_n$ is a cyclic group of automorphisms of V. Then one believes that V^G has exactly n^2 irreducible modules, indexed by the elements of $G \times G$ in a natural way. Moreover the indexing should be such that if $V(x, y)$ is the irreducible V^G-module corresponding to $(x, y) \in G \times G$ then the direct sum $\oplus_y V(x, y)$ for fixed x should be the unique irreducible x-twisted module for V.

Now it is well known that if p is a prime and $p-1$ divides 24 then Λ admits an automorphism of order p, call it g_p, such that g_p fixes no non-zero elements of Λ. We are interested in the theory of the VOA $V_\Lambda^{\langle g_p \rangle}$. In the case of a general theory based on a lattice, one knows [16] how to construct a family of twisted modules, though their uniqueness if the lattice is self dual remains an open question. In particular, we can construct x-twisted irreducible modules for V_Λ and each $x \in \langle g_p \rangle$. Denote them by $V_\Lambda^{(x)}$. Moreover from the construction there is an action of a group, naturally identified with $\langle g_p \rangle$, on $V_\Lambda^{(x)}$. Then the eigenspaces of $\langle g_p \rangle$ on $V_\Lambda^{(x)}$ become irreducible $V_\Lambda^{\langle g_p \rangle}$-modules which may be identified with $V_\Lambda(x, y)$ above. Thus $V_\Lambda^{(x)} = \oplus_y V_\Lambda(x, y)$.

Consider, for example, the case $p = 2$, so that g_p is just the -1 automorphism of Λ (in this case, by the way, we know by [9] that the spaces $V_\Lambda(x, y)$ are the *only* irreducible $V_\Lambda^{\langle g_2 \rangle}$-modules). With an obvious notational change we have

$$\begin{aligned} V_\Lambda = V_\Lambda^{(+)} &= V_\Lambda(+,+) \oplus V_\Lambda(+,-) \\ V_\Lambda^{(-)} &= V_\Lambda(-,+) \oplus V_\Lambda(-,-). \end{aligned} \tag{1}$$

As a graded vector space, the Moonshine Module may be defined as

$$V_{\Lambda/\mathbb{Z}_2} = V_\Lambda(+,+) \oplus V_\Lambda(-,-).$$

($V_{\Lambda/\mathbb{Z}_2}$ is denoted by $V^\natural$ in [13].) Thus it is a priori the sum of a VOA $V_\Lambda(+,+) = V_\Lambda^{\langle g_2 \rangle}$ together with one of its modules $V_\Lambda(-,-)$. The main point in [13] concerning $V_{\Lambda/\mathbb{Z}_2}$ is that it can itself be made into a VOA. That the Monster M acts on $V_{\Lambda/\mathbb{Z}_2}$ is, in their approach, not merely an extrinsic property of $V_{\Lambda/\mathbb{Z}_2}$ but something inextricably bound up with the construction of $V_{\Lambda/\mathbb{Z}_2}$ as a VOA.

Now consider the more general case where $p-1$ divides 24. One defines

$$V_{\Lambda/\mathbb{Z}_p} = V_\Lambda^{\langle g_p \rangle} \oplus \left(\oplus_{k=1}^{p-1} V_\Lambda^{T,k} \right) \tag{2}$$

where $V_\Lambda^{T,k}$ $(k = 1, \ldots, p-1)$ are a certain set of $V_\Lambda^{\langle g_p \rangle}$-modules described more carefully below. We refer to them loosely as the "twisted sectors." Then we expect that $V_{\Lambda/\mathbb{Z}_p}$ can be made into a VOA, and we can prove this in case $p = 3$.

This is achieved by showing that $V_{\Lambda/\mathbb{Z}_p}$ also admits the Monster M as a group of automorphisms, and it will be worthwhile to stop and consider M for a while in this context. It is well known (cf. [17], [3]) that for each of the primes p, the Monster has a subgroup C_p which can be described as the middle term of a short exact sequence

$$1 \to p^{1+2d} \to C_p \to G_p \to 1. \tag{3}$$

Here we have set $24 = (p-1)2d$ $(d \in \mathbb{Z})$; p^{1+2d} is a so-called extra-special p-group of the indicated order – essentially a finite Heisenberg group defined over $GF(p)$ and corresponding to a $GF(p)$-symplectic space of dimension $2d$; and G_p is the group of elements of $\mathrm{Aut}(\Lambda)$ which commute with g_p, modulo $\langle g_p\rangle$. Note that (3) is a non-split extension if $p = 2$ or 3. These subgroups of M were discovered by group theorists even before M was actually constructed by Griess [14], and our work can be thought of as ascribing a geometric meaning to C_p. Namely, we expect that C_p is the subgroup of M which preserves each of the twisted sectors in (2). Again for $p = 3$, this is a theorem.

One proceeds by first carefully constructing a faithful, degree-preserving action of C_p on the space $V_{\Lambda/\mathbb{Z}_p}$ in such a way that the center of C_p, which is exactly the center of p^{1+2d}, and hence isomorphic to $\mathbb{Z}_p$ and identifiable with $\langle g_p\rangle$, acts faithfully on $V_\Lambda^{\langle g_p\rangle}$ and has the twisted sectors as its non-trivial eigenspaces.

Next (this is the hard part!) we want to define a further symmetry σ_p of $V_{\Lambda/\mathbb{Z}_p}$ which mixes the twisted sectors and the "untwisted" sector $V_\Lambda^{\langle g_p\rangle}$. Of course, there are no end of ways to do this if we consider $V_{\Lambda/\mathbb{Z}_p}$ only as a vector space, but we demand much more of σ_p : remembering that $V_\Lambda^{\langle g_p\rangle}$ is a VOA in its own right and that $V_{\Lambda/\mathbb{Z}_p}$ is a $V_\Lambda^{\langle g_p\rangle}$-module, we use the symmetry σ_p to transport this structure to the image of $V_\Lambda^{\langle g_p\rangle}$ under σ_p. Together with C_p, which by dint of its construction is a group of automorphisms of the VOA $V_\Lambda^{\langle g_p\rangle}$, this allows us to manufacture vertex operators $Y(v, z)$ for all v in $V_{\Lambda/\mathbb{Z}_p}$. It is a stringent demand upon σ_p that now the "Jacobi identity" [13] should hold, which is the main ingredient for establishing that $V_{\Lambda/\mathbb{Z}_p}$ is a VOA.

Remarkably, it seems that there is a *canonical* choice of σ_p which makes this procedure work. In his original work [14], Griess constructed M by adjoining to C_2 an astutely chosen additional element σ. Here $\langle C_2, \sigma\rangle$ is a group of transformations of the Griess algebra B, essentially the subspace of $V_{\Lambda/\mathbb{Z}_2}$ of conformal weight 2, which was the only part of $V_{\Lambda/\mathbb{Z}_2}$ that Griess had available. He succeeded in proving that $\langle C_2, \sigma\rangle \simeq M$. Then FLM showed essentially that σ can be extended to a symmetry σ_2 of $V_{\Lambda/\mathbb{Z}_2}$, whence $M = \langle C_2, \sigma\rangle$ acts on $V_{\Lambda/\mathbb{Z}_2}$ by Griess' work.

If p is odd, the element σ_p, although canonical, is not an extension of σ per se, but is merely analogous to it. For one thing σ_p has order 4 if p is odd, whereas σ has order 2. The ultimate reason for this is that the Weyl element $w = \begin{pmatrix} 0 & -1 \\ 1 & 0 \end{pmatrix}$ of $SL_2(p)$ has order 4 if p is odd, and order 2 if $p = 2$. We outline below the tortuous passage by which w can be souped-up to a suitable operator σ_p on $V_{\Lambda/\mathbb{Z}_p}$. At least, we expect that this is true. Unfortunately, the amount of work involved in achieving this increases with p. While no conceptual barriers remain, we have only completed the calculation in case $p = 3$.

At this point, we will have a group $\langle C_p, \sigma_p \rangle$ of linear automorphisms of $V_{\Lambda/\mathbb{Z}_p}$ (although we do not yet know that it is the Monster) and $V_{\Lambda/\mathbb{Z}_p}$ will be known to be a VOA, although we do not know yet that it is the same VOA as as the Moonshine Module $V_{\Lambda/\mathbb{Z}_2}$.

Now we can calculate without too much difficulty the graded character of $V_{\Lambda/\mathbb{Z}_p}$, and it is as expected, namely the modular function $J(q)$ with constant term 0 :

$$J(q) = q^{-1} + 196884q + \cdots . \tag{4}$$

As is well known [13], since $V_{\Lambda/\mathbb{Z}_p}$ is VOA it confers on the conformal weight 2 subspace B_p (of dimension 196884) the structure of a commutative, not necessarily associative algebra of which $\langle C_p, \sigma_p \rangle$ is a group of algebra automorphisms. But for odd p, the structure constants for the algebra B_p are so convoluted that it appears hopeless to attempt a direct identification of B_p with the Griess algebra B. If we could do this then one can prove easily that $\langle C_p, \sigma_p \rangle \simeq M$, but in practice we must turn the tables. Taking $p = 3$, we first employ some nontrivial group theory (the classification of finite simple groups, as well as some of its methodology) to prove that indeed $\langle C_3, \sigma_3 \rangle \simeq M$. Note that even the finiteness of $\langle C_3, \sigma_3 \rangle$ is in doubt, though we pass over this issue here.

Having identified $\langle C_3, \sigma_3 \rangle$ with M, this *forces* $B_3 \simeq B$ since there is essentially only one nontrivial M-invariant structure of commutative algebra possible on a faithful M-module of dimension 196884.

How can we prove that the VOAs $V_{\Lambda/\mathbb{Z}_3}$ and $V_{\Lambda/\mathbb{Z}_2}$ are isomorphic? FLM have made a general conjecture that says that any holomorphic VOA with graded character (4) is necessarily the Moonshine Module. In lieu of having this result available (if we did then we would not need to construct M acting on $V_{\Lambda/\mathbb{Z}_3}$), we use the following result:

Theorem *Let $\hat{B} = B \otimes \mathbb{C}[t, t^{-1}] \otimes \mathbb{C}e$ be the affinization of the Griess algebra B (see [13]). Then the Moonshine Module is the unique irreducible $\hat{B}$-module $\coprod_{n \geq -1} V_n$ for which $\dim V_{-1} = 1$.*

Given this result, the proof of which will appear elsewhere, we can immediately identify $V_{\Lambda/\mathbb{Z}_2}$ with $V_{\Lambda/\mathbb{Z}_3}$ since we have gone to some lengths to show that in both VOAs the weight 2 subspaces can be identified with B, and the hypotheses of the theorem apply to both VOAs, as can be easily seen. Hence

Corollary *There is an M-equivariant isomorphism of VOAs*

$$V_{\Lambda/\mathbb{Z}_2} \simeq V_{\Lambda/\mathbb{Z}_3}.$$

Vafa has discussed in [20] the philosophy of isomorphisms of conformal field theories and its relation to mirror geometry. The corollary is a non-trivial example of such a phenomenon. Moreover, there should be analogues of these results for many elements of M, not just those of order p. A heuristic discussion of what one may expect, from a string-theoretic point of view, may be found in [19].

In the rest of the paper we will enlarge on some of the main ideas which arise in the course of establishing these results.

2 p-Golay codes and the group C_p

We continue with previous notation, in particular, p is a prime such that $p-1$ divides 24. So $p = 2, 3, 5, 7$ or 13. Also let $24 = (p-1)2d$, so that $d = 12, 6, 3, 2$, or 1 respectively.

The rôle that certain binary codes play in the construction of both lattices and conformal field theories is well known [4], [13]. In particular, the binary Golay code is crucial in studying V_Λ. In our case certain codes define over $GF(p)$ are important – they are the p-analogues of the binary Golay code.

Proposition 1 *Up to isomorphism there is a unique d-dimensional subspace $\mathcal{C}(p) \subset GF(p)^{2d}$ such that*

(a) $\mathcal{C}(p)$ is totally isotropic with respect to the usual dot product;

(b) $\mathcal{C}(p)$ has minimal weight $m = 8$ if $p = 2$, d if $p > 2$;

(c) $\mathcal{C}(p)$ contains a vector w of weight $2d$.

(Every element of $\mathcal{C}(p)$ is a $2d$-tuple $x = (x_1, \cdots, x_{2d})$ with $x_i \in GF(p)$. The weight $\omega(x)$ of x is the number of x_i not equal to zero; the weight of $\mathcal{C}(p)$ is minimum of $\omega(x)$ for $x \in \mathcal{C}(p) - \{0\}$.) $\mathcal{C}(2)$ is just the binary Golay code, and $\mathcal{C}(3)$ the ternary Golay code.

It is convenient now to use the language of algebraic number theory [2]. Fix a primitive pth root of unity λ, and let $K = \mathbb{Q}(\lambda)$, $D = \mathbb{Z}[\lambda]$ the ring of algebraic integers in K. Then $P = D\theta$ is the unique prime ideal in D which contains p, where $\theta = 1 - \lambda$, and p is totally ramified in D. Thus $D/P \simeq GF(p)$.

A real algebraic integer in D has a unique expression of the form

$$x = n_0 + \sum_{i=2}^{\frac{p-1}{2}} n_i(\lambda^i + \bar{\lambda}^i), \quad n_i \in \mathbb{Z}.$$

We define $f : D \cap \mathbb{R} \to \mathbb{Z}$ via

$$f(x) = n_0.$$

Then for $v = (v_1, \cdots, v_{2d}) \in D^{2d}$ we define a rational quadratic form $q : D^{2d} \to \mathbb{Q}$ via

$$q(v) = 2p^{-kd/6} \sum_{i=1}^{2d} f(v_i \bar{v}_i), \tag{5}$$

where $k = 2$ if $p \leq 5$ and $\frac{p-1}{2}$ if $p > 5$. Because $D/P \simeq GF(p)$, $GF(p)$ is naturally a D-module and there is a D-module epimorphism $D^{2d} \to GF(p)^{2d}$. Fix a section α of this map, so that in particular $\alpha(C) \in D^{2d}$ for $C \in \mathcal{C}$. Define D-submodules of D^{2d} as follows:

$$N_p = D - \text{span}\langle (P^k)^{2d} \cup \{\theta^{k-1}\alpha(C), C \in \mathcal{C}(p)\}\rangle$$

$$\Lambda_p = D - \text{span}\langle \{x \in (P^k)^{2d} | x \cdot \alpha(w) \in P^{k+1}\}$$
$$\cup \{\theta^{k-1}\alpha(C), C \in \mathcal{C}(p), \theta^{k-2}\alpha'(w)\}\rangle.$$

Here we have set

$$\alpha'(w) = \begin{cases} (-3, 1^{23}) & \text{if } p = 2 \\ (4, 1^{11}) & \text{if } p = 3 \\ \alpha(w) & \text{if } p > 3. \end{cases}$$

If we consider N_p and Λ_p as Z-modules (of rank 24) equipped with the quadratic form q (5) by restriction then we have

Proposition 2 *(N_p, q) is isometric to the Niemeier lattice of type A_{p-1}^{2d}*

Proposition 3 *(Λ_p, q) is isometric to the Leech lattice.*

Proposition 2 means that N_p is the (unique) self dual, positive definite integral lattice of rank 24 whose short vectors x (i.e., those satisfying $q(x) = 2$) form a root system of type A_{p-1}^{2d}. See Venkov's paper [21] reproduced in [4]. In fact $Q = (P^k)^{2d} \subset N_p$ is the root lattice of type A_{p-1}^{2d}. This result facilitates our entré into Lie algebras, and since $|\Lambda_p : \Lambda_p \cap N_p| = |N_p : \Lambda_p \cap N_p| = p$, Λ_p and N_p are intrinstely related. Observe also that since both are D-modules, multiplication by λ is an automorphism of each lattice of order p (λ leaves q invariant) leaving only the zero vector fixed. Denote this automorphism by t_p.

We turn now to a description of the group C_p, where for the sake of uniformity it is convenient to take p odd. Let β be the bilinear form corresponding to the quadratic from q (5), and following [16] define a map $\epsilon : \Lambda_p \times \Lambda_p \to GF(p)$ via

$$\epsilon(x, y) = \beta(\sum_{p/2<n<p} n\lambda^{-n}x, y) \pmod{p}.$$

As ϵ is $\mathbb{Z}$-bilinear it defines ipso facto a 2-cocycle on Λ_p with values in the trivial Λ_p-module $GF(p)$, and thereby a central extension

$$1 \to GF(p) \to \hat{\Lambda}_p \bar{\to} \Lambda_p \to 1.$$

Let $H_p = \mathrm{Aut}_D(\Lambda_p)$, the group of automorphisms of the D-module Λ_p, also equal to the subgroup of $\mathrm{Aut}_{\mathbb{Z}}(\Lambda)$ which commutes with t_p. Then H_p preserves ϵ and induces a natural group of automorphisms of $\hat{\Lambda}_p$, so we can form the split extension $H_p \cdot \hat{\Lambda}_p$. (This is where we need p odd – such a group does not exist if $p = 2$!) Now the center of $\hat{\Lambda}_p$ can be identified with $GF(p) \times P\Lambda_p$, and contains the subgroup $K_p = [\hat{\Lambda}_p, t_p]$. So $K_p \trianglelefteq \hat{\Lambda}_p$ and K_p is stable under the action of H_p, that is $K_p \trianglelefteq H_p \cdot \hat{\Lambda}_p$. We set

$$\hat{C}_p = H_p \cdot \hat{\Lambda}_p / K_p.$$

Now $E_p = \hat{\Lambda}_p / K_p \simeq p^{1+2d}$ and the center of $\hat{C}_p$ can be identified with $Z(E_p) \times \langle t_p \rangle \simeq \mathbb{Z}_p \times \mathbb{Z}_p$. We chose a subgroup of order p in this group distinct from $\langle t_p \rangle$ and $Z(E_p)$. Then C_p is the quotient of $\hat{C}_p$ by this central subgroup of order p. We have the following diagram giving rise to (3):

$$\begin{array}{ccccccccccc}
 & & & & 1 & & & & & & \\
 & & & & \downarrow & & & & & & \\
 & & & & K_p & & & & & & \\
 & & & & \downarrow & & & & & & \\
1 & \to & GF(p) & \to & \hat{\Lambda}_p & \to & \Lambda & \to & 1 & & \\
 & & & & \downarrow & & & & & & \\
 & & 1 & \to & E_p & \to & \hat{C}_p & \to & H_p & \to & 1 \\
 & & & & \downarrow & \searrow & \downarrow & & \downarrow & & \\
 & & & & 1 & & C_p & \to & G_p & \to & 1
\end{array}$$

3 Construction of the p-Moonshine Module as C_p-module

Let $A_0 = \langle x \in (P^k)^{2d}, \theta^{k-1}\alpha(w) | x \cdot \alpha(w) \in P^{k+1} \rangle$, a subgroup of Λ_p, and A the inverse image of A_0 under the map $\hat{\Lambda}_p \bar{\to} \Lambda_p$. Then A is a maximal abelian subgroup of $\hat{\Lambda}_p$ which projects onto a maximal abelian subgroup of E_p. For $j = 1, ..., p-1$ we choose (very carefully!) a 1-dimensional A-module L_j such that the restriction of L_j to a fixed generator f of $GF(p) \subset Z(\hat{\Lambda}_p)$ is multiplication by λ^j. Then the spaces

$$T_j = \mathbb{C}[\hat{\Lambda}_p] \otimes_{\mathbb{C}[A]} L_j$$

for $j = 1, ..., p-1$ are a complete set of inequivalent, faithful irreducible E_p-modules, since K_p acts trivially by the construction. One can show that each T_j prolongs to a $\hat{C}_p$-module on which t_p acts trivially.

We now write down a large graded space, then proceed to explain its meaning (q is a formal variable):

$$W_p = q^{-1} S\left(\sum_{n=1}^{\infty} \mathbf{h} q^n\right) \otimes \mathbb{C}[\Lambda_p] + \sum_{j=1}^{p-1} q^{1/p} S\left(\sum_{n=1}^{\infty} \mathbf{h}_{nj} q^{n/p}\right) \otimes T_j. \qquad (6)$$

Here $\mathbb{C}[\Lambda_p]$ is the group algebra of Λ_p with a basis $\{e^y q^{\beta(y,y)/2} | y \in \Lambda_p\}$. $\mathbb{C}[\Lambda_p]$ is a graded $\hat{C}_p$-module in which f acts trivially and the action of $\hat{C}_p$ factors through the action of $H_p \cdot \Lambda_p$ where $(h, x) : e^y \mapsto \lambda^{\epsilon(x,y)-\epsilon(y,x)} e^{h(y)}$ for $x, y \in \Lambda_p$ and $h \in H_p$. Next we have $\mathbf{h} = \mathbb{C} \otimes_{\mathbb{Z}} \Lambda_p$ of dimension 24, $S(\sum_{n=1}^{\infty} \mathbf{h} q^n)$ the symmetric algebra on the direct sum of copies of $\mathbf{h}$, one in each dimension. Then $\hat{C}_p$ acts on $\mathbf{h}$ naturally (with kernel E_p), and hence on the symmetric algebra too. Let $\mathbf{h}_r$ be the λ^r-eigenspace of t_p on $\mathbf{h}$, which is also a $\hat{C}_p$-module. Since T_j is too, we have shown how to turn W_p into a $\hat{C}_p$-module graded by $q^{1/p}$ – that is a formal sum of $\hat{C}_p$-modules considered as an element of $R\hat{C}_p[[q^{1/p}, q^{-1/p}]]$ where $R\hat{C}_p$ is the Grothendieck group of $\hat{C}_p$-modules. The desired space $V_{\Lambda_p/\mathbb{Z}_p}$ is defined to be the subspace of W_p of the fixed-points of $\langle f t_p \rangle = \ker(\hat{C}_p \to C_p)$. So if we define

$$V_{\Lambda/\mathbb{Z}_p} = W_p^{\langle f t_p \rangle} \qquad (7)$$

then $V_{\Lambda/\mathbb{Z}_p}$ is a q-graded C_p-module. If we set

$$V_{\Lambda/\mathbb{Z}_p} = \sum_{n \in \mathbb{Z}} V_n q^n$$

then we have

Proposition 4 *The graded character* $\sum_{n \in \mathbb{Z}} \dim V_n q^n$ *is the modular function* $J(q) = q^{-1} + 196884q + \cdots$.

4 Vertex operator algebras and modules

The general definition of vertex (operator) algebra can be found in [1] and [13]. We have a $\mathbb{Z}$-graded vector space $V = \coprod_{n \in \mathbb{Z}} V_n$, $\dim V_n$ finite, equipped with a linear map $V \to (\operatorname{End} V)[[z, z^{-1}]]$. The image of $v \in V$ is a *vertex operator* $Y(v, z) = \sum_{n \in \mathbb{Z}} v_n z^{-n-1}$, $v_n \in \operatorname{End} V$. The main axiom (the *Jacobi identity*) is a complicated identity essentially measuring deviation from commutivity of the operators u_m, v_n for $u, v \in V$. We will not reproduce the identity here. An automorphism g of V is a linear automorphism such that $gY(v,z)g^{-1} = Y(gv,z)$ for all $v \in V$ and such that g also leaves certain distinguished vectors invariant. Similarly, for an automorphism g of the VOA V of finite order r, a g-twisted module for V is a $\mathbb{Q}$-graded vector space W with linear map $V \to (\operatorname{End} W)[[z^{1/r}, z^{-1/r}]]$, $v \mapsto Y(v,z)$ which is required to satisfy a g-twisted Jacobi identity (see [8] and [12]). If we take $g = 1$, we obtain the definition of a V-module.

Given an even lattice L, it is shown in [13] and [1] how to construct a vertex algebra V_L, moreover if L is positive definite and self dual then V_L has a unique irreducible module, namely itself (see [7]). We briefly recall the construction relevant for us, namely the VOA V_{Λ_p}.

With $\mathbf{h} = \mathbb{C} \otimes_{\mathbb{Z}} \Lambda_p$ as before, extend the bilinear form β to $\mathbf{h}$ by $\mathbb{C}$-linearity. Then $\hat{\mathbf{h}} = \mathbf{h} \otimes \mathbb{C}[t,t^{-1}] \oplus \mathbb{C}c$ is a Lie algebra with $[c, \hat{\mathbf{h}}] = 0$ and $[x \otimes t^m, y \otimes t^n] = m\langle x,y\rangle \delta_{m+n,0}c$ for $x, y \in \mathbf{h}$ and $m, n \in \mathbb{Z}$. Set $\hat{\mathbf{h}}^{\pm} = \mathbf{h} \otimes t^{\pm}\mathbb{C}[t^{\pm}]$ and make $\mathbb{C}$ a $\hat{\mathbf{h}}^{+} \oplus \mathbb{C}c$-module by $\hat{\mathbf{h}}^{+} \cdot 1 = 0$ and $c \cdot 1 = 1$. If $U(\cdot)$ denotes universal enveloping algebra there is an identification between the induced $\hat{\mathbf{h}}$-module $U(\hat{\mathbf{h}}) \otimes_{U(\hat{\mathbf{h}}^{+} \oplus \mathbb{C}c)} \mathbb{C}$ with the symmetric algebra $S(\hat{\mathbf{h}}^{-})$. Set

$$V_{\Lambda_p} = S(\hat{\mathbf{h}}^{-}) \otimes \mathbb{C}[\Lambda_p]. \tag{8}$$

We have

Proposition 5 *([13]) Then there exists a linear map*

$$\begin{array}{rccl} Y: & V_{\Lambda_p} & \to & (\operatorname{End} V_{\Lambda_p})[[z, z^{-1}]] \\ & v & \mapsto & Y(v,z) = \sum_{n \in \mathbb{Z}} v_n z^{-n-1} \end{array} \tag{9}$$

$(v_n \in \operatorname{End} V_{\Lambda_p})$ *such that* V_{Λ_p} *equipped with the map* Y *is a VOA.*

Note that V_{Λ_p} is the first summand of W_p in (6). Similarly, the other $p-1$ summands of W_p – the twisted spaces – are t_p^j-twisted modules for V_{Λ_p} as described below. Let $1 \neq g \in \langle t_p \rangle$. Then $\mathbf{h}$ has a decomposition $\mathbf{h} = \mathbf{h}_0 \oplus \cdots \oplus \mathbf{h}_{p-1}$ where $\mathbf{h}_n = \{h \in \mathbf{h} | g \cdot h = e^{2\pi i n/p} h\}$. The g-twisted affine algebra associated with $\mathbf{h}$ is $\hat{\mathbf{h}}[g] = \coprod_{n \in \mathbb{Z}} \mathbf{h}_n \otimes t^{n/p} \oplus \mathbb{C}c$ with the obvious bracket. Similarly, we define $\hat{\mathbf{h}}[g]^{\pm}$ and the irreducible $\hat{\mathbf{h}}[g]$-module $S(\hat{\mathbf{h}}[g]^{-})$ on which c acts as 1. Set $V_{\Lambda_p}^j = S(\hat{\mathbf{h}}[t_p^{-j}]^{-}) \otimes T_j$ for $j = 1, ..., p-1$.

Proposition 6 ([16], [10]) *There exists a linear map*

$$\begin{array}{rccl} Y_j: & V_{\Lambda_p} & \to & (\operatorname{End} V_{\Lambda_p}^j)[[z^{1/p}, z^{-1/p}]] \\ & v & \mapsto & Y_j(v,z) = \sum_{n \in \frac{1}{p}\mathbb{Z}} v_n^j z^{-n-1} \end{array} \tag{10}$$

$(v_n \in \operatorname{End} V_{\Lambda_p}^j)$ *such that* $V_{\Lambda_p}^j$ *is an irreducible* t_p^{-j}*-twisted module for* V_{Λ_p}.

Then we have $W_p = V_{\Lambda_p} \oplus V^1_{\Lambda_p} \oplus \cdots \oplus V^{p-1}_{\Lambda_p}$ (as vector spaces) as graded in (6) and $Y(v,z) = Y(v,z) \oplus Y_1(v,z) \oplus \cdots \oplus Y_{p-1}(v,z)$ defines a linear map Y from V_{Λ_p} to $(\operatorname{End} W_p)[[z^{1/p}, z^{-1/p}]]$. Because all of these constructions are "natural," and likewise since the action of $\hat{C}_p$ on W_p is natural, the next result should not be surprising:

Proposition 7 *The actions of $\hat{C}_p$ and the vertex operators $Y(v,z)$ for $v \in V_{\Lambda_p}$, on W_p are compatible in the sense that*

$$gY(v,z)g^{-1} = Y(gv, z,), \quad g \in \hat{C}_p.$$

It follows from this that on our version $V_{\Lambda_p/\mathbb{Z}_p}$ of the Moonshine Module (7), the group C_p operates in a way that is compatible with the vertex operators $Y(v,z)$ for $v \in V_{\Lambda_p}^{\langle ft_p \rangle}$, which is the *untwisted* subspace of $V_{\Lambda_p/\mathbb{Z}_p}$.

5 An additional symmetry σ

Just as we defined $V_{\Lambda_p/\mathbb{Z}_p}$ to be a certain subspace of W_p, so too we will define σ_p first on a space which contains $V_{\Lambda_p/\mathbb{Z}_p}$. To explain this we start with some results from [18]. It is shown there that the simple Lie algebra $\mathbf{sl}(p,\mathbb{C})$ of type A_{p-1} admits a canonical group of automorphisms $\Gamma \simeq SL_2(p) \cdot (\mathbb{Z}_p \times \mathbb{Z}_p)$ with the natural action of $SL_2(p)$ on the normal subgroup. This action extends to one on the affine Lie algebra $\mathcal{L} = A^{(1)}_{p-1}$. Let $H \simeq \mathbb{Z}_p$ be a subgroup of the normal subgroup of Γ. Then again $\mathcal{L}^H$ (the fixed point subalgebra) is isomorphic to $\mathcal{L}$ (see [15]), hence has just p level one irreducible integrable modules. They are naturally indexed by elements $c \in \mathbb{Z}/p\mathbb{Z}$, and we denote them by $M[a,b,c]$ where $H = \langle (a,b) \rangle$.

It is convenient to embed $\mathbb{Z}/p\mathbb{Z}$ in the projective line $P^1(GF(p)) = \{0, 1, ..., p-1, \infty\}$ where $SL_2(p)$ acts in the natural way. For $j \in P^1(GF(p))$ set $j^* = j^{-1}$ if j is finite, $\infty^* = 0$, etc. Now let $w = (w_1, ..., w_{2d})$ be a codeword in $\mathcal{C}(p)$ satisfying the conditions of Proposition 1(c); w plays a crucial, but perhaps somewhat mysterious, rôle in everything that follows. Define elements $d_i \in \Gamma$ via

$$d_i = \left(\begin{pmatrix} (w_i^*)^{1/2} & 0 \\ 0 & w_i^{1/2} \end{pmatrix}, (0, w_i^{1/2}) \right), \quad 1 \le i \le 2d.$$

Now the action of Γ on $\mathcal{L}$ induces a permutation of the modules $M[a,b,c]$ (in a rather complicated way – see [18]). Let Γ_i denote the action on the $M[a,b,c]$ obtained by composing the original action together with conjugation by d_i. In this way we get an action $\Gamma_1 \times \cdots \times \Gamma_{2d}$ on tensor product modules $\otimes_{i=1}^{2d} M[a_i, b_i, c_i]$. The latter modules can be regarded as irreducible modules for $(A_{p-1}^{(1)})^{2d}$.

Now we define further spaces as follows. It is convenient to take p odd, which we assume from now on. For $1 \leq j \leq p-1$ and $t \in P^1(GF(p))$ define

$$V_t^j = \begin{cases} \oplus_C \otimes_{r=1}^{2d} M[-jt^* w_r, j, -j^* a_r + j^*/2] & \text{if } t \neq 0 \\ \oplus_C \otimes_{r=1}^{2d} M[jw_r, 0, a_r] & \text{if } t = 0 \end{cases} \tag{11}$$

where the sum runs over all codewords $(a_1, ..., a_{2d})$ in $\mathcal{C}(p)$. So each V_t^j is a direct sum of p^d irreducible modules for $(A_{p-1}^{(1)})^{2d}$. Set

$$V_t = \oplus_{j=1}^{p-1} V_t^j, \quad t \in P^1(GF(p)). \tag{12}$$

Proposition 8 *Let $\Sigma \leq \Gamma$ be the natural diagonal subgroup of $\Gamma_1 \times \cdots \times \Gamma_{2d}$. Then Σ permutes the modules V_t, and in fact*

$$g: V_t \to V_{\frac{at+b}{ct+d}}$$

for $g = \begin{pmatrix} a & b \\ c & d \end{pmatrix} \in \Sigma.$

The extra symmetry σ_p is now defined via

$$\sigma_p = (\sigma_{w_1}, ..., \sigma_{w_{2d}}) \in \Sigma \tag{13}$$

where $\sigma = \begin{pmatrix} 0 & -1 \\ 1 & 0 \end{pmatrix}$ and $\sigma_{w_i} = \left(\begin{pmatrix} 0 & -w_i^* \\ w_i & 0 \end{pmatrix}, (w_i/2, -1/2)\right) \in \Gamma$ is the action of σ on the ith coordinate. We can establish that

$$\sigma_p : V_t^j \to \begin{cases} V_{-t^*}^{-jt^*} & \text{if } t \neq 0, \infty \\ V_\infty^j & \text{if } t = 0 \\ V_0^{-j} & \text{if } t = \infty. \end{cases} \tag{14}$$

Now let $L_p = N_p + \Lambda_p$, the sum of the lattices being in D^{2d} as in Section 2. Of course L_p is no longer an integral lattice, nevertheless we can construct a space U_p analogous to W_p (6) but based instead on the lattice L_p. We form the space V_{L_p} analogous to V_{Λ_p} in Proposition 5. Note that V_{L_p} contains V_{Λ_p} and we have

$$V_{L_p} = \oplus_{j=0}^{p-1} V_{L_p^j}$$

where $L_p^j = N_p + j\theta^{k-2}\alpha'(w)$ is a coset of N_p in L_p and $V_{L_p^j}$ is the obvious subspace of V_{L_p} corresponding L_p^j. The root lattice $Q = (P^k)^{2d} \subset N_p$ has index p^d in N_p, so each L_p^j is a direct sum of p^d irreducible modules for $(A_{p-1}^{(1)})^{2d}$.

There is also a central extension $\hat{L}_p$ of L_p which contains $\hat{\Lambda}_p$ and such that the pullback $\hat{Q}$ of Q is a maximal abelian subgroup of $\hat{L}_p$. We form analogues of the spaces T_j namely $T_j^{L_p} = \mathbb{C}[\hat{L}_p] \otimes_{\mathbb{C}[\hat{Q}]} L'_j$ for a suitable one dimension $\hat{Q}$-module L'_j and as in Proposition 6 set $V_{L_p}^j = S(\hat{\mathbf{h}}[t_p^{-j}]^-) \otimes T_j^{L_p}$. Choose $a \in \hat{L}_p$ such that $\bar{a} = \theta^{k-2}\alpha'(w)$. Then $T_{L_p}^j$ contains the p^d-dimensional module $\{v \in T_j^{L_p} | a^{-1}t_p(a)v = v\}$ for $\hat{\Lambda}_p$, which is isomorphic to T_j as defined in Section 4. We view $V_{\Lambda_p}^j$, defined in Section 4, as a subspace of $V_{L_p}^j$. The analogue of Proposition 6 holds viz., the existence of a linear map

$$\begin{array}{rccl} Y_j : & V_{L_p} & \to & (\operatorname{End} V_{L_p}^j)[[z^{1/p}, z^{-1/p}]] \\ & v & \mapsto & Y_j(v,z) = \sum_{n \in \frac{1}{p}\mathbb{Z}} v_n^j z^{-n-1} \end{array}$$

such that the restriction of $Y_j(v,z)$ to $V_{\Lambda_p}^j$ is the operator of Proposition 6 (see [10]). Now set

$$U_p = V_{L_p} \oplus \left(\oplus_{j=1}^{p-1} V_{L_p}^j\right).$$

So by construction, U_p contains a canonical subspace isomorphic to $V_{\Lambda/\mathbb{Z}_p}$ and we have vertex operators $Y(v,z)$ on U_p for $v \in V_{L_p}$. Note also that if V_Q denotes the subspace of V_{L_p} corresponding to the root lattice Q, then the component operators of $Y_j(v,z)$ for v in the weight one subspace of V_Q generate a copy of the affine algebra $(A_{p-1}^{(1)})^{2d}$, which therefore acts on U_p. Set, for $0 \le i, j \le p-1$,

$$U_i^j = \left\{ \begin{array}{ll} V_{L_p^i} & \text{if } j = 0 \\ V_{L_p^i}^j & \text{if } j > 0 \end{array} \right.$$

where $V_{L_p^i}^j = S(\hat{\mathbf{h}}[t_p^{-j}]^-) \otimes \mathbb{C}[\hat{L}_p^i] \otimes L'_j \subset V_{L_p}^j$. Now we can at last forge the connection between these spaces and the V_t^j. Each U_i^j is also an $(A_{p-1}^{(1)})^{2d}$-module. We have

Proposition 9 *There are explicit isomorphisms of $(A_{p-1}^{(1)})^{2d}$-modules*

$$V_t^j = \left\{ \begin{array}{ll} U_{-t^*j}^j & \text{if } 0 \neq t \in P^1(GF(p)) \\ U_0^j & \text{if } t = 0. \end{array} \right.$$

In view of Proposition 8, the group Σ acts on the sum $\oplus_{(i,j)\neq(0,0)} U_i^j$. It is shown that the group also acts on $U_0^0 = V_{N_p}$. So Σ acts on U_p.

Form (14) we deduce the action of σ_p on the space U_j^i. Moreover

Proposition 10 *The operators $Y(v,z)$ for $v \in V_Q$ preserve each U_i^j and*

$$\sigma_p Y(v,z)\sigma_p^{-1} = Y(\sigma_p v, z).$$

We have already pointed out that our Moonshine Module $V_{\Lambda_p/\mathbb{Z}_p}$ is a subspace of U_p. We expect that σ_p leaves it invariant, but we have only worked out the details for $p = 3$. Thus

Proposition 11 *If $p = 3$ then σ_p leaves $V_{\Lambda_p/\mathbb{Z}_3}$ invariant.*

6 Completion of the proofs

Define τ to be the element corresponding to the element of $E_p = \hat{\Lambda}_p/K_p$ which is the image of $\theta^{k-2}\alpha'(w)$. Then τ acts on $V_{\Lambda_p/\mathbb{Z}_p}$ as follows:

$$\tau : V_{\Lambda_p/\mathbb{Z}_p} \cap U_i^j \to V_{\Lambda_p/\mathbb{Z}_p} \cap U_{i+1}^j.$$

We can then define vertex operator $Y(v,z)$ for $v \in V_{\Lambda_p/\mathbb{Z}_p} \cap U_i^j$ as follows:

$$Y(v,z) = \tau^i \sigma_p Y((\tau^i \sigma_p)^{-1} v, z)(\tau^i \sigma_p)^{-1}. \tag{15}$$

The point is that from the previous propositions, $(\tau^i\sigma_p)^{-1}v$ lies in the untwisted space $V_{\Lambda_p}^{\langle f t_p \rangle}$ of $V_{\Lambda_p/\mathbb{Z}_p}$ (cf. Section 4). So the vertex operator in (15) is the vertex operator guaranteed to exist for such v (see the discussion following Proposition 7). Thus (15) "propagates" vertex operators from the untwisted space to the twisted sectors of $V_{\Lambda_p/\mathbb{Z}_p}$.

We should caution that in fact, this analysis is only meaningful in case σ_p leaves $V_{\Lambda_p/\mathbb{Z}_p}$ invariant. So we have it for $p = 3$ courtesy of Proposition 11. Let us take $p = 3$ from now on.

We have gone to some lengths to get the group $\langle C_3, \sigma_3 \rangle$ (yet to be identified with the Monster) acting on $V_{\Lambda_p/\mathbb{Z}_p}$. By construction, σ_p satisfies

$$\sigma_p Y(v,z)\sigma_p^{-1} = Y(\sigma_p v, z) \tag{16}$$

for the vertex operators (15). Moreover, this also holds for the subgroup $D_3 \subset C_3$ which normalizes $\langle t_p, \tau \rangle$. We have $C_3 = \langle D_3, \mu \rangle$ for a certain canonical element (described first by Conway; see Chapter 10 of [4]).

Proposition 12 $\langle \sigma_3, \mu \rangle \simeq S_4$.

With this result, we can prove (16) for all $g \in \langle C_3, \sigma_3 \rangle = \langle D_3, \sigma_3, \mu \rangle$. The argument is similar to the analogous result in [13]. The proof of Proposition 12 itself is a long calculation, and is one of the main technical points of our work.

Now a great deal more effort yields

Proposition 13 *The vertex operators (15) satisfy the Jacobi identity and* $V_{\Lambda_3/\mathbb{Z}_3}$ *is a VOA.*

The remainder of the proof is as discussed in the introduction.

REFERENCES

1. R. E. Borcherds, Vertex algebras, Kac-Moody algebras, and the Monster, *Proc. Natl. Acad. Sci. USA* **83** (1986), 3068-3071.

2. J. W. S. Cassels and A. Fröhlich, Algebraic number theory, Academic Press, London, 1988.

3. J. H. Conway and S. P. Norton, Monstrous Moonshine, *Bull. London. Math. Soc.* **11** (1979), 308-339.

4. J. H. Conway and N. J. A. Sloane, Sphere packings, lattices and groups, Springer Verlag, New York, 1988.

5. R. Dijkgraaf, C. Vafa, E. Verlinde and H. Verlinde, The operator algebra of orbifold models, *Comm. Math. Phys.* **123** (1989), 485-526.

6. L. Dixon, J. A. Harvey, C. Vafa and E. Witten, Strings on obifolds, *Nucl. Phys.* **B261** (1985), 651, II *Nucl. Phys.* **B282** (1987), 13.

7. C. Dong, Vertex algebras associated with even lattices, *J. Algebra* **160** (1993), 245-265.

8. C. Dong, Twisted modules for vertex algebras associated with even lattices, *J. Algebra*, **165** (1994), 90-112.

9. C. Dong, Representations of the Moonshine Module vertex operator algebra, in this volume, *Contemporary Math.*

10. C. Dong and J. Lepowsky, The algebraic structure of relative twisted vertex operators, preprint.

11. C. Dong and G. Mason, On the construction of the Moonshine Module as a $\mathbb{Z}_p$-orbifold, preprint.

12. A. J. Feingold, I. B. Frenkel and J. F. X. Ries, Spinor construction of vertex operator algebras, triality and $E_8^{(1)}$, *Contemporary Math.* **121,** 1991.

13. I. B. Frenkel, J. Lepowsky and A. Meurman, Vertex Operator Algebras and the Monster, *Pure and Applied Math.*, Vol. **134**, Academic Press, 1988.

14. R. L. Griess Jr., The Friendly Giant, *Invent. Math.* **69** (1982), 1-102.

15. V. G. Kac, Infinite dimensional Lie algebras, 3rd ed., Cambridge Univ. Press, Cambridge, 1990.

16. J. Lepowsky, Calculus of twisted vertex operators, *Proc. Natl. Acad Sci. USA* **82** (1985), 8295-8299.

17. G. Mason, Finite groups and modular functions, *Proc. Symp. Pure Math., American Math. Soc.* **47** (1987), 181-210.

18. G. Mason, Vertex operator representations of $\hat{A}_N$, *J. Algebra* **157** (1993), 128-160.

19. M. Tuite, Monstrous Moonshine from orbifolds, preprint.

20. C. Vafa, Topological mirrors and quantum rings, preprint.

21. B. B. Venkov, On the classification of integral even unimodular 24-dimensional quadratic forms, *Proc. Steklov Inst. Math.* **4** (1980), 63-74.

Department of Mathematics, University of California, Santa Cruz, CA 95064

E-mail address: dong@cats.ucsc.edu

Department of Mathematics, University of California, Santa Cruz, CA 95064

E-mail address: gem@cats.ucsc.edu

Contemporary Mathematics
Volume **175**, 1994

Star Products, Quantum Groups, Cyclic Cohomology and Pseudodifferential Calculus.

MOSHE FLATO and DANIEL STERNHEIMER

ABSTRACT. We start with a short historical overview of the developments of deformation (star) quantization on symplectic manifolds and of its relations with quantum groups. Then we briefly review the main points in the deformation-quantization approach, including the question of covariance (and related star-representations) and describe its relevance for a cohomological interpretation of renormalization in quantum field theory. We concentrate on the newly introduced notion of closed star product, for which a trace can be defined (by integration over the manifold) and is classified by cyclic (instead of Hochschild) cohomology ; this allows to define a character (the cohomology class of cocycle in the cyclic cohomology bicomplex). In particular we show that the star product of symbols of pseudodifferential operators on a compact Riemannian manifold is closed and that its character coincides with that given by the trace, thus is given by the Todd class, while in general not satisfying the integrality condition.

In the last section we discuss the relations between star products and quantum groups, showing in particular that "quantized universal enveloping algebras" (QUEAs) can be realized, essentially in a unique way (using a strong star-invariance condition), as star product algebras with a different quantization parameter. Finally we show (in the $sl(2)$ case) that these QUEAs are dense in a model Fréchet-Hopf algebra, stable under bialgebra deformations, containing all of them (for different parameter values) and that they have the same product and equivalent coproducts with the original algebra.

1991 Mathematics Subject Classification. Primary 81R50, 17B37, 81S10, 58G15, 58B30.

Lectures presented at the 10th annual Joint Summer Research Conference (AMS - IMS - SIAM) on Conformal Field Theory, Topological Field Theory and Quantum Groups (Mount Holyoke College, South Hadley, MA; June 13-19, 1992), delivered by D. Sternheimer.

1. Introduction.

1.1. Historical background. The mathematical theory of deformations is inherent to the development of modern physics. However this fact has essentially been recognized a posteriori [1]. For example Newtonian mechanics is invariant under the 10-parameter Galilei group of transformations of space-time, while relativistic mechanics brings in its deformation, the Poincaré group, obtained by attributing a non-zero value to the inverse of the velocity of light. If in addition one gives a (non-zero) constant curvature to space-time one gets the De Sitter groups and (since they are simple) the buck stops there - as far as Lie groups are concerned. Considering the Lie bialgebra structure permits, as we shall see below, to get "quantum groups" by deforming the coproduct - but then the "transformation group" point of view is lost, or at least needs to be seriously revised (which wasn't done yet). Had the theory of deformations of Lie groups been developed a century ago, one could have already then found mathematically relativistic theories, before any experimental fact.

Quantum mechanics can in a similar way be considered as a deformation of classical mechanics. However this fact was made explicit and rigorous only 15 years ago [2], while again it might have been developed mathematically a century ago (had the mathematical tools been developed then). In 1927, H. Weyl [3] gave an integral formula (a Fourier transform with operatorial kernel) for a passage from classical to quantum observables, and E. Wigner gave in 1931 an inverse formula [4]. At the end of the 40's, Moyal and Groenewold [5] used the latter to write the symbols of a bracket and a product (respectively) of quantum observables, in a form from where the structure of deformation could have been read off, had the theory of deformation of algebras been developed and had they looked for its physical interpretation (instead of trying to interpret somehow the symbols as probability densities, which cannot be done). Many other works followed along similar lines including some that relied implicitely on the idea of deformation (e.g. the so-called semi-classical approximations), but none "hit the nail on the head" until the mid 70's.

Around 1970, coming from the representation theory of Lie groups, a theory called "geometric quantization" became quite popular [6]. The idea is to do for more general Lie groups (than the Heisenberg groups) and more general symplectic manifolds (than the flat ones) what the Weyl transform does, map functions on the symplectic manifold to operators on a Hilbert space of functions on "half" of the variables. While very interesting for representation theory, this approach proved not very effective physically; in particular polarizations (which in many cases could be found only in the complex domain) had to be introduced to get "half" of the variables, and relatively few classical observables could be thus quantized.

1.2. Star-products and quantum groups. Our approach [2], sometimes called "deformation quantization", was different - though starting from similar premises. We did not look for polarizations or operators, but had the quantum theory "built in" the algebra of classical observables by deforming the composition laws, product and Poisson bracket of functions, to what we called star-products and the associated commutators. We could develop in an autonomous manner, even on general symplectic manifolds, a quantum theory that can be mapped into the usual one by a Weyl mapping (when there exists one). And this has naturally lead to a parallel development of representation theory of Lie groups, without operators (cf. a partial summary in [7]). Furthermore the same approach can be developed for infinite-dimensional phase-spaces, giving rise to a star-product approach to quantum field theories and in particular to a cohomological interpretation of cancellation of infinities [8].

Around 1980, a new mathematical notion appeared in the quantization of 2-dimensional integrable models [9], and has gained tremendous popularity after Drinfeld [10] coined the (somewhat misleading but very effective) term "quantum group" to qualify it. The basic point is that, on a Lie group with a compatible Poisson structure, the functions have a natural Hopf algebra structure that can be deformed by replacing the usual product by a star-product; in the dual approach of Jimbo [11] one deforms the coproduct on a completion of the enveloping algebra of the Lie algebra (which has a bialgebra structure) and gets in particular the strange commutation relations obtained in [9]. A realization of these structures was given by Woronowicz [12] using the basic representations of compact Lie groups while taking coefficients in C^*-algebras (with some relations); this was inspired by the "non-commutative geometry" of A. Connes [13].

1.3. Main new points. In this paper we shall first briefly review the main points of the star-product theory, including the new developments of star-quantization in field theory. Then we shall develop the new notion of closed star-product [14], for which a trace can be defined by integration over phase-space, show its relation to cyclic cohomology and apply it to the pseudodifferential calculus on a compact Riemannian manifold : for the corresponding star-product a character can be defined (in the cyclic cohomology) that is given by the Todd class of the manifold (via the index theorem).

Then we shall concentrate on the relations between star-products and quantum groups. In particular we shall show, by the example of $sl(2)$, that "quantum groups" are neither quantum nor groups, but examples of star-products requiring another star-product in the background (with another parameter $\hbar$) in order to realize the quantized universal enveloping algebras (QUEAs) as star-product algebras and that this realization is then essentially unique. Finally, we shall indicate that there exists a universal Fréchet-Hopf algebra (containing densely the QUEAs), rigid as bialgebra, the dual of which contains the original (simple compact) group as a "hidden group".

2. Star-products, star quantization and closed star-products.

2.1. The framework. *a. Phase-space.* The phase-space is a Poisson manifold, i.e. a manifold where a Poisson bracket can be defined. In the case of a finite-dimensional manifold W the Poisson structure is given by a contravariant skew-symmetric 2-tensor Λ satisfying $[\Lambda, \Lambda] = 0$ in the sense of the supersymmetric Schouten brackets; the latter condition is equivalent [2] to the fact that the Poisson bracket :

$$P(u, v) = i(\Lambda)(du \wedge dv) \quad ; \quad u, v \in C^\infty(W) = N \tag{1}$$

satisfies the Jacobi identity. When Λ is everywhere non degenerate, W is a symplectic manifold, of even dimension $2l$ with a closed 2-form ω. Poisson structures can however be defined also for infinite-dimensional (e.g. Banach or Fréchet) manifolds, for instance on the space of initial conditions of a wave equation such as Klein-Gordon ([15, 16]).

b. Star-products, deformations of algebras and cohomologies. The relation between associative (resp. Lie) algebra deformations and Hochschild (resp. Chevalley) cohomology is well known [17], and can be made more precise (at each step) both for the existence question (determined by the 3-cohomology of the algebra valued in itself) and the equivalence question (classified by the 2-cohomology) [2]. For the associative (resp. Lie) algebra N it is consistent [2] to restrict oneself to differentiable cohomologies and to a formal series of differential operators for the equivalence.

DEFINITION. *A* **star-product** *is an associative deformation of N with a complex parameter ν :*

$$u * v = \sum_{r=0}^{\infty} \nu^r \, C_r(u,v) \quad ; \quad u, v \in N \tag{2}$$

$$C_0(u,v) = uv, \quad C_1(u,v) - C_1(v,u) = 2P(u,v)$$

where the C_r are bidifferential operators.

We thus get a deformation of the Lie algebra (N, P) by the "star-commutator" :

$$\frac{1}{2\nu}(u * v - v * u) \equiv [u,v]_\nu = P(u,v) + \sum_{r=1}^{\infty} \nu^r \, C'_r(u,v) \tag{3}$$

for which the relevant cohomologies are finite-dimensional (the Chevalley 2-cohomology has dimension $1 + b_2(W)$, where $b_2(W)$ is the second Betti number of W). Equivalence between two star-products $*$ and $*'$ (resp. brackets) is given by a formal series $T = I + \sum_{s=1}^{\infty} \nu^s \, T_s$ (where the T_s are necessarily differential operators [2]) such that $T(u *' v) = Tu * Tv$ (and similarly for the brackets). All this extends naturally from the algebra N of functions to $\mathcal{A} = N[[\nu]]$, the formal series in ν with coefficients in N.

For quantum groups one needs to consider Hopf algebras [10] where the essential ingredients are the product and coproduct, that together define a bialgebra structure for which a similar theory can be developed [18,19].

c. Example. The typical example is $W = \mathbb{R}^{2l}$ with the Moyal star-product for which one takes the rth powers of the bidifferential operator P :

$$r! C_r(u,v) = P^r(u,v) = \Lambda^{i_1 j_1} \ldots \Lambda^{i_r j_r} (\partial_{i_1 \ldots i_r} u)(\partial_{j_1 \ldots j_r} v) \tag{4}$$

so that the star-product and bracket can be written :

$$u * v = \exp(\nu P)(u,v) \tag{5}$$

$$M(u,v) = \nu^{-1} \sinh(\nu P)(u,v) \tag{6}$$

The Weyl maps can be defined, for $u \in N$, by

$$\Omega_w(u) = \int_{\mathbb{R}^{2l}} \mathcal{F}(u)(\xi,\eta)w(\xi,\eta)\ \exp\ ((\xi.P + \eta.Q)/2\nu)d^l\xi\ d^l\eta \tag{7}$$

where $\mathcal{F}$ is the inverse Fourier transform, (P_α, Q_α, I) are generators of the Heisenberg Lie algebra $\mathcal{H}_l$ in the Von Neumann representation on $L^2(\mathbb{R}^l)$ satisfying $[P_\alpha, Q_\beta] = 2\nu\delta_{\alpha\beta}I$ $(\alpha,\beta = 1,...,l)$, and (in the Moyal case) the weight function $w = 1$ and $2\nu = i\hbar$. Then Ω_1 will map functions in $C^\infty(\mathbb{R}^{2l})$ into operators in $L^2(\mathbb{R}^l)$, star-products into operator product, and the Moyal bracket M into commutators. The inverse map can be defined by a trace :

$$u = (2\pi\hbar)^{-1}Tr(\Omega_1(u)e^{(\xi.P+\eta.Q)/i\hbar}) \tag{8}$$

is one-to-one between $L^2(\mathbb{R}^{2l})$ and Hilbert-Schmidt operators on $L^2(\mathbb{R}^l)$, and can be defined on larger spaces of functions (cf.e.g.[20]). One then has, for $\Omega_1(u)$ trace-class :

$$Tr(\Omega_1(u)) = (2\pi\hbar)^{-l}\int_{\mathbb{R}^{2l}} u\ \omega^l \equiv Tr_M(u) \tag{9}$$

where $\omega = \sum_{\alpha=1}^{l} dp_\alpha \wedge dq_\alpha$ is the usual symplectic form on $\mathbb{R}^{2l}$, so that

$$Tr_M(u * v) = Tr_M(v * u).$$

The latter property can be seen directly since (due to the skew-symmetry of Λ, by integration by parts), for all r and C_r defined by (4) :

$$\int_W C_r(u,v)\omega^l = \int_W C_r(v,u)\omega^l. \tag{10}$$

For other orderings (other weight functions w in (7)), formula (9) is only approximate, i.e. valid modulo higher powers of $\hbar$. This is in particular true of the so-called standard ordering (all q's on the left, as in the usual way of writing differential operators) for which [21] $w(\xi,\eta) = exp(\frac{1}{2}i\xi\eta)$, that corresponds to the pseudodifferential calculus; if we denote by Ω_S the corresponding Weyl map, then [14], whenever defined :

$$Tr(\Omega_S(u)) = (2\pi\hbar)^{-l}\int_{\mathbb{R}^{2l}} u\omega^l + O(\hbar^{1-l}). \tag{11}$$

To define a trace on $\mathcal{A}$ with integration on W one is thus lead to look at the coefficient of $\hbar^l$ in $u * v$ for $u, v \in \mathcal{A}$.

This will motivate our definition of closed star products.

2.2. Closed star-products and cyclic cohomology.

a. DEFINITION. *Let W be a symplectic manifold of dimension $2l$, and $*$ be a star-product on N defined by (2). The star-product is said* **closed** *if (10) holds for all $u, v \in N$ and $r \leq l$, and* **strongly closed** *if (10) holds for all r.*

Equivalently, if $u = \sum_{k=0}^{\infty} \nu^k u_k \in \mathcal{A}$ and $v = \sum_{j=0}^{\infty} \nu^j v_j \in \mathcal{A}$, then the coefficient $a_l(u, v)$ of ν^l in $u * v$ is $\sum_{r+j+k=l} C_r(u_k, v_j)$ and the condition for closedness is

$$\int_W a_l(u,v)\, \omega^l = \int_W a_l(v,u)\, \omega^l. \tag{12}$$

Note that (10) is always true for $r = 0$ and 1 (thus all star-products on 2-dimensional symplectic manifolds are closed). If the star-product is closed, then the map τ defined by

$$\mathcal{A} \ni u = \sum_{k=0}^{\infty} \nu^k\, u_k \mapsto \tau(u) = \int_W u_l\, \omega^l \tag{13}$$

has the properties of a trace for the algebra $\mathcal{A}$ with the product $*$:

$$\tau(u * v) = \tau(v * u)$$

whenever both sides are defined.

b. Cyclic cohomology [13,14]. If M is a N-module, the coboundary operation b in the Hochschild cohomology can be defined on a p-cochain C by :

$$bC(f_0, ..., f_p) = f_0\; C(f_1, ..., f_p) - C(f_0 f_1, ..., f_p) + ...$$

$$+(-1)^p\; C(f_0, ..., f_{p-1} f_p) + (-1)^{p+1} C(f_0, ..., f_{p-1}) f_p. \tag{14}$$

To every $f \in \mathcal{A}$ one can associate an element $\tilde{f}$ in the dual $\mathcal{A}^*$ by :

$$\tilde{f} : g \mapsto \int fg\, \omega^l.$$

The action of $\mathcal{A}$ on $\mathcal{A}^*$ is given by $(x\varphi y)(a) = \varphi(yax)$ when $\varphi \in \mathcal{A}^*$ and $a, x, y \in \mathcal{A}$. The map $f \mapsto \tilde{f}$ gives then a map of p-cochains :

$$C^p(\mathcal{A}, \mathcal{A}) \to C^p(\mathcal{A}, \mathcal{A}^*)$$

that is compatible with the coboundary operation : if $\tilde{C}_r$ denotes the image in $C^p(\mathcal{A}, \mathcal{A}^*)$ of $C_r \in C^p(\mathcal{A}, \mathcal{A})$, then one has $b\tilde{C}_r = \widetilde{bC}_r$. If we define a bicomplex by $\{0\}$ for $n < m$ and, for $n \geq m$:

$$C^{n,m} = C^{n-m}(\mathcal{A}, \mathcal{A}^*) \tag{15}$$

the Hochschild coboundary b is of degree 1. We can, in addition, define another operation B of degree -1 that anticommutes with b ($bB = -Bb$, with $B^2 = 0 = b^2$) as $B = A_S B_0$ where A_S is the cyclic antisymmetrization and, for $\tilde{C} \in C^{n+1}(\mathcal{A}, \mathcal{A}^*)$:

$$B_0 \tilde{C}(f_0, ..., f_{n-1}) = \tilde{C}(1, f_0, ..., f_{n-1}) + (-1)^n\, \tilde{C}(f_0, ..., f_{n-1}, 1). \tag{16}$$

Let us denote by $C^n_\lambda \subset C^n(\mathcal{A},\mathcal{A}^*)$ the space of cochains $\tilde{C}$ satisfying the cyclicity condition:

$$\tilde{C}(f_1,...,f_n)(f_0) = (-1)^n\, C(f_2,...,f_n,f_0)(f_1). \tag{17}$$

DEFINITION. *The* **cyclic cohomology** *of $\mathcal{A}$, denoted by $HC^n(\mathcal{A})$, is the cohomology of the complex (C^n_λ, b).*

A fundamental (and non trivial) property of cyclic cohomology is :

PROPOSITION. *At each level n, one has*

$$HC^n(\mathcal{A}) = (kerb \cap kerB)\ /\ b(kerB). \tag{18}$$

c. Classification of closed star-products. For $r = 2$, the closedness condition (10) can be written $B\tilde{C}_2 = 0$. (This condition is necessary; it is sufficient if $dimW = 4$). If we start with a Hochschild 2-cocycle C_1 (e.g. $C_1 = P$), standard deformation theory [2, 14, 17] gives a 3-cocycle E_2 (determined by C_1) that has to be equal to bC_2. Therefore $b\tilde{E}_2 = 0 = B\tilde{E}_2$ and $b\tilde{C}_2 = \tilde{E}_2 \in kerb \cap kerB$. From (18) we therefore get, since the same can be done [14] successively at each order of deformation, and in one degree less for the equivalence operators T :

PROPOSITION. *At each order, the obstructions to the existence of closed star-products are classified by $HC^3(\mathcal{A})$, and the obstructions to equivalence by $HC^2(\mathcal{A})$.*

If C is a non-closed current on W ($dC \neq 0$), $\tilde{C}^2(f_1,f_2)(f_0) =< C, f_0 df_1 \wedge df_2 >$ will give $B\tilde{C}_2 \neq 0$ and therefore we get in this way a non-closed star-product.

2.3. Existence, uniqueness and rigidity of star-products.

a. THEOREM [22]. *On any symplectic manifold W there exists a strongly closed star-product.*

The Moyal product is closed, and can be defined on any canonical chart of W. The problem is to "glue" together these products, and to do it in a way that preserves the integrals of functions. The authors of [22] do both things together by defining globally on W a locally trivial algebra bundle (called Weyl manifold) giving a (globally defined) formal completion of the Heisenberg enveloping algebra; functions in $\mathcal{A}$ are then extended to sections of this bundle (called Weyl functions) in an integration-preserving way.

Previous existence proofs of star-products and brackets were first done by supposing $b_3(W) = 0$, to control the multiple intersections of charts. Then, after other generalizations, Lecomte and De Wilde gave a very abstract existence proof in the general case, and later simplified a first existence proof by Omori, Maeda and Yoshioka by showing that Moyal products on canonical charts can be transformed by equivalences so as to coincide on intersections [23] - but these equivalences were not constructed in a way that obviously preserved closedness. All these proofs extend to regular Poisson manifolds (where the symplectic leaves have all the same dimension).

b. Uniqueness. The Hochschild differentiable p-cohomology spaces $\tilde{H}^p(N)$ for the associative algebra $N = C^\infty(W)$ are isomorphic [24] to $\Lambda_p(W)$, the space of skew-symmetric contravariant p-tensors on W. More precisely [25], any differentiable p-cocycle can be written as such a tensor plus the coboundary of a differentiable $(p-1)$ cochain. Therefore they are huge; except for *dim* $W = 2$, the obstructions to existence (classified by $\tilde{H}^3(N)$) make the previous theorem highly non trivial.

However consideration of the Lie algebra (N, P) reduces the choice. We shall call [2] **Vey product** a star-product (2) for which the cochains C_r are bidifferential operators that vanish on constants (i.e. have no constants terms), have the same parity as r (thus $C_1 = P$) and have the same principal part as P^r in any canonical chart (a star-product satisfying all these conditions except the last one can always be brought into the Vey form by equivalence [26]).

We then get associated brackets where only odd cochains C_{2r+1} appear. The relevant Chevalley cohomology spaces $H^p(N)$ are finite-dimensional. H^2 and H^3 can be explicitely computed [25], and in particular *dim* $H^2(N) = 1 + b_2(W)$. Explicit expressions for the cochains up to C_4 in a general Vey product can also be given ([25, 26]), in terms of a symplectic connection Γ and some tensors (the general expressions for C_5 and above could not be written in such terms; however the existence proof of Fedosov [23], which remained obscure until his MIT preprint of December 1992, gives a recurrence algorithm showing that a star-product can be constructed globally in terms of a symplectic connection and its covariant derivatives). In particular one has, for a Vey product [26] :

$$C_2 = P_\Gamma^2 + bH \quad , \quad C_3 = S_\Gamma^3 + T + 3\partial H \tag{19}$$

where H is a differential operator of maximal order 2, T a 2-tensor corresponding to a closed 2-form, ∂ the Chevalley coboundary operator, and P_Γ^2 and S_Γ^3 are given (in canonical coordinates) by expressions similar to (4) where, in P^2, usual derivatives are replaced by covariant derivatives and, in P^3, by the relevant components of the Lie derivative of Γ in the direction of the vector field associated to the function (u or v).

It is then possible [2] to work by steps of 2 and to reduce the possible choices of Vey products from the infinity suggested by the Hochschild cohomology to the finite alternatives of Vey brackets. In particular :

PROPOSITION. *If* $b_2(W) = 0$, *the Moyal-Vey bracket is unique up to (mathematical) equivalence.*

The more general definition of a star-product given here is required in particular because normal and standard orderings are star-products in this sense, equivalent to Moyal (on flat space) but not Vey products (the cochains C_r are not of the same parity as r). Consideration of the associated Lie algebra is however still possible, and will give information on the skew-symmetric parts of the cochains C_r. The analysis developed in [2] for Vey products can however be carried over to this more general definition.

c. Rigidity. A natural question then arises : is it the end of the story, and in particular can the Moyal-Vey product be deformed ?

The answer to the latter is : essentially no. More precisely one can show [27] that any (Vey) star-product of the form

$$u *_{\nu\rho} v = \sum_{r,s=0}^{\infty} \nu^r \rho^s \, C_{rs}(u,v), \quad C_{rs}(u,v) = (-1)^r C_r(u,v) \tag{20}$$

can be transformed by equivalence to $*_\nu$, by deforming the 2-tensor Λ with the parameter ρ.

However if the requirement that $C_1 = P$ is relaxed that result is no longer true. Examples can be given where an intertwinning operator will still exist, but will transform the Poisson into a Jacobi bracket (associated with a conformal symplectic structure) [28]: one writes

$$u *_{\nu\rho} v = u *_\nu f_{\nu\rho} *_\nu v \tag{21}$$

where $f_{\nu\rho} = \sum_{r=0}^{\infty} \nu^{2r} \, f_{2r,\rho} \in (N[[\nu]])[[\rho]]$ with $f_{0,\rho}$ invertible and $f_{0,0} = 1$. In particular one can take for $f_{\nu\rho}$ the $*_\nu$ exponential of ρH, $H \in N$ (and ρ proportional to the inverse temperature, in applications) : the so-called KMS conditions in statistical mechanics take then a simple expression (the condition for a state to be KMS is similar to a trace).

Quantum groups are also, in a sense, deformations of the usual star-products where the covariance enveloping algebra is deformed. We shall now study more closely that notion of covariance of a star-product.

2.4. Invariance and covariance of star-products. Star representations. *a. Invariance.* The Poisson bracket P is (by definition) invariant under all symplectomorphisms. However its powers P^r $(r \geq 2)$ given by (4) are invariant only under the vector fields generated by the polynomials of degree at most 2. For general Vey products the cochains C_2 and C_3 given by (19) are invariant under a subgroup of the finite-dimensional Lie group of symplectomorphisms preserving the symplectic connection Γ. That is also the case in $\mathbb{R}^{2l}$ for orderings other than Moyal (only the linear polynomials remain, possibly supplemented - e.g. for standard and normal orderings - by one second-degree polynomial). We thus have [2] a finite-dimensional Lie algebra $\mathcal{O}$ of "**preferred observables**", under which the star-product will be **invariant** :

$$\mathcal{O} = \{a \in N \; ; \; [a,u]_\nu = P(a,u) \quad \forall u \in N\}. \tag{22}$$

b. Covariance. A star-product will be said **covariant** under a Lie algebra $\mathcal{O}$ of functions if $[a,b]_\nu = P(a,b)$ for all $a, b \in \mathcal{O}$. It can then be shown [29] that $*$ is $\mathcal{O}$-covariant iff there exists a representation τ of the Lie group G whose Lie algebra is $\mathcal{O}$ into $Aut\ (\mathcal{A}; *)$ such that

$$\tau_g u = (Id_N + \sum_{r=1}^{\infty} \nu^r \, \tau_g^r)(g.u) \tag{23}$$

where $g \in G, u \in N$, G acts on N by the natural action induced by the vector fields associated with $\mathcal{O}$, $(g.u)(x) = u(g^{-1}x)$, and where the τ_g^r are differential operators on W. Invariance of course means that the geometric action preserves the star-product : $g.u * g.v = g(u * v)$.

c. Star representations. Let G be a Lie group (connected and simply connected), acting by symplectomorphisms on a symplectic manifold W (e.g. coadjoint orbits in the dual of the Lie algebra $\mathcal{L}$ of G). The elements $x, y \in \mathcal{L}$ will be supposed realized by functions u_x, u_y in N so that their Lie bracket $[x, y]_{\mathcal{L}}$ is realized by $P(u_x, u_y)$. Let us take a G-covariant star-product $*$, so that $P(u_x, u_y) = [u_x, u_y]_\nu$. We can now define the **star exponential** :

$$E(e^x) = Exp(x) \equiv \sum_{n=0}^{\infty} (n!)^{-1} (u_x/2\nu)^{*n} \tag{24}$$

where $x \in \mathcal{L}$, $e^x \in G$ and the power $*n$ denotes the nth star-power of the corresponding function. By the Campbell-Hausdorff formula one can extend E to a **group homomorphism** :

$$E : G \to (N[[\nu, \nu^{-1}]], *) \tag{25}$$

where, in the formal series, ν and ν^{-1} are treated as independent parameters for the time being. Alternatively, the values of E can be taken in the algebra $(\mathcal{P}[[\nu^{-1}]], *)$, where $\mathcal{P}$ is the algebra generated by $\mathcal{L}$ with the $*$-product (a representation of the enveloping algebra).

We now call **star representation** [2] of G a distribution $\mathcal{E}$ (valued in $Im\ E$) on W defined by

$$D \ni f \mapsto \mathcal{E}(f) = \int_G f(g)\ E(g^{-1}) dg \tag{26}$$

where D is some space of test-functions on G. The corresponding **character** χ is the (scalar-valued) distribution defined by

$$D \ni f \mapsto \chi(f) = \int_W \mathcal{E}(f)\ d\mu \tag{27}$$

where $d\mu$ is a quasi-invariant measure on W. The character permit a comparison with usual representation theory.

This theory is now very developed, and parallels in many ways the usual (operatorial) representation theory. For a review of developments (and of star-product theory) until 1986, see e.g. [30] and references quoted therein. Among notable results one may quote :

i) An exhaustive treatment of **nilpotent** groups and of solvable groups of exponential type [31]. The coadjoint orbits are there symplectomorphic to $\mathbb{R}^{2l}$, and one can lift the Moyal product to the orbits in a way that is adapted to the Plancherel formula. Polarizations are not required, and "star-polarizations" can always be introduced to compare with usual theory.

ii) For **semi-simple** Lie groups an array of results is already available, including a complete treatment of the holomorphic discrete series [32] (that includes the case of compact Lie groups) and scattered results for specific examples.

iii) For semi-direct products, and in particular the Poincaré and Euclidean groups, an autonomous theory has also been developed (see e.g.[33]).

Comparison with the usual results of "operatorial" theory of Lie group representations can be performed in several ways, in particular by constructing an invariant Weyl transform generalizing (7), finding "star-polarizations" that always exist, in contradistinction with the geometric quantization approach (where at best one can find complex polarizations), study of spectra (of elements in the center of the enveloping algebra and of compact generators) in the sense of the next subsection, comparison of characters, etc. But our main insistence is that the theory of star-representations is an **autonomous** one that can be formulated completely within this framework, based on coadjoint orbits (and some additional ingredients when required).

2.5. Star-quantization.

a. DEFINITION. *Let W be a symplectic (or Poisson) manifold and N an algebra of classical observables (functions, possibly including distributions if proper care is taken for the product). We shall call* **star-quantization** *a star-product on N invariant under some Lie algebra $\mathcal{O}$ of "preferred observables".*

Invariance of the star-product ensures that the classical and quantum evolutions of observables under a Hamiltonian $H \in \mathcal{O}$ will coincide [2]. The typical example is the case described in (2.1.c), the Moyal product on $W = \mathbb{R}^{2l}$. Physicists often prefer to work with the so-called **normal ordering**, for which the Weyl map (7) is taken [21] with a weight $w(\xi, \eta) = exp(-\frac{1}{4}(\xi^2 + \eta^2))$.

b. Spectrality. Physicists want to get numbers that match experimental results, e.g. for energy levels of a system. That is usually achieved by describing the spectrum of a given Hamiltonian $\hat{H}$ supposed to be a self-adjoint operator so as to get a real spectrum and so that the evolution operator (the exponential of $it\hat{H}$) is unitary (thus preserves probability). A similar spectral theory can be done here, in an autonomous manner. The most efficient way to achieve it is to consider [2] the star exponential (corresponding to the evolution operator)

$$Exp(Ht) \equiv \sum_{n=0}^{\infty} (n!)^{-1} \, (t/i\hbar)^n \, (H*)^n \tag{28}$$

where $(H*)^n$ means the nth star power of $H \in N$ (or $\mathcal{A}$). Then one writes its Fourier-Stieltjes transform $d\mu$ (in the distribution sense), formally

$$Exp(Ht) = \int e^{\lambda t/i\hbar} d\mu(\lambda). \tag{29}$$

DEFINITION. *The spectrum of $(H/\hbar)$ is the support S of $d\mu$.*

In the particular case that H has discrete spectrum, the integral can be written as a sum : the distribution $d\mu$ is a sum of "delta functions" supported at the points of S, multiplied by the symbols of the corresponding eigenprojectors.

In different "orderings" given by (7) with various weight funtions w, one gets in general different operators for different classical observables H, thus different spectra. For $W = \mathbb{R}^{2l}$ all those are mathematically equivalent (to Moyal under the Fourier transform T_w of the weight function w of (7)). This means that every observable H will have the same spectrum under Moyal ordering as $T_w H$

under the equivalent ordering. But this does not imply "physical equivalence", i.e. the fact that H will have the same spectrum under both orderings. In fact, the opposite is true :

PROPOSITION [34]. *If two equivalent star-products are isospectral (give the same spectrum for a "large family" of observables and all ν) they are identical.*

It is worth mentioning that our definition of spectrum permits to define a spectrum even for symbols of non-spectrable operators, such as the derivative on a half-line (that has different deficiency indices). That is one of the many advantages of our autonomous approach to quantization.

c. Applications. i) In **quantum mechanics** it is preferable to work (for $W = \mathbb{R}^{2l}$) with the star-product that has maximal symmetry, i.e. $sp(\mathbb{R}^{2l}).\mathcal{H}_l$ as algebra of preferred observables : the Moyal product (5). One indeed finds [2] that the star exponential of these observables (polynomials of order ≤ 2) is proportional to the usual exponential. More precisely, if $X = \alpha p^2 + \beta pq + \gamma q^2 \in sl(2)$ $(p, q \in \mathbb{R}^l\ ;\ \alpha, \beta, \gamma \in \mathbb{R})$, setting $d = \alpha\gamma - \beta^2$ and $\delta = |d|^{1/2}$ one gets (as distributions)

$$Exp(Xt) = \begin{cases} (\cos\,\delta t)^{-l}\exp((X/i\hbar\delta)\,\tan(\delta t)) & \text{for } d > 0 \\ \exp(Xt/i\hbar) & \text{for } d = 0 \\ (\cosh\,\delta t)^{-l}\exp((X/i\hbar\delta)\,\tanh(\delta t)) & \text{for } d < 0 \end{cases} \tag{30}$$

hence the Fourier decompositions

$$Exp(Xt) = \begin{cases} \sum_{n=0}^{\infty} \Pi_n^{(l)}\, e^{(n+\frac{l}{2})t} & \text{for } d > 0 \\ \int_{-\infty}^{\infty} e^{\lambda t/i\hbar}\Pi(\lambda, X)d\lambda & \text{for } d < 0. \end{cases} \tag{31}$$

We thus get the discrete spectrum $(n+\frac{l}{2})\hbar$ of the **harmonic oscillator** and the continuous spectrum $\mathbb{R}$ for the dilation generator pq. The "eigenprojectors" $\Pi_n^{(l)}$ and $\Pi(\lambda, X)$ are given by some special functions [2].

Other examples can be brought to this case by some functional manipulations [2]. For instance the Casimir element C of $so(l)$ representing angular momentum, which can be written $C = p^2q^2 - (pq)^2 - l(l-1)\frac{\hbar^2}{4}$, has $n(n+(l-2))\hbar^2$ for spectrum. For the **hydrogen atom,** with Hamiltonian $H = \frac{1}{2}p^2 - |q|^{-1}$, the Moyal product on $\mathbb{R}^{2l+2}$ ($l = 3$ in the physical case) induces a star product on $W = T^*S^l$ and the energy levels, solutions of $(H - E) * \phi = 0$, are found to be (from (31), for $l = 3$) :

$E = \frac{1}{2}(n+1)^{-2}\hbar^{-2}$ for the discrete spectrum

$E \in \mathbb{R}_+$ for the continuous spectrum.

It is worth noting that the term $\frac{l}{2}$ in the harmonic oscillator spectrum, obvious source of divergences in the infinite-dimensional case, disappears if the normal star-product is used instead of Moyal - which is one of the reasons it is preferred in field theory.

ii) **Path integrals** are intimately connected to star exponentials. In fact, in quantum mechanics the path integral of the action is nothing but the partial Fourier transform of the star exponential (28) with respect to the momentum variables, for $W = \mathbb{R}^{2l}$ as phase space with the Moyal star-product [35]. For compact groups the star exponential E given by (25) can be expressed in terms of unitary characters using a global coherent state formalism [36] based on the Berezin dequantization of compact group representation theory used in [32] (that gives star-products somewhat similar to normal ordering); the star exponential of any Hamiltonian on G/T (where T is a maximal torus in the compact group G) is then equal to the path integral for this Hamiltonian.

iii) For **field theory** similar results hold. In particular [8] the star exponential of the Hamiltonian of the free scalar field (for the normal star-product) is equal to a path integral. That is only part of the results obtained for field theory [8]. It has indeed been found that other star-products "close to normal" enjoy similar properties and permit nonstandard quantizations of the Klein-Gordon equation (not necessarily leading to a free field theory). And for interacting fields one can show that by transforming the normal star-product by a suitable equivalence one can remove some of the divergences occurring in $\lambda\phi_2^4$ theory, which indicates that **the processus of renormalization is cohomological in essence** (removing an infinite coboundary to get a finite cocycle).

2.6. Pseudodifferential calculus and character of a star-product.

a. Closedness of the star-product arising in pseudodifferential calculus.

Let $g \in \mathcal{A}$, $g = \sum_{r=0}^{\infty} \nu^r\, g_r$ ($g_r \in N, \nu = \frac{1}{2}i\hbar$). Let us define

$$\tau(g) = \int g_l\, \omega^l. \tag{32}$$

From (11) we see that, in the pseudodifferential calculus on $\mathbb{R}^{2l}$, $(2\pi)^l\, Tr(\Omega_S(g))$ is equal to $\tau(g)$ modulo multiples of $\hbar$ (and $\hbar^{-1}$).

Therefore τ enjoys the same properties as a trace with respect to the standard star product, which is thus closed. More precisely, for a pseudodifferential operator D (with compact support) on $\mathbb{R}^l$ one can define its symbol (on $W = T^*\mathbb{R}^l$) by [37]

$$\sigma_D(q_0, p) = D\, e^{i(q-q_0).p}|_{q=q_0} \tag{33}$$

so that one has for a product :

$$\sigma_{BD} = \sigma_B *_S \sigma_D = \sum_{r=0}^{\infty} \frac{i^{-r}}{r!} \frac{\partial^r \sigma_B}{\partial p^r} \frac{\partial^r \sigma_D}{\partial q^r}. \tag{34}$$

These formulas extend [37] to $W = T^*M, M$ Riemannian compact manifold, using a globally defined function $L \in C^\infty(T^*M \times M)$ that can, on canonical charts, be written $L(q_0; p, q) = p.(q - q_0)$. As for standard ordering one can then show (by integration by parts e.g.) that the corresponding star-product on $W = T^*M$ is closed.

b. The character. The deviation from algebra homomorphism of the identity map on $\mathcal{A}$ between the algebras $\mathcal{A}_0$ ($\mathcal{A}$ endowed with commutative product) and $\mathcal{A}_\nu$ ($\mathcal{A}$ endowed with $*_\nu$ product) is expressed by

$$\sigma(f, g) = f * g - fg \quad (f, g \in \mathcal{A}). \tag{35}$$

As for operator algebras [13], since the map τ behaves like a trace, one is thus lead to define (for $f_i \in \mathcal{A}$) :

$$(36) \qquad \varphi_{2k}(f_0, f_1, ..., f_{2k}) = \tau(f_0 * \sigma(f_1, f_2) * ... * \sigma(f_{2k-1}, f_{2k}))$$

and $\varphi_{2k} = 0$ except for $l \leq 2k \leq 2l$. This makes sense for a general closed star-product.

DEFINITION. *The character of a closed star-product $*$ is the cohomology class of the cocycle φ with components φ_{2k} defined by (38) in the cyclic cohomology bicomplex.*

The previous considerations then imply :

PROPOSITION. *On $W = T^*M$, M Riemannian compact, the character φ defined by (36) (with standard ordering) coincides with the character defined by the trace on pseudodifferential operators.*

c. *Examples.* The top component φ_{2l} of φ is given by [14]

$$\varphi_{2l}(f_0, f_1, ..., f_{2l}) = \int f_0 P(f_1, f_2) ... P(f_{2l-1}, f_{2l})\omega^l$$

$$= \int f_0 \, df_1 \wedge ... \wedge df_{2l}.$$

If *dim* $W = 4$, the computation of φ can be carried out easily and one gets in addition $\varphi_2 = \tilde{C}_2$ (in the notations of (2.2.b)) and thus $b\varphi_2 = -\frac{1}{2}B\varphi_4$. Now, since $HC^2(N) = Z^2(W,\mathbb{C}) \oplus \mathbb{C}$ where Z^2 stands for the closed 2-currents, a change in C_2 will affect the class in HC^4 thus the class of φ, which has then no reason to be an integer - as is the case in the traditional approaches to quantization.

d. *Consequences.* Combining the above proposition with the algebraic index theory of A. Connes et al. [13] one gets [14] :

THEOREM. *The character φ of the star product of the pseudodifferential calculus belongs to $HC^{ev}(T^*M)$ and is given by the Todd class $Td(T^*M)$ as a current over T^*M.*

Since the trace on compact operators is unique, whenever a star product has an associated Weyl map that includes them in the image, the cyclic cocycle φ will necessarily be proportional to an integral one. But we have seen above that the character is not necessarily integral. Therefore on one hand star quantization (with a closed star-product) can be applied in cases where there are obstructions to the traditional (Bohr-Sommerfeld) quantization; and on the other hand it paves the way for a **generalized index theory where the index is no more necessarily an integer.**

For finite-dimensional manifolds, because the trace formula (9) is exact (not modulo $\hbar$ terms) in the Moyal case, the corresponding "symmetric pseudodifferential calculus" should be of special interest. And in the infinite-dimensional case since orderings "close to normal" permit a cohomological interpretation of cancellation of infinities [8], the corresponding "complex pseudodifferential calculus" appears to be an appropriate framework.

3. Quantum groups and star-products.

3.1. The background. *a. Yang-Baxter equations and Hopf algebras.*

As mentioned in the introduction, "quantum groups" appeared around 1980 in the quantization of 2-dimensional integrable models [9]. A basic ingredient there is an equation which the LOMI group called **Yang-Baxter equation.** If $R \in End(V \otimes V)$, where V is a vector space, and if we define $R^{12} = R \otimes I \in End(V \otimes V \otimes V)$ and similarly for R^{13}, R^{23}, that equation writes

$$R^{12}\, R^{13}\, R^{23} = R^{23}\, R^{13}\, R^{12} \text{ with } R^{12}\, R^{21} = I. \tag{37}$$

This is the quantum Yang-Baxter equation (QYBE). Writing

$$R = I + \sum_{j=1}^{\infty} \nu^j\, r_j \in End(V \otimes V)[[\nu]] \tag{38}$$

one gets (at order 2) the classical Yang-Baxter equation (CYBE) :

$$[r_1^{12}, r_1^{13}] + [r_1^{12}, r_1^{23}] + [r_1^{13}, r_1^{23}] = 0 = r_1^{12} + r_1^{21} \tag{39}$$

Looking for solutions lead to a strange modification of the $sl(2)$ commutation relations, which later was extended to a wide array of Lie groups and became systematized by Drinfeld in the framework of **Hopf algebras** [10]. These are **bialgebras**, i.e. associative algebras $\mathcal{A}$ with a coproduct $\Delta : \mathcal{A} \to \mathcal{A} \otimes \mathcal{A}$, that posess in addition a counit $\varepsilon : \mathcal{A} \to \mathbb{C}$ and antipode map $S : \mathcal{A} \to \mathcal{A}$, satisfying the expected (and well-known) relations and commutative diagrams.

b. Poisson-Lie groups and Lie bialgebras. The typical example is the associative algebra N of functions over a Lie group G with pointwise multiplication, comultiplication $\Delta : N \to N \otimes N$ defined (by duality) from the group multiplication $G \times G \to G$, counit defined by the value at the identity e of G and antipode by the value at the inverse (in G). The group is said to be a **Poisson-Lie group** when it has a Poisson structure Λ for which the above Δ is a Poisson morphism, i.e.

$$\Delta P(u, v) = P(\Delta u, \Delta v) \tag{40}$$

where P denotes the Poisson bracket in G and $G \times G$, defined by Λ.

The infinitesimal version (dual to it in the sense that the enveloping algebra $\mathcal{U}(g)$ can be considered as a space of distributions on G with support at the identity) is the notion of **Lie bialgebras.** It is a Lie algebra g with a "compatible" bracket φ^* on its dual g^* (i.e. $\varphi : g \to g \otimes g$ is a 1-cocycle for the adjoint action of g on $g \otimes g$).

A case of special interest (triangular Poisson-Lie group [10]) is when φ is the coboundary of $r \in g \wedge g$ such that the Poisson-Lie structure Λ on the corresponding group G is the difference of the left and right invariant skew-symmetric 2-tensors defined by r : this r is solution of the CYBE (39).

c. Their quantization. It is now natural to try and "quantize" (40). This can indeed be done. More precisely [38,39] on $N = C^\infty(G)$ where G is a triangular Poisson Lie group there exists an invariant star product defined by invariant bidifferential operators $F_k(x,y)$, i.e. by a formal series $F = I + \sum_{k=1}^{\infty} \nu^k F_k$ such that the associativity condition for the star product gives a solution $S(x,y) = F^{-1}(y,x)F(x,y)$ of the QYBE (37) on $\mathcal{U}(g)[[\nu]]$.

Moreover, from that invariant star product one can build [38] an equivalent one ($*'$, non invariant) such that (with the same coproduct Δ on N) :

$$\Delta(u *' v) = \Delta u *' \Delta v. \tag{41}$$

These results were more or less implicit in Drinfeld's work but precise statements and proofs can be found in [39], together with related results.

In the dual approach [11] it is the coproduct Δ, on a completion $\mathcal{U}_\nu(g)$ of the enveloping algebra $\mathcal{U}(g)$, that is deformed to Δ_ν. The simplest example is when $g = sl(2)$ with generators H and $E^\pm$ (or more generally H_α and $E_\alpha^\pm$ for simple roots α on a simple Lie algebra g) where

$$\Delta_\nu(H) = H \otimes 1 + 1 \otimes H$$

$$\Delta_\nu(E^\pm) = E^\pm \otimes \exp(\frac{1}{2}\nu H) + \exp(-\frac{1}{2}\nu H) \otimes E^\pm$$

(see [40] for a very detailed exposition).

A quantum group can therefore be seen as a non commutative star-product deformation of the Hopf algebra of functions on a Poisson-Lie group, or as a non cocommutative deformation of its dual.

3.2. Quantized Universal Enveloping Algebras as star-product algebras. *a. Example* [41]. First let us remark that the following functions on $\mathbb{R}^2$:

$$H = pq, \quad J_+ = q, \quad J_-^\eta = (1 - \cos \eta pq)/\eta^2 q \tag{42}$$

satisfy, with respect to Poisson brackets, the same commutation relations as those of the QUEA $\mathcal{U}_\eta(sl(2))$. Since H and J_+ are preferred observables this is also true for Moyal brackets and in particular

$$M(J_+, J_-^\eta) = -\eta^{-1}\sin(\eta pq) \tag{43}$$

Now from (30) one gets that a strong invariance condition

$$f(*tX/i\hbar) = \alpha(t)\, f(\beta(t)X/i\hbar) \tag{44}$$

holds when f is an exponential or a trigonometric function (sine or cosine), where on the left-hand side we mean that we take (Moyal) star powers in the power series expansion of f. Therefore (after a slight rescaling) (43) can be rewritten with ($\sin *$) instead of sin, and the defining QUEA relations can be expressed entirely in the $*$-product algebra generated by the Lie generators.

b. Generalizations. The same can be done for all higher rank simple algebras, e.g. the A_n series [42]. The cubic Serre relations are also expressed in the $*$-product algebra.

One can show [41] that **for Moyal product, the only power series for which (44) holds, are sine, cosine and exponential** (with exception of these series truncated at order two). Therefore if one requires that the "quantum" commutation relations be a deformation of the "classical" ones, the strong invariance condition (44) alone shows that the algebra found in the litterature is unique (except for the truncated sine $f(x) = a_1 x + a_3 x^3$).

For other star-products than Moyal there are even less functions f satisfying (44). For instance [42] for a family of star-products interpolating between standard and antistandard ordering, the only function is the exponential.

In conclusion, we have seen that the QUEAs can be realized essentially uniquely, as star-product algebras with a quantization parameter $\hbar$ (in the star-product) different from the QUEA parameter. We have in fact a **2-parameter deformation**.

3.3. A universal rigid model for quantum groups [43].

a. The model. Let us start with $G = SU(2)$. Denote by (π_n, V_n) its $2n+1$-dimensional representation ($2n$ an integer) by matrices in $\mathcal{L}(V_n)$. The direct product $A = \Pi_{n=0}^{\infty} \mathcal{L}(V_n)$, endowed with the product topology and algebra law, becomes a Fréchet algebra into which both G and the enveloping algebra $\mathcal{U}(g)$ can be imbedded by $x \mapsto (\pi_n(x))$. These imbeddings are total because A is the bicommutant of the direct sum representation $\pi = \sum \pi_n$ on $V = \sum V_n$, and every π_n extends by continuity to a representation of A.

b. Universality. We have :

PROPOSITION. *If $t \notin 2\pi\mathbf{Q}$, the QUE algebras $\mathcal{U}_t(g)$ can be imbedded into a dense subalgebra A_t of A; π_n are still a complete set of representations of A_t, and the Hopf structures (coproduct Δ_t, counit ϵ_t, and antipode S_t) on $\mathcal{U}_t(g)$ have unique extensions to topological Hopf structures (with equivalent coproducts) on A, when $A \otimes A$ is endowed with the completed projective tensor topology so that*

$$A \otimes A \approx \Pi_{n,p}\, \mathcal{L}(V_n \otimes V_p).$$

The classical limit is $\mathcal{U}_0(g) \approx \mathcal{U}(g) \otimes \theta$ where $\theta^2 = 1$ (a parity), with coproduct Δ_0 equivalent to Δ_t : there exists $P(t) \in A \otimes A$ such that $\Delta_t = P(t)\, \Delta_0\, P(t)^{-1}$. The R-matrix is $R(t) = P(-t)\, P(t)^{-1}$ and can be chosen so that A_t is a quasitriangular Hopf algebra ($q = e^{it}$ in the usual notations).

c. The star-product. The topological dual A^* is isomorphic to $\oplus \mathcal{L}(V_n) = H$, endowed with the inductive topology, and can thus be viewed as polynomial functions on the complex extension $G_{\mathbf{C}}$ of G; as Hopf algebra,

$$H \approx \mathbf{C}[a,b,c,d]/(ad - bc = 1)$$

where the product is that of functions and the coproduct is defined by the product in G. Each co-associative coproduct Δ_t induces an associative star-product on H such that $(a+d)^{*k} = (a+d)^k$, $\forall k \in \mathbf{N}$, and conversely if this is true for a $*$-product compatible with the usual coproduct Δ_0, this $*$ is induced by a coproduct $\Delta_t = P(t) \Delta_0 P(t)^{-1}$ on A : *we recover Manin's description of functions of non-commutative arguments.*

d. Rigidity as bialgebra. Let B be a bialgebra, an associative algebra B with coproduct $\Delta : B \to B \otimes B$. We then define [18] in a natural way the notion of isomorphism of bialgebras (associative algebra isomorphism with equivalence of coproducts), and of deformations of such structures, where we denote by $H(B,\Delta)$ the relevant 2-cohomology.

It can be shown [43] that if (for Hochschild cohomologies)

$$\tilde{H}^2(B) = 0 = \tilde{H}^1(B, B \otimes B)$$

then $H(B,\Delta) = 0$. Whitehead's lemmas therefore prove that (A, Δ_t) **is rigid for all** $t \notin 2\pi\mathbf{Q}$.

REMARK. These results can be extended to all simple compact groups. From this follows that (in the generic case) the QUEAs associated with such groups have a hidden group structure, that of the original group, included in a universal (reflexive) topological Hopf algebra completion.

REFERENCES

1. M. Flato, Deformation view of physical theories, Czechoslovak J. Phys. **B32** (1982), 472-475.

2. F. Bayen, M. Flato, C. Fronsdal, A. Lichnerowicz and D. Sternheimer, Quantum mechanics as a deformation of classical mechanics. Lett. Math. Phys. **1** (1977), 521-530; Deformation theory and quantization, I and II,Ann. Physics **111** (1978), 61-110 and 111-151. C. Fronsdal, Some ideas about quantization. Reports On Math. Phys. **15** (1978), 111-145.

3. H. Weyl, *The theory of groups and quantum mechanics* Dover, New-York, 1931. (Translated from *Gruppentheorie and Quantenmechanik*, Hirzel Verlag, Leipzig (1928); see also Quantenmechanik und Gruppentheorie, Z. Physik **46** (1927), 1-46).

4. E. Wigner, Quantum corrections for thermodynamic equilibrium, Phys. Rev. **40** (1932), 749-759.

5. J.E. Moyal, Quantum mechanics as a statistical theory, Proc. Cambridge Phil. Soc. **45** (1949), 99-124; A. Groenewold, On the principles of elementary quantum mechanics, Physica **12** (1946), 405-460.

6. B. Kostant, Quantization and unitary representations, in : Lecture Notes in Math. vol. 170, Springer Verlag, Berlin, (1970), 87-208; J.M. Souriau, *Structure des sytèmes dynamiques*, Dunod, Paris 1970.

7. D. Sternheimer, Phase-space representations, *Lectures in Applied Math.* vol. 21, Amer. Math. Soc., Providence, 1985, p. 255-267.

8. J. Dito, Star-product approach to quantum field theory : the free scalar field, Lett. Math. Phys. **20** (1992), 125-134 ; Star-products and nonstandard quantization for Klein-Gordon equation, J. Math. Phys. **33** (1992), 791-801; An example of cancellation of infinities in star-quantization of fields, Lett. Math. Phys. **27** (1993), 73-80.

9. P.P. Kulish and N. Yu. Reshetikhin, Quantum linear problem for the sine-Gordon equation and higher representations. Zap. Nauch. Sem. LOMI **101** (1981), 101-110 (in Russian; English translation in Jour. Sov. Math. **23** (1983), 24-35).

10. V.G. Drinfeld, Quantum groups, in : Proc. ICM86, Berkeley, vol. 1, Amer. Math. Soc. 1987, 798-820.

11. M. Jimbo, A q-difference algebra of $U(g)$ and the Yang-Baxter equation, Lett. Math. Phys. **10** (1985), 63-69.

12. S.L. Woronowicz, Compact matrix pseudogroups, Commun. Math. Phys. **111** (1987), 613-665.

13. A. Connes, Non-commutative differential geometry, Publ. Math. IHES **62** (1986), 41-144; *Géométrie non commutative*, Interéditions, Paris, 1990 (English expanded version to be published by Academic Press).

14. A. Connes, M. Flato and D. Sternheimer, Closed star-products and cyclic cohomology, Lett. Math. Phys. **24** (1992), 1-12.

15. I.E. Segal, Symplectic structures and the quantization problem for wave equations, Symposia Mathematica **14** (1974), 79-117.

16. D. Arnal, J.C. Cortet, M. Flato and D. Sternheimer, Star products : quantization and representations without operators, in : *Field theory, quantization and statistical physics* (E. Tirapegui, Ed.), D. Reidel, Dordrecht, 1981, p. 85-111.

17. M. Gerstenhaber, On the deformation of rings and algebras, Ann. of Math. (2) **79** (1964), 59-103.

18. P. Bonneau, Cohomology and associated deformations for not necessarily coassociative bialgebras, Lett. Math. Phys. **26** (1992), 277-283.

19. M. Gerstenhaber and S.D. Schack, Bialgebra cohomology, deformations and quantum groups, Proc. Nat. Acad. Sci. USA **87** (1990), 478-481.

20. J.M. Maillard, On the twisted convolution product and the Weyl transform of tempered distributions, J. Geom. Phys. **3** (1986), 231-261; J.M. Kammerer, Analysis of the Moyal product in a flat space, J. Math. Phys. **27** (1986), 529-535; F. Hansen, The Moyal product and spectral theory for a class of infinite-dimensional matrices, Publ. RIMS, Kyoto Univ., **26** (1990), 885-933.

21. G.S. Agarwal and E. Wolf, Calculus for functions of noncommuting operators and general phase-space methods in quantum mechanics I, II, III, Phys. Rev. D(3) **2** (1970), 2161-2186, 2187-2205, 2206-2225.

22. H. Omori, Y. Maeda and A. Yoshioka, Existence of a closed star-product, Lett. Math. Phys. **26** (1992), 285-294.

23. H. Omori, Y. Maeda and A. Yoshioka, Weyl manifolds and deformation quantization, Adv. in Math. **85** (1991), 225-255; M. De Wilde and P. Lecomte, Existence of star-products revisited, Note di Mat. vol. X (1992); Existence of star-products and of formal deformations of the Poisson Lie algebra of arbitrary symplectic manifolds, Lett. Math. Phys. **7** (1983), 487-496; see also B.V. Fedosov, A simple geometrical construction of deformation quantization, J. Diff. Geom. (1994) (in press) and "Formal quantization" in *Some topics of modern mathematics and their applications to problems of mathematical physics* (in Russian) Moscow 1985, p. 129-139.

24. J. Vey, Déformation du crochet de Poisson sur une variété symplectique, Comment. Math. Helv. **50** (1975), 421-454.

25. S. Gutt, *Déformations formelles de l'algèbre des fonctions différentielles sur une variété symplectique*, Thesis, Bruxelles, 1980.

26. A. Lichnerowicz, Déformations d'algèbres associées à une variété symplectique (les $*_\nu$-produits), Ann. Inst. Fourier, Grenoble, **32** (1982), 157-209.

27. H. Basart and A. Lichnerowicz, Déformations d'un star-produit sur une variété symplectique. C. R. Acad. Sci. Paris **293** I (1981), 347-350.

28. H. Basart, M. Flato, A. Lichnerowicz and D. Sternheimer, Deformation theory applied to quantization and statistical mechanics, Lett. Math. Phys. **8** (1984), 483-494.

29. D. Arnal, J.C. Cortet, P. Molin and G. Pinczon, Covariance and geometrical invariance in star-quantization, J. Math. Phys. **24** (1983), 276-283.

30. D. Sternheimer, Approche invariante de l'analyse non linéaire et de la quantification, Sem. Math. Sup. Montréal **102** (1986), 260-293.

31. D. Arnal and J.C. Cortet, Nilpotent Fourier transform and applications, Lett. Math. Phys. **9** (1985), 25-34; Star-products in the method of orbits for nilpotent Lie groups, J. Geom. Phys. **2**

(1985), 83-116; Représentations star des groupes exponentiels, J. Funct. Anal. **92** (1990), 103-135.

32. D. Arnal, M. Cahen and S. Gutt, Representations of compact Lie groups and quantization by deformation, Bull. Acad. Royale Belg. **74** (1988), 123-141; Star exponential and holomorphic discrete series, Bull. Soc. Math. Belg. **41** (1989), 207-227; C. Moreno, Invariant star-products and representations of compact semi-simple Lie groups, Lett. Math. Phys. **12** (1986), 217-229.

33. D. Arnal, J.C. Cortet and P. Molin, Star-produit et représentation de masse nulle du groupe de Poincaré, C.R. Acad. Sci. Paris A **293**(1981), 309-312; P. Molin, Existence de produits star fortement invariants sur les orbites massives de la coadjointe du groupe de Poincaré, C.R. Acad. Sci. Paris A **293** (1981), 309-312.

34. M. Cahen, M. Flato, S. Gutt and D. Sternheimer, Do different deformations lead to the same spectrum ? J. Geom. Phys. **2** (1985), 35-48.

35. P. Sharan, Star-product representation of path integrals, Phys. Rev. D **20** (1979), 414-418.

36. C. Cadavid and M. Nakashima, The star-exponential and path integrals on compact groups, Lett. Math. Phys. **23** (1991), 111-115.

37. H. Widom, A complete symbolic calculus for pseudodifferential operators, Bull. Sci. Math. (2), **104** (1980), 19-63.

38. L. Takhtajan, Introduction to quantum groups, in : Lecture Notes in Physics, vol. 370, Springer Verlag, Berlin, 1990, p. 3-28; Lectures on quantum groups in : Nankai Lectures on Math. Phys. (M. Ge and B. Zhao eds.), World Scientific, Singapore 1990, p. 69.

39. C. Moreno and L. Valero, Star-products and quantization of Poisson-Lie groups, J. Geom. Phys. **9** (1992), 369-402.

40. M. Rosso, Représentations des groupes quantiques, Séminaire Bourbaki, exposé 744 (Juin 1991).

41. M. Flato and D. Sternheimer, A possible origin of quantum groups, Lett. Math. Phys. **22** (1991), 155-160; M. Flato and Z.C. Lu, Remarks on quantum groups, Lett. Math. Phys. **21** (1991), 85-88.

42. Z.C. Lu, Quantum algebras on phase-space, J. Math. Phys. **33** (1992), 446-453; M. Flato, Z.C. Lu and D. Sternheimer, From where do quantum groups come? Foundations of Physics **23** (1993), 587-598.

43. P. Bonneau, M. Flato and G. Pinczon, A natural and rigid model of quantum groups, Lett. Math. Phys. **25** (1992), 75-84; P. Bonneau, M. Flato, M. Gerstenhaber and G. Pinczon, The hidden group structure of quantum groups: strong duality, rigidity and preferred deformations, Commun. Math. Phys. **161** (1994), 125-156.

PHYSIQUE MATHEMATIQUE, UNIVERSITE DE BOURGOGNE
B.P. 138, F-21004 DIJON Cedex, FRANCE.
e-mail addresses: flato@satie.u-bourgogne.fr, daste@ccr.jussieu.fr

Contemporary Mathematics
Volume 175, 1994

The Universal T-Matrix.

C.Fronsdal A. Galindo

ABSTRACT. The universal T-matrix of a quantum group is the Hopf algebra dual form, expressed in terms of the generators of the algebra and the generators of its dual. In the physical applications it is the familiar T-matrix of integrable models, here calculated in the structure, without specialization to a representation of the algebra of physical variables, nor to a representation of the auxiliary algebra. This article deals with some rather surprising facts that were discovered by examination of the formula for the universal T-matrix for $U_q(sl_2)$, among them the existence of a new series of quantum deformations of $U(gl_n)$ and a generalization of the quantum double. The new quantum groups have physical applications with essentially new features, principally arising from the fact that the dual Lie algebra (algebra of physical variables) is not solvable.

1 Introduction

The principal object that defines an integrable classical field theory is its "L-matrix", the spatial component of a flat Lax connection. In Drinfeld's interpretation it is a Lie bialgebra dual form. If $(l_i), i = 1, ..., n$, is a basis for a Lie bialgebra G, and (x^i) is the dual basis for the vector space dual G^*, then it is

$$L = \sum_i x^i l_i. \tag{1}$$

In practice, the l_i are matrices and the x^i are dynamical variables, but nothing so far depends on the choice of a representation; duality, in this context, is a concept with a clear structural content. After quantization, which is understood as a Hopf algebra deformation of the universal enveloping algebra $U(g)$ of G, the attention of the physicist shifts to the "T-matrix". Classically, the T-matrix is the monodromy associated with the Lax connection; after quantization one tries to preserve this interpretation. Unlike the situation in the classical theory, the structural meaning of the quantum T-matrix is not always clear; it is normally studied in a particular representation of G (the Woronowicz picture [1]) or in special representations of G^* (the Kulish-Reshetikhin picture [2]).

1991 Mathematical Subjects Classification. Primary 81R50. Secondary 70G50.
Supported by National Science Foundation Grant No. NSF/Phys 89-15286.
This paper is in final form and no version of it will be submitted for publication elsewhere.
Presented to the Conference by Christian Fronsdal.

Section 2 examines the Hopf algebra dual of the quantum group $U_{q,q'}(gl_2)$. It is a deformation $U_{q,q'}(gl_2{}^*)$ of the enveloping algebra of the bialgebra dual $gl(2)^*$ of $gl(2)$, with the structure of Lie algebra preserved, the deformation affecting only the coproduct. Both Hopf algebras are thus finitely generated. The Hopf algebra dual form can be obtained explicitly in terms of the generators x^i and l_i of $gl_2{}^*$ and of gl_2:

$$T = e_{1/k}^{x^- l_-} e^{x^0 l_0} e^{x^1 l_1} e_k^{x^+ l_+}. \tag{2}$$

Here $k = qq'$ and e_k^x is the usual twisted exponential. The result was obtained from the known structure of $U_{q,q'}(gl_2)$, both product and coproduct, using the relations that characterize and in fact determine the dual form, namely

$$T_{x,l}T_{x,l'} = T_{x,\Delta(l)}, \quad T_{x,l}T_{x',l} = T_{\Delta(x),l}. \tag{3}$$

Here x and x', resp. l and l', refer to two identical copies of $U_{q,q'}(gl_2{}^*)$, resp. of $U_{q,q'}(gl_2)$. The second relation is the multiplicative property that is the essence of complete integrability. The coproducts are denoted $x^i \mapsto \Delta(x^i)$ and $l_i \mapsto \Delta(l_i)$. The calculation yields the expression for T, as well as the structure of the dual. Since Eq.(2) is structural, independent of any reference to representations of gl_2 or of $gl_2{}^*$, the term Universal T-matrix is justified. Eqs.(3) express Hopf algebra duality in a particularly succinct form.

Section 3 reports the discovery of a new quantum deformation of $U(gl_3)$. To demonstrate the practical utility of the result (2) we investigated the Hopf algebra that results from taking the x^i (not, as is usual, the l_i) in a small, faithful representation. We are thus led to

$$T = \begin{pmatrix} a & b & 0 \\ 0 & d & 0 \\ 0 & c & \hat{a} \end{pmatrix}, \tag{4}$$

where the entries are expressed in terms of the l_i and satisfy the relations

$$\begin{gathered} ab = q^{-1}ba, \quad ac = qca, \quad [a,\hat{a}] = 0, \\ [a,d] = 0 = [\hat{a},d], \quad \hat{a}b = q^{-1}b\hat{a}, \quad \hat{a}c = qc\hat{a}, \\ bd = q^{-1}db, \quad cd = qdc, \quad [b,c] = \lambda(a\hat{a} - dd). \end{gathered} \tag{5}$$

[The associated compatible coproduct is of course given by multiplication of matrices, this being an application of Eq.(3).] The respective roles of gl_2 and $gl_2{}^*$ have thus been reversed as compared to the usual situation. The physical significance of this is that the physical variables a, b, c, d now have the structure of quantum $gl(2)$, instead of that of its solvable dual. Applications in the form of ice models and new types of Toda field theories and nonlinear sigma models are envisaged. Because of (3), the matrix elements of (4) generate a Hopf algebra. However, since it is gl_2 and not $gl_2{}^*$ that is a coboundary Lie algebra, this Hopf algebra is not of the coboundary type. Nevertheless, a Yang-Baxter matrix for the T-matrix (4) does exist. The explanation for this lies in the fact that $U_{q,q'}(gl_2{}^*)$ is a sub-Hopf algebra of a new quantum deformation $U_{q,q'}(gl_3)$ of the enveloping algebra of $gl(3)$, and this one *is* of the coboundary type. We determined the

R-matrix using Manin's technique and from this the new structure of "esoteric quantum $gl(3)$".

In Section 4 the generalization of the formula (2) for the universal T-matrix, to standard quantum $gl(n)$, is obtained.

In Section 5 we generalize the results of Section 3 and present the structure of the new quantum groups, "esoteric quantum $gl(n)$". The most important feature of these new quantum groups is that the dual algebras are not solvable; this has a profound effect on the physical applications. The Hopf algebra dual form of these quantum groups can also be calculated; a start is made here. The essential step is a factorization that generalizes the famililar quantum double.

Section 6. Belavin and Drinfeld [3] have classified the coboundary Lie bialgebra structures associated with the simple Lie algebras. Analogous results for the quantum deformations are not yet available. To the simplest r-matrices for $gl(n)$ correspond multi-parameter deformations $U_q(gl_n)$; here q stands for a point in a multi-dimensional parameter space. The "roots of unity" points are of course exceptional, but that is not all. Investigating the rigidity of $U_q(gl_n)$ under deformations within the category of coboundary Hopf algebras, one finds rigidity in general (including the standard, one-parameter case), but nontrivial deformations in special cases, among them the new quantizations of $gl(n)$ obtained in Section 5.

2 The Hopf Algebra Dual Form for $U_{q,q'}(gl_2)$

With Woronowicz, we consider the matrix

$$T = \begin{pmatrix} a & b \\ c & d \end{pmatrix} \tag{6}$$

with matrix elements that satisfy the relations

$$\begin{gathered} ab = q'ba, \;\; ac = qca, \\ bd = qdb, \;\; cd = q'dc, \\ q'bc = qcb, \;\; ad - da = (q' - q^{-1})bc, \end{gathered} \tag{7}$$

involving two independent parameters q, q'. The algebra generated by a, b, c, d will be augmented by the addition of an inverse to a, to authorize the factorization

$$T = \begin{pmatrix} 1 & 0 \\ x^- & 1 \end{pmatrix} \begin{pmatrix} a & 0 \\ 0 & \hat{d} \end{pmatrix} \begin{pmatrix} 1 & x^+ \\ 0 & 1 \end{pmatrix}. \tag{8}$$

The algebra is finally completed by including logarithms of a and d, to allow the representation

$$a = e^{x^0}, \;\; \hat{d} = e^{-x^1}, \;\; q = e^h, \;\; q' = e^{h'}. \tag{9}$$

Proposition. The relations (7) are equivalent to the following structure of Lie algebra

$$\begin{gathered} [x^0, x^+] = h'x^+, \;\; [x^0, x^-] = hx^-, \;\; [x^+, x^-] = 0, \\ [x^1, x^+] = hx^+, \;\; [x^1, x^-] = h'x^-, \;\; [x^0, x^1] = 0. \end{gathered} \tag{10}$$

The compatible coproduct, defined by matrix multiplication, takes the form

$$(11)\quad \begin{aligned}\Delta(x^-) &= x^- \otimes 1 + (e^{-x^1} \otimes x^-)(1 + x^+ \otimes x^-)^{-1}(e^{-x^0} \otimes 1),\\ \Delta(x^0) &= x^0 \otimes 1 + 1 \otimes x^0 - (h+h')\textstyle\sum_n(-x^+ \otimes x^-)^n/(e^{nh} - e^{-nh'}),\\ \Delta(x^1) &= x^1 \otimes 1 + 1 \otimes x^1 - (h+h')\textstyle\sum_n(-x^+ \otimes x^-)^n/(e^{nh} - e^{-nh'}),\\ \Delta(x^+) &= 1 \otimes x^+ + (1 \otimes e^{-x^0})(1 + x^+ \otimes x^-)^{-1}(x^+ \otimes e^{-x^1}).\end{aligned}$$

There is a sense in which the Hopf algebra generated by a, b, c, d is dual to the 2-parameter quantum group $U_{q,q'}(gl_2)$. A precise concept of duality can be formulated as follows:

Definition. Let (A, B) be an algebra A with countable basis B. The topological dual (A^*, B^*) is the vector space of linear functions on A, containing all those vanishing on all but a finite subset of B, (with B^* as the dual basis), and certain formal series [4] of elements of B^*. [When, as in the case that A is finitely generated, the choice of B is more or less natural, we speak of the dual A^* of A without referring to B.]

Proposition. The bialgebra dual of the deformed enveloping algebra $U_{q,q'}(gl_2)$ is the deformed enveloping algebra $U_{q,q'}(gl_2{}^*)$ generated by $(x^i), i = -, 0, 1, +$, with relations (10) and coproduct (11).

Proof. In $U_{q,q'}(gl_2{}^*)$ take the basis $(x^-)^m(x^0)^n(x^1)^r(x^+)^s, m, n, r, s = 0, 1, 2, \ldots$, and let (P_{mnrs}) be the dual basis, so that the dual form is

$$(12)\qquad T = \sum (x^-)^m(x^0)^n(x^1)^r(x^+)^s\, P_{mnrs}.$$

Direct calculation using the fundamental relations (3), the relations (10) and the coproduct (11), immediately yields the result that the dual is generated by

$$(13)\qquad p_- = P_{1000},\ p_0 = P_{0100},\ p_1 = P_{0010},\ p_+ = P_{0001},$$

with the relations and the coproduct that characterize $U_{q,q'}(gl_2)$, namely

$$(14)\qquad \begin{aligned}&[p_0, p_\pm] = \pm p_\pm = [p_1, p_\pm],\ [p_0, p_1] = 0,\\ &[p_+, p_-] = (q - q'^{-1})^{-1}[e^{hp_0}e^{h'p_1} - e^{-h'p_0}e^{-hp_1}].\end{aligned}$$

and

$$(15)\qquad \begin{aligned}\Delta(p_0) = p_0 \otimes 1 + 1 \otimes p_0,\ &\Delta(p_1) = p_1 \otimes 1 + 1 \otimes p_1,\\ \Delta(p_+) &= p_- \otimes e^{-h'p_0 - hp_1} + 1 \otimes p_+,\\ \Delta(p_-) &= p_- \otimes 1 + e^{hp_0 + h'p_1} \otimes p_-,\end{aligned}$$

Proposition. The dual form (12) can be expressed in terms of the generators, as follows

$$(16)\qquad T = e_{1/k}^{x^-p_-} e^{x^0p_0} e^{x^1p_1} e_k^{x^+p_+}.$$

Here $k = qq'$ and e_k^x is the usual twisted exponential

$$(17)\qquad e_k^x := \sum_n \frac{x^n}{[n!]_k},\ [n!]_k := [1]_k[2]_k \ldots [n]_k,\ [n]_k = \frac{k^{n+1} - 1}{k - 1}.$$

To summarize. The bialgebra dual of $U_{q,q'}(gl_2)$, with structure relations (14) and coproduct (15), is the solvable quantum group $U_{q,q'}(gl_2)$ with structure (10) and coproduct (11). Both are thus finitely generated and the Hopf algebra dual form can be expressed, as in (16), in terms of the generators.

3 Esoteric Quantum gl_3

All recent activity in the field of integrable models makes essential use of the Yang-Baxter matrix, in its original quantum mechanical context or in the classical form developed later [5]. The existence of the classical r-matrix was interpreted by Drinfeld [6] in terms of coboundary Lie bialgebras, and the construction of a related quantum R-matrix could then be viewed as arising from quantization, especially when quantization is interpreted as a deformation of classical structures [7]. A principal new result reported here is that Drinfeld's restriction to coboundary Lie bialgebras can be relaxed.

Basic to all applications is the construction of a Lax pair, a matrix valued, flat connection. Two Lie algebras are involved. In this section, G_x will denote the Lie algebra of basic dynamical variables. This is not the infinite dimensional Poisson algebra on phase space, but some finite subalgebra of it that is the focus of interest and that forms the basis for quantization. In the context of integrable models, G_x is generated by the dynamical variables that appear as matrix elements of the Lax connection. Typically, G_x is the Heisenberg algebra (as in the nonlinear Schroedinger model) or the Euclidean group (as in the sine-Gordon theory), and sometimes it is a simple Lie algebra (as in the case of spin systems). On the other hand, G_l will denote the auxiliary matrix algebra where the Lax connection lives. There is no *a priori* known principle that commands a particular choice of G_l, except that a preference for simplicity usually leads to the lowest possible dimension; this explains why this algebra is often simple; in the familiar examples it is $gl(2)$.

A Lax pair does not automatically imply the existence of an r-matrix. According to the insight of Drinfeld [6], the r-matrix is intimately related to the fact that G_x and G_l can be promoted to a pair of mutually dual Lie bialgebras, and more especially to the fact that G_l (but not G_x!) is a coboundary Lie bialgebra. The connection is provided by the relation $f = dr$, where f is the G_l one-form that determines the coproduct on G_l (and the Lie product on G_x), and dr is the differential of r. This then implies very strong conditions of compatibility between the two algebras. For example, if G_l is $sl(2)$, then the structure of G_x is fixed; it is not the Euclidean algebra $E(2)$, though it is similar to it (one of the signs in the commutation relations is different). Therefore, sl_2 cannot be used as the auxiliary algebra to solve the sine-Gordon model, for example, for the dynamical algebra of this model is $E(2)$. The impasse is resolved by the device of the spectral parameter. Instead of $sl(2)$, one introduces the infinite dimensional Lie algebra $sl(2)[\lambda]$, of matrices of power series in a parameter λ, the spectral parameter. This infinite dimensional Lie algebra can be endowed with a coproduct that turns it into a coboundary Lie algebra. Also, and this is the point, its (infinite dimensional) dual contains $E(2)$ as a quotient. Here are the details.

The r-matrix that was used to solve the sine-Gordon model is [8]

$$r = (l_3 \otimes l_3 \cosh u + l_+ \otimes l_- + l_- \otimes l_+) / \sinh u, \quad u = \lambda - \mu \tag{18}$$

in which (l_i) is the usual Weyl basis for sl_2. The algebra $sl_2[\lambda]$ has the basis $(l_{ni} = \lambda^n l_i), i = 3, +, -$, with n running over an infinite set of integers to be specified. The dual form is

$$\Gamma = \sum_{n,i} \gamma^{ni} l_{ni}, \tag{19}$$

and the Lie structure of $sl(2)[\lambda]^*$ (expressed as a Poisson bracket) is determined by

$$\{\Gamma, \Gamma'\} = [r, \Gamma + \Gamma'], \quad \Gamma' = \sum_{n,i} \gamma^{ni} l'_{ni}, \tag{20}$$

where l'_{ni} is the basis for another copy of $sl(2)[\lambda]$. One finds that

$$\{\gamma^{n+}, \gamma^{m-}\} = \tau_n^{n+m} \gamma^{n+m,3}, \quad \{\gamma^{m,\pm}, \gamma^{n,3}\} = \sigma_m^{n+m} \gamma^{n+m,\pm}, \tag{21}$$

with the following non-zero coefficients:

$$\begin{aligned} \tau_k^n &= 1, \quad k = 1-n, \dots -3, -1, 1, 3, \dots, n-1, \\ \sigma_k^m &= \begin{cases} -1, & k = m = 1, 3, \dots, \\ +1, & k = m = -1, -3, \dots, \\ -2, & k = 1, 3, 5, \dots, m-2, \\ +2, & k = -1, -3, \dots, 2-m. \end{cases} \end{aligned} \tag{22}$$

This dual algebra contains $E(2)$ as the quotient by the ideal generated by all γ^{ni} except $\gamma^{0,3}, \gamma^{1,+}$, and $\gamma^{-1,-}$.

Thus it seems as if, to incorporate a physically relevant dynamical algebra - $E(2)$, $sl_2, \dots$ - into the scheme, the use of a spectral parameter is essential, so that one has to deal with infinite dimensional Lie algebras and infinite dimensional quantum groups. This does not annihilate the subject; but it is interesting to note that all recent work on quantum groups (this paper included) deals with finite dimensional Lie algebras, although this makes the connection to solvable models a little distant. It would certainly be interesting to see some more examples (besides the Liouville model and the conformal Toda models that generalize it [10]) that do not require infinite dimensional Lie algebras but instead provide direct application of q-deformations of finite dimensional Lie bialgebras to integrable field theories.

Our inspiration comes from the high degree of symmetry that exists in the dual relationship between G_x and G_l. The quantum group $U_{q,q'}(gl_2)$ is a deformation of the enveloping algebra $U(gl_2)$ and its bialgebra dual is also a deformation of an enveloping algebra. It is not too much to conjecture that, quite generally, the bialgebra dual of $U_q(G_l)$ is a deformation $U_q(G_x)$ of the enveloping algebra of G_x. This has interesting implications. Namely, by stressing the symmetry between the two Lie bialgebras, rather than the relation of duality between them, one should be able to use them interchangeably; in spite of the fact that only one of them is a coboundary. In the simplest interesting case, that of the quantum group $U_{q,q'}(gl_2)$, the bialgebra dual form is known explicitly in terms of q-exponentials, Eq.(2). The two Lie bialgebras appear quite symmetrically in this "universal T-matrix."

Let us see what kinds of models can be built on this structure, without introducing a spectral parameter. Only $gl(2)$ is a coboundary Lie bialgebra, its dual is not; therefore, according to Drinfeld, it is $gl(2)$ that is suitable for use as matrix algebra; and $gl(2)^*$ must play the role of the dynamical algebra. But this is quite contrary to the requirements of the applications, so it is natural to persist in trying to reverse the situation. Thus, we identify $gl(2)$ with the dynamical algebra, and the solvable dual $gl(2)^*$ with the matrix algebra.

If in the formula (2) one takes the l's in the 2-dimensional representation of $gl(2)$ (which is what one normally does), then the matrix elements T_i^j of T satisfy the algebraic conditions (7) of Woronowicz's matrix pseudogroup. Instead, we shall now take the x's in a faithful representation of $gl_2{}^*$, then the matrix elements of T satisfy another set of algebraic relations. This new algebra of matrix elements is related to gl_2 and to $U_{q,q'}(gl_2)$ in the same way that the algebra of Woronowicz is related to $gl(2)^*$ and to $U_{q,q'}(gl_2{}^*)$.

The dual Lie algebra $gl(2)^*$ is solvable; the faithful representations of smallest dimension are 3-dimensional. In one such representation the universal T-matrix (2) reduces to

$$T = \begin{pmatrix} a & b & 0 \\ 0 & d & 0 \\ 0 & c & \hat{a} \end{pmatrix}, \tag{23}$$

with matrix elements

$$\begin{gathered} a = e^{h[p_1 - \eta(p_1+p_2)]}, \; \hat{a} = e^{h[p_1+\eta(p_1+p_2)]} \\ b = hp_- d, \; c = h\hat{a}p_+, \; d = e^{hp_2} \end{gathered} \tag{24}$$

satisfying the relations

$$\begin{gathered} ab = q^{-1}ba, \; ac = qca, \; [a,\hat{a}] = 0, \; [a,d] = 0 = [\hat{a},d], \\ \hat{a}b = q^{-1}b\hat{a}, \; \hat{a}c = qc\hat{a}, \; bd = q^{-1}db, \; cd = qdc, \\ [b,c] = \lambda(a\hat{a} - dd). \end{gathered} \tag{25}$$

The two parameters q and q' of $U_{q,q'}(gl_2)$ have been replaced by $e^{h(1-\eta)}$ and $e^{h(1+\eta)}$, respectively. One regards η as a fixed parameter of the undeformed theory and h as a single deformation parameter. From now on this will uniformly be the significance of η, h and $q = e^h$.

The roles of the two underlying Lie algebras have thus been reversed. Recall that, in the applications, the matrices are used to construct the Lax connection, while the matrix elements are the dynamical variables. Our example is therefore appropriate for systems whose principal dynamical variables are those of gl_2 or sl_2 (after restriction) or $E(2)$ (after contraction); for example, spin systems and the sine-Gordon model.

In order for this to be useful one needs (according to present wisdom) an R-matrix. The existence of an R-matrix satisfying

$$R(T \otimes 1)(1 \otimes T) = (1 \otimes T)(T \otimes 1)R \tag{26}$$

would imply the "multiplication property", but the converse is not true. The multiplication property is built into (2) and is expressed by (3), but the existence of an R-matrix satisfying (26), (the "coboundary" property) must be checked directly. A matrix satisfying this condition will be constructed explicitly below, after some preparation.

Once more according to current wisdom, this matrix should have to satisfy the Yang-Baxter or braid relation

$$R_{12}R_{13}R_{23} = R_{23}R_{13}R_{12} \tag{27}$$

That this relation turns out to be satisfied in the case at hand is a surprise. Recall that the infinitesimal form of this relation is the classical Yang-Baxter relation, and that this is related to the coboundary property of a Lie bialgebra, the matrix algebra G_l. In our case that would be the solvable dual of $gl(2)$. But this Lie bialgebra is not coboundary! Nevertheless, we verify that the R-matrix does have an expansion around unity, and that the first order term satisfies a modified classical Yang-Baxter relation. This surprising and, we believe important result does not contradict what was known previously. The point is that our r-matrix does not live in $gl(2) \otimes gl(2)$, but in $M_3 \otimes M_3$.

The wider implication of this result is that, to find a useful Lax pair for integrable models in which the dynamical algebra is $E(2)$ or $sl(2)$, it may be unnecessary to introduce a spectral parameter and, with it, an infinite dimensional auxiliary algebra. A flat sl(2) connection does not exist, but one can be found that is valued in a larger, finite dimensional matrix algebra.

We try to interpret the matrix (23), with matrix elements satisfying the relations (25), in terms of automorphisms of a 3-dimensional quantum plane, an algebra with generators ("coordinates") $(\xi^i), i = 1, 2$, and quadratic relations [9]. That is, we require that the mapping given by

$$\xi \mapsto \xi T = \xi' = (\xi^1 a, \xi^1 b + \xi^2 d + \xi^3 c, \xi^3 \hat{a}), \tag{28}$$

preserve the relations. It is easy to see that this condition is satisfied if the relations among the coordinates are

$$q\xi^1\xi^2 = \xi^2\xi^1, \; q\xi^2\xi^3 = \xi^3\xi^2, \; q^2\xi^1\xi^3 = \xi^3\xi^1, \tag{29}$$

and that these are the only relations of the type $\xi^i\xi^j = q_{ij}\xi^j\xi^i$, $i < j$, that are so preserved.

Now our strategy for constructing the R-matrix calls for a similar statement about an associated exterior algebra. Let $(\theta^i), i = 1, 2, 3$, be the generators of this exterior algebra; what is preserved is the unique structure

$$\begin{gathered} \theta^1\theta^2 + q\theta^2\theta^1 = 0, \; \theta^2\theta^3 + q\theta^3\theta^2 = 0, \; \theta^1\theta^3 + \theta^3\theta^1 = 0, \\ \theta^1\theta^1 = \theta^2\theta^2 = 0, \; \theta^2\theta^2 + \lambda\theta^3\theta^1 = 0. \end{gathered} \tag{30}$$

There is nothing canonical about the relations (29-30), except for their number, 3^2; but they are all preserved when T acts from the right.

This property of the relations (29-30) has the important implication that there is a numerical 9×9 matrix U such that

$$\theta^i\theta^j U_{ij}^{kl} = 0, \; \xi^i\xi^j(U - 1 - q^2)_{ij}^{kl} = 0, \tag{31}$$

which is unique if $1 + q^2 \neq 0$, namely

$$(32)\quad U = (1+q^2) \otimes \begin{pmatrix} 1 & q \\ q & q^2 \end{pmatrix} \oplus \begin{pmatrix} 1 & -\lambda q^2 & q^2 \\ 0 & 1+q^2 & 0 \\ 1 & \lambda & q^2 \end{pmatrix} \oplus \begin{pmatrix} 1 & q \\ q & q^2 \end{pmatrix} \oplus (1+q^2).$$

[The subspaces are in correspondence with $i + j = 2, 3, ..., 6$.]

The fact that U preserves both structures is equivalent to the statement that U commutes with $(T \otimes T)$; hence U is, up to the addition of a scalar, the matrix that is conventionally denoted $\hat{R}$:

$$(33)\qquad \hat{R} = U - q^2.$$

This turns (31) into

$$(34)\qquad \theta\theta(\hat{R} + q^2) = 0,\ \xi\xi(\hat{R} - 1) = 0,$$

so that $\xi^i \xi^j$ is "symmetric" (the deformed permutation $\hat{R}$ is unity on $\xi\xi$). Also, this makes $R \mapsto 1$ when $h \mapsto 0$. This construction of R does not guarantee that it satisfies the Yang-Baxter relation; nevertheless it does. In other words, we have discovered a new coboundary bialgebra deformation of $U(gl_3)$. It is related to a known but somewhat esoteric classical r-matrix [3].

4 The Dual Form for Standard Quantum $gl(n)$

Following Woronowicz [1] and Manin [9], we investigate a bialgebra A, generated by elements $(z_i^j), i, j = 1, ..., n$. The relations are

$$(35)\qquad \begin{aligned} z_i^j z_i^k &= q z_i^k z_i^j,\ j < k, \\ z_i^k z_j^k &= q z_j^k z_i^k,\ i < j, \\ [z_i^j, z_k^l] &= 0, \qquad j < l,\ i > k, \\ [z_i^j, z_k^l] &= \lambda z_k^j z_i^l,\ j < l,\ i < k, \end{aligned}$$

with $\lambda := q - q^{-1}$, and the coproduct is defined by matrix multiplication,

$$(36)\qquad \Delta(z_i^j) = z_i^k \otimes z_k^j.$$

We do not fix the quantum determinant.

Define three "bialgebra quotients" of A. As algebras, $(A_i), i = -, 0, +$, are quotients of A by ideals I_i that are generated respectively by

$$(37)\qquad I_- : \{z_i^j, i < j\},\ I_0 : \{z_i^j, i \neq j\},\ I_+ : \{z_i^j, i > j\}\ ;$$

the relations of A_i are obtained from A by setting the generators of I_i to zero.

The reason why these algebraic quotients deserve to be called bialgebra quotients is that a compatible coproduct is naturally induced on each from that of A; namely, the coproduct (36) has the property

$$(38)\qquad \Delta(I) \subset I \otimes A + A \otimes I,$$

for $I = I_-, I_0$ or I_+. For example, taking z_i^j in I_-, thus $i < j$, it is evident that $\Delta(z_i^j)$ has one or the other factor in I_-.

We introduce copies of A_i, with the correspondence

$$A_- : z_i^j \mapsto \hat{X}_i^j, \ A_0 : z_i^i \mapsto z_i, \ A_+ : z_i^j \mapsto \hat{Y}_i^j. \tag{39}$$

The formula

$$z_i^j = \hat{X}_i^k \hat{Y}_k^j \tag{40}$$

is an imbedding of A into $A_- \otimes A_+$, but to obtain a unique decomposition of A we need to take a factor A_0 out of A_- and A_+. The generators z_i^i of A are assumed to be invertible. Now there are unique factorizations (x_i and y_i being invertible),

$$\begin{aligned} \hat{X}_i^j &= X_i^j x_j, \ j \leq i, \quad X_i^i = 1, \\ \hat{Y}_i^j &= y_i Y_i^j, \ i \leq j, \quad Y_i^i = 1. \end{aligned} \tag{41}$$

The elements x_i generate a copy A_{0-} of A_0, an Abelian bialgebra (commutative and co-commutative), and the y_i's generate another copy, A_{0+}. The relations among the X_i^j (the Y_i^j) are of the same form as those that hold for the z_i^j, with the obvious specializations. These quantities generate bialgebras A_{--} and A_{++}, with the usual formula for the coproducts, but they are not bialgebra quotients of A_- and A_+.

We imbed A_0 into $A_{0-} \otimes A_{0+}$ by setting

$$z_i = x_i y_i \tag{42}$$

and notice that the factorization (40) takes the form

$$z_i^j = \sum_{k \leq i,j} X_i^k z_k Y_k^j. \tag{43}$$

Proposition. This factorization is unique, $A = A_- \otimes_{A_0} A_+$.

Proof. One begins by noticing that $z_1^1 = z_1$ and proceeds by induction.

The relations satisfied by the X_i^k, the z_k and the Y_k^j are easily determined from the formulas given; note in particular that X_i^j and x_i commute with Y_k^l and y_k while for example,

$$X_i^j z_i = q z_i X_i^j, \ i > j, \quad z_i Y_i^j = q Y_i^j z_i, \ i < j. \tag{44}$$

We analyze the algebraic structure of A_{--}. Among the generators $X_i^j, i < j$, we distinguish the "simple generators"

$$X_i = X_{i+1}^i, \ i = 1, \ldots, N-1. \tag{45}$$

They satisfy the following relations, the commutation relations

$$[X_i, X_j] = 0, \ |i - j| > 1 \tag{46}$$

$$X_iX_{i+1} - q^{-1}X_{i+1}X_i =: [X_i, X_{i+1}]_{1/q} = \lambda X^i_{i+2}, \tag{47}$$

and the "Serre relations"

$$\begin{aligned} [X_i, [X_i, X_{i+1}]_{1/q}]_q &= 0, \\ [[X_i, X_{i+1}]_{1/q}, X_{i+1}]_q &= 0. \end{aligned} \tag{48}$$

Recall that the generators x_i are supposed to be invertible; set $q = e^h$ and introduce new generators ρ_i by

$$x_i = e^{\rho_i}. \tag{49}$$

The relations, expressed in terms of the X^j_i and the ρ_i, reduce in the limit $h \mapsto 0$ to the commutation relations of a parabolic subgroup of $gl(n)$. After replacing the x_i by the ρ_i we find a deformation of the enveloping algebra of this Lie algebra; by abuse of notation we still call this algebra A_-. We now calculate its dual, and the bialgebra dual form.

Consider the formal series

$$T_{x,p} = \sum \prod_{i<j,m} (X^j_i)^{a_{ij}} (\rho_m)^{\alpha_m} P_{[a][\alpha]}. \tag{50}$$

The first two groups of factors run over a basis for A_-, the last over a basis for $A_-{}^*$. The subscript $[a]$ stands for the set of indices $(a_{ij}, i < j)$ and $[\alpha]$ for the set (α_m). What defines the bialgebra dual form is the set of relations

$$T_{x,p}T_{x',p} = T_{\Delta(x),p}, \quad T_{x,p}T_{x,p'} = T_{x,\Delta(p)}. \tag{51}$$

In the first relation, x and x' refer to two copies of A_-; the coproduct is known and the equation determines the relations of $A_-{}^*$. In the second relation, p and p' refer to two copies of $A_-{}^*$; the relations of A_- are known and the equation determines the coproduct of $A_-{}^*$.

The calculation is quite painless provided that the most convenient ordering of the X^j_i is chosen:

$$X^1_2, X^1_3, \ldots, X^1_n, X^2_3, \ldots, X^2_n, X^3_n, \ldots \tag{52}$$

We find that $A_-{}^*$ is generated by

$$P^j_i = P_{[a][0]}, \quad a^j_i = 1, \text{ all others zero,} \tag{53}$$

and

$$P_m = P_{[0][\alpha]}, \quad \alpha_m = 1, \text{ all others zero.} \tag{54}$$

The relations satisfied by the generators are

$$\begin{aligned} [P_i, P_j] &= 0, \\ [P_i, P^k_j] &= (\delta_{ij} - \delta_{ik})P^k_j, \ j > k, \\ [P^j_i, P^l_j]_q &= P^l_i, \ i > j > l, \\ [P^j_i, P^l_k]_q &= 0, \ i > j \neq k > l. \end{aligned} \tag{55}$$

The expression for $T_{x,p}$ is simply

$$T_{x,p} = \prod_{i<j,m} e_k^{X_i^j P_i^j} e^{\rho_m P_m}, \ k := q^{-2}, \tag{56}$$

the factors to be ordered as above. This completes the calculation of the universal T-matrix for A_-. Similarly, for A_+, set $y_i = e^{\sigma_i}$ and

$$T_{x,p} = \prod e^{\sigma_m P_m} e_k^{Y_i^j \hat{P}_i^j}, \tag{57}$$

to discover the relations of $A_+{}^*$:

$$\begin{aligned} [P_i, P_j] &= 0, \\ [P_i, \hat{P}_j^k] &= (\delta_{ij} - \delta_{ik})\hat{P}_j^k, \ j < k, \\ [\hat{P}_i^j, \hat{P}_j^l]_q &= P_i^l, \ i < j < l, \\ [\hat{P}_i^j, \hat{P}_k^l]_q &= 0, \ i < j \neq k < l. \end{aligned} \tag{58}$$

For the complete algebra A we combine (56) and (57).

Theorem. The bialgebra dual of the standard quantum deformation of $U(gl_n)$ is generated by (X_i^j, τ_i, Y_i^j), $\tau_i = \rho_i + \sigma_i$, and with relations and coproduct inherited from (35) and (36). The bialgebra dual form can be expressed in terms of the generators,

$$T_{x,p} = \prod e_{1/k}^{X_i^j P_i^j} e^{\tau_i P_i} e_k^{Y_k^l \hat{P}_k^l}, \ k := q^2, \tag{59}$$

(The factors involving the X_i^j have to be ordered as in (52) and those involving the Y_i^j as in $Y_1^2, Y_1^3, \ldots$.)

5 Esoteric Quantum $gl(2n-1)$

We shall describe a generalization of esoteric quantum gl_3. The R-matrix takes the form

$$R = R_0 + R_1, \tag{60}$$

with

$$\begin{aligned} R_0 = \sum_{i=1}^{2n-1} M_i^i \otimes M_i^i + \sum_{i<j<2n} (1-q^2) M_j^i \otimes M_i^j + \sum_{1\neq j, i+j\neq 2n} q M_i^i \otimes M_j^j \\ + \sum_{i<n} M_i^i \otimes M_{i'}^{i'} + \sum_{i<n} q^2 M_{i'}^{i'} \otimes M_i^i, \end{aligned} \tag{61}$$

and R_1 equal to

$$\sum_{k<i<n} (\mu_i' M_i^n \otimes M_{i'}^n + \mu_i M_{i'}^n \otimes M_i^n) + \sum_{k<i<j<n} (\lambda_{ij} M_i^j \otimes M_{i'}^{j'} + q^2 \lambda_{ij}' M_{i'}^{j'} \otimes M_i^j). \tag{62}$$

The coefficients $(\mu_i \neq 0), k < i < n$ determine the others,

$$
\begin{aligned}
\mu'_i &= -q^{2(i-n)}\mu_i, \quad k < i < n, \\
\lambda_{ij} &= (1-q^2)q^{2(i-j)}(\mu_i/\mu_j), \\
\lambda'_{ij} &= (1-q^{-2})(\mu_i/\mu_j), \quad i < j < n.
\end{aligned} \tag{63}
$$

Here $i' := 2n - i$ and k is fixed, $0 \leq k \leq n-2$.

The bialgebra A is the algebra generated by the matrix elements $(T_i^j), i,j = 1,...,2n-1$ of a matrix T, with relations

$$R_{12}T_2T_1 = T_1T_2R_{12}, \tag{64}$$

and with the coproduct

$$\Delta(T_i^j) = T_i^k \otimes T_k^j. \tag{65}$$

The factorization used in the standard case does not work here.

We shall describe three bialgebra quotients, denoted A_-, A_+ and A_0; they will play the same roles as their homologues in the standard case, but their construction and their properties will be different. In particular, A_0 will not be Abelian.

We need some descriptive terms for matrices. If $m = (m_i^j)$ is any matrix we denote by (i,j) the position of m_i^j in it. Two regions are important,

$$D_+ = \{(i,j), k < i < j \leq n\}, \quad D_- = \{(i,j), n \leq j < i < 2n-k\}. \tag{66}$$

Definition. A matrix (m_i^j) is called almost lower triangular if $m_i^j = 0$ whenever $i \ll j$, which means that $i < j$ and $(i,j) \notin D_+$; it is called almost upper triangular if $m_i^j = 0$ whenever $i \gg j$, by which is meant that $i > j$ and $(i,j) \notin D_-$.

As matrices, A_- is a space of almost lower triangular matrices and A_+ is a space of almost upper triangular matrices, while A_0 consists of matrices that are almost upper triangular as well as almost lower triangular; they may be said to be almost diagonal:

$$
\begin{pmatrix}
z & 0 & 0 & 0 & 0 & 0 & 0 \\
0 & z & z & z & 0 & 0 & 0 \\
0 & 0 & z & z & 0 & 0 & 0 \\
0 & 0 & 0 & z & z & 0 & 0 \\
0 & 0 & 0 & z & z & z & 0 \\
0 & 0 & 0 & 0 & 0 & 0 & z
\end{pmatrix} \tag{67}
$$

In this illustration $n = 4$ and $k = 1$. As algebras, $(A_i), i = -,+,0$, are quotients of A by ideals I_i that are generated respectively by

$$I_- : \{z_i^j, i \ll j\}, \quad I_+ : \{z_i^j, i \gg j\}, \quad I_0 : \{z_i^j, i \ll j \text{ or } i \gg j\}. \tag{68}$$

The algebras A_i are generated by the complimentary sets of generators of A.

We now explain what motivated the above definitions.

It is well known, and an immediate consequence of the Yang-Baxter relation, that the mapping π_-, from A to the space of matrices of dimension $2n-1$, generated by

$$T_i^j \mapsto \pi_-(T_i^j) = R_i^j = \{(R_i^j)_a^b\} = \{R_{ia}^{jb}\} \tag{69}$$

is a representation of A as an algebra. As for the mapping

$$T_i^j \mapsto \pi_+(T_i^j) = S_i^j = \{(S_i^j)_a^b\} = \{R_{ai}^{bj}\}, \tag{70}$$

it is (almost) an anti-homomorphism. Now it is seen by inspection of R that the kernels of π_- and π_+ contain the ideals I_- and I_+, respectively, that is,

$$R_i^j = 0,\ i \ll j,\ S_i^j = 0,\ i \gg j. \tag{71}$$

Furthermore, the representatives R_i^j and S_i^j of the generators z_i^j are invertible. This shows that these generators can be taken to be invertible, as in the standard case, as generators of A_-, of A_+ or A.

There is a unique decomposition of A, similar to the decomposition (43) of the standard case, into three factors,

$$A = A_- \otimes_{A_0} A_+. \tag{72}$$

The factors are defined in the same way as before, starting with the new ideals defined in (67). This decomposition is useful for understanding the structure of esoteric quantum $gl(2n-1)$ and suitable for evaluating the associated universal T-matrix. Complete results will be published elsewhere.

We list all the relations Eq.(65). When $a < i, b < j$, and $a \neq i'$, $b \neq j'$,

$$\begin{aligned}
[z_a^j, z_a^b]_q &= 0,\\
[z_a^{b'}, z_a^b]_{q^2} &= q^2 \textstyle\sum_l \lambda_{lb} z_a^l z_a^{l'} = -\sum_l \lambda'_{lb} z_a^{l'} z_a^l,\\
[z_i^b, z_a^b]_q &= \delta_n^b \textstyle\sum_l(-\mu'_l z_i^{l'} z_a^l - \mu_l z_i^l z_a^{l'}) = \delta_{bn} q^{-1} \sum_l(\mu_l z_a^l z_i^{l'} + \mu'_l z_a^{l'} z_i^l),\\
[z_a^b, z_{a'}^b] &= \delta_n^b q^{-2} \textstyle\sum_l(-\mu'_l z_a^{l'} z_{a'}^l - \mu_l z_a^l z_{a'}^{l'}) + \sum_l \lambda_{al} z_{l'}^b z_l^b + q^{-2}\mu_a z_n^b z_n^b\\
&= \delta_n^b \textstyle\sum_l(\mu_l z_{a'}^l z_a^{l'} + \mu'_l z_{a'}^{l'} z_a^l) - \mu'_a z_n^b z_n^b - \sum_l \lambda'_{al} z_l^b z_{l'}^b,\\
[z_a^b, z_i^j] &= 0,\\
[z_a^{b'}, z_i^b]_q &= -\textstyle\sum_l \lambda'_{lb} z_a^{l'} z_i^l,\\
[z_a^j, z_{a'}^b]_q &= q^{-1}\mu_a z_n^b z_n^j + q \textstyle\sum_l \lambda_{al} z_{l'}^b, z_l^j,\\
[z_a^{b'}, z_{a'}^b]_{q^2} &= q^2 \textstyle\sum_l \lambda_{al} z_{l'}^b z_l^{b'} - \sum_l \lambda'_{ib} z_a^{l'} z_{a'}^l + \mu_a z_n^b z_n^{b'},\\
[z_a^b, z_i^j] &= (q^{-1} - q) z_a^j z_i^b,\\
[z_i^{b'}, z_a^b]_q &= (q - q^{-1}) z_a^{b'} z_i^b + q \textstyle\sum_l \lambda_{lb} z_a^l z_i^{l'},\\
[z_a^b, z_{a'}^j]_q &= (q^{-1} - q) z_a^j z_{a'}^b + q^{-1}\mu_a z_n^j z_n^b + q \textstyle\sum_l \lambda_{al} z_{l'}^j z_l^b,\\
[z_a^b, z_{a'}^{b'}] &= (q^{-2} - 1) z_a^{b'} z_{a'}^b - \textstyle\sum_l \lambda_{lb} z_a^l z_{a'}^{l'} + q^{-2}\mu_a z_n^{b'} z_n^{b'} + \sum_l \lambda_{al} z_{l'}^{b'} z_l^b.
\end{aligned} \tag{73}$$

It is easy to see that the relations are just enough to allow every element of the algebra to be expressed in the form of a standard ordered polynomial.

Now we pass to the quotient A_0 by setting

$$z_i^j = 0, \text{ unless } k < i \leq j \leq n \text{ or } n \leq j \leq i < 2n - k. \tag{74}$$

Again one verifies that the standard ordered monomials span the algebra. The relations of A_0 are a little simpler; with $a < i, b < j$ and $a' \neq i, b' \neq j$,

$$
\begin{aligned}
&[z_a^j, z_a^b]_q &&= 0,\\
&[z_i^b, z_a^b]_q &&= \delta_n^b \textstyle\sum_l(-\mu_l' z_i^{l'} z_a^l - \mu_l z_i^l z_a^{l'}) = \delta_n^b q^{-1} \textstyle\sum_l(\mu_l z_a^l z_i^{l'} + \mu_l' z_a^{l'} z_i^l),\\
&[z_a^b, z_{a'}^b] &&= \delta_n^b(-q^{-2} \textstyle\sum_l \mu_l z_a^l z_{a'}^{l'} + \textstyle\sum_l \lambda_{al} z_{l'}^n z_l^n + q^{-2}\mu_a z_n^n z_n^n)\\
& &&= \delta_n^b(\textstyle\sum_l \mu_l' z_{a'}^{l'} z_a^l - \mu_a' z_n^n z_n^n - \textstyle\sum_l \lambda_{al}' z_l^n z_{l'}^n),\\
(75)\quad &[z_a^j, z_i^b] &&= 0,\\
&[z_a^b, z_i^j] &&= (q^{-1} - q) z_a^j z_i^b,\\
&[z_i^{b'}, z_a^b]_q &&= q \textstyle\sum_l \lambda_{lb} z_a^l z_i^{l'},\\
&[z_a^b, z_{a'}^j]_q &&= q \textstyle\sum_l \lambda_{al} z_{l'}^j z_l^b,\\
&[z_a^b, z_{a'}^{b'}] &&= -\textstyle\sum_l \lambda_{lb} z_a^l z_{a'}^{l'} + \textstyle\sum_l \lambda_{al} z_{l'}^{b'} z_l^b.
\end{aligned}
$$

The positive and negative simple roots are

$$
(76)\qquad e_i = z_i^{i+1}, \text{ and } f_i = z_{i'}^{i'-1}, \; k < i < n;
$$

the former commute with the latter except that

$$
(77)\qquad
\begin{aligned}
[e_{n-1}, f_{n-1}] &= q^{-2}\mu_{n-1}(h_n h_n - h_{n-1} h_{n+1}), \; h_i = z_i^i,\\
[e_i, f_i] &= \lambda_{i,i+1}(h_{i+1} h_{i'-1} - h_i h_{i'}), \; i < n-1.
\end{aligned}
$$

This algebra has a nontrivial center that includes $\{h_a - h_{a'}, a = 1, ..., n-1\}$.

6 Deformations of Standard Multiparameter $U(gl_n)$.

The structure of esoteric quantum $gl(3)$ was found by accident, the generalization to $gl(n)$ by making a general ansatz for the R-matrix and demanding that the quantum Yang-Baxter relation be satisfied. A more satisfactory approach is through the use of deformation theory [11]. Deformations around the classical theory gives nothing in the first order; the classical Yang-Baxter relation appears only in the second order. Perturbations around multiparameter $U_q(gl_n)$, $q \in \mathbf{R}^N$, are more rewarding. We find rigidity at general position in parameter space and interesting, essential deformations on certain algebraic surfaces. We have determined the equivalence classes of essential deformations, and we have related these results to the classification, by Belavin and Drinfeld [3], of classical Yang-Baxter matrices. We describe, briefly, a very interesting special case.

Taking $n = 3$, for simplicity, we note that standard multiparameter quantum $gl(3)$ [12] is characterized by three q-parameters and, in addition, the Hecke parameter; that is, the second eigenvalue of $\hat{R}$, the first one being unity. A very special kind of deformation exists under some conditions on the q's, provided the fourth parameter is a cubic root of -1. The R-matrix for this unusual case has the form (61), with

$$
(78)\qquad R_1 = \mu M_1^3 M_2^3.
$$

We thank Moshe Flato, Georges Pinczon and V.S. Varadarajan for consultation.

REFERENCES

1. S.L. Woronowicz, Twisted $SU(2)$ group. An example of noncommutative differential geometry. Publ. Res. Inst. Math. Sci., Kyoto University **23**(1987) 117-181.
2. P.P.Kulish and N.Y. Reshetikhin, Quantum linear problem for the sine-Gordon equation and higher representations. J.Sov.Math. **23**(1983) 2435-2445.
3. A.A. Belavin and V.G Drinfeld, Triangle equations and simple Lie algebras. Sov.Sci.Rev.Math. **4**(1984) 93-165.
4. P. Truini and V.S. Varadarajan, Universal Deformations of Simple Lie Algebras. Lett.Math.Phys.**24**(1992) 63-72.
5. C.N.Yang, Some exact results for the many-body problem in one dimension with repulsive delta-function interaction. Phys.Rev.Lett. **19**(1967) 1312-1315. R.J. Baxter, Partition Function of the Eight-vertex Lattice Model. Ann.Phys. **70**(1971) 193-228. E.K.Sklyanin, Quantum version of the method of inverse scattering problem. J.Sov.Math. **19**(1980) 1546-1596.
6. V.G. Drinfeld, Quantum Groups. Proc. Int.Congr.Math., Berkeley, CA 1986 798-820.
7. F. Bayen, M. Flato, A. Lichnerowicz, C. Fronsdal and D. Sternheimer, Deformation Theory and Quantization. I. Deformations of Symplectic Structures; II.Physical Applications. Ann.Phys. **111**(1978) 61-110 and 111-151.
8. L.A. Takhtajan and L.D. Faddeev, Essentially nonlinear one-dimensional model of classical field theory. Theor.Math.Phys. **21**(1974) 1046-1057.
9. Y.I. Manin, *Topics in noncommutative geometry.* Princeton University Press, Princeton, NJ 1991.
10. A.N. Leznov and M.V. Saveliev, Exact monopole solutions in gauge theories for an arbitrary semisimple compact group. Lett.Math.Phys. **3**(1979) 207-211.
11. M. Gerstenhaber and S.D. Schack, Bialgebra Cohomology, Deformations, and Quantum Groups. Proc.Nat.Acad.Sci. **87**(1990) 478-481.
12. A. Sudbery, Consistent multiparameter quantization of GL(n), J.Phys. A.Math.Gen. **23**(1990) L697. N.Y. Reshetikhin, Multiparameter quantum groups and twisted quasitriangular Hopf algebras. Lett.Math.Phys. **20**(1990) 331-336. A.Schirrmacher, Multiparameter R-matrices and their quantum groups. J.Phys A:Math.Gen. **24**(1991) L1249-L1258. D.B. Fairlie and C.K.Zachos, Multiparameter associative generalizations of canonical commutation relations and quantized planes. Phys.Lett.B **256**(1991 43-49.

Physics Department, University of California, Los Angeles CA 90024.
Departamento de Física Teórica, Universidad Complutense, Madrid.
E-mail address: Bitnet fronsdal@uclaph.

Contemporary Mathematics
Volume **175**, 1994

Fusion rings for modular representations of Chevalley groups

Galin GEORGIEV and Olivier MATHIEU

Introduction. First recall the definition of the classical tensor product multiplicities. Let $G(\mathbf{C})$ be a simply connected, simple algebraic group over $\mathbf{C}$ associated to some Dynkin diagram I. Let P^+ be the set of dominant weights and for any $\lambda \in P^+$ denote by $L(\lambda)$ the simple representation with highest weight λ. For $\lambda, \mu \in P^+$ we have $L(\lambda) \otimes L(\mu) = \Sigma_{\nu \in P^+} K^{\nu}_{\lambda,\mu}.L(\nu)$ where the integers $K^{\nu}_{\lambda,\mu}$ are called the tensor product multiplicities. Note that these multiplicities $K^{\nu}_{\lambda,\mu}$ are explicitly given by a formula of B. Kostant (see **[H]**).

Now define some modified tensor product multiplicities. Let k be an algebraically closed field of characteristic p and let $G = G(k)$ be the simply connected Chevalley group over k associated to I. For any $\lambda \in P^+$ let $P(\lambda)$ be the unique indecomposable tilting module with highest weight λ. These modules $P(\lambda)$ have been discovered recently by C.M. Ringel **[R]** (see section 3 for more details; see also **[D2]**). A theorem of Ringel states that any tilting module is a direct sum of such indecomposable modules $P(\lambda)$. Moreover in view of **[D1][M]** the tensor product of two tilting modules is a tilting module. Thus we can define some new tensor product multiplicities $V^{\nu}_{\lambda,\mu}$ by the requirement $P(\lambda) \otimes P(\mu) = \Sigma_{\nu \in P^+} V^{\nu}_{\lambda,\mu}.P(\nu)$, where $\lambda, \mu \in P^+$. It would be very interesting to compute all these multiplicities. For example such a formula for groups of type A would imply a computation of the dimension of modular simple representations of symmetric groups (via Schur-Weyl duality).

The main result of our paper is a computation of $V^{\nu}_{\lambda,\mu}$ when ν belongs to the fundamental alcove C_p, $\lambda, \mu, \nu \in P^+$. More precisely, there is a constant $n(I)$, which depends only on the Dynkin diagram I such that for $p > n(I)$ we prove the following statement :

THEOREM 4.10 ($p > n(I)$) *Let $\lambda, \mu, \nu \in P^+$. Assume that ν belongs to the fundamental alcove C_p.*

(i) If $\lambda + \rho$ or $\mu + \rho$ is not in the fundamental alcove, then $V^{\nu}_{\lambda,\mu} = 0$.

(ii) If $\lambda + \rho$ and $\mu + \rho$ are in the fundamental alcove, then

$$V^{\nu}_{\lambda,\mu} = \sum_{\substack{w \in W_p \\ w^{-1}(\nu+\rho)-\rho \in P^+}} \epsilon(w).K^{w^{-1}(\nu+\rho)-\rho}_{\lambda,\mu}.$$

1991 *Mathematics Subject Classification.* Primary 17B10; Secondary 17B81, 20C20, 16E20.

This paper is in final form.

The proof of the theorem is essentially based on the fact that the (positive integer) constants $\{ V^{\nu}_{\lambda,\mu} \}_{\lambda,\mu,\nu \in C_p \cap P^+}$ are the structure constants of an associative algebra. The associativity trick follows from this very simple (but apparently new) lemma :

LEMMA 2.7 *Let K be an algebraic group over k, and let M and N be two rational representations of K. Assume that M is indecomposable and of dimension divisible by p. Then the dimension of any direct summand of $M \otimes N$ is divisible by p.*

In the remaining part of the introduction we will explain the relationship with Moore and Seiberg tensor products for affine Kac-Moody Lie algebras.

Let $\mathcal{L}$ be a complex affine Kac-Moody Lie algebra, let l be a positive integer and let $\mathcal{O}^l_{int}$ be the category of integrable modules in the category $\mathcal{O}$ of level l. For M, $N \in \mathcal{O}^l_{int}$, the module $M \otimes N$ has level $2l$ and usually the multiplicities of simple components in $M \otimes N$ are infinite. However Moore and Seiberg **[MS]** defined a new tensor product $M \otimes^{\cdot} N$ in a such way that $M \otimes^{\cdot} N$ has level l and the mutiplicities of simple components are finite (for a rigorous mathematical treatment of this modified product see **[KL0 $\div$ 5]**;**[F]** for positive integer level and **[HL]** for general Vertex Operator Algebra modules). Moreover the multiplicities are given by a particular case of Verlinde's formula **[V]**. The Grothendiek ring $K(\mathcal{O}^p_{int})$ (with product induced by $\otimes^{\cdot}$) is called the fusion ring. Thus the fusion ring is the Grothendieck ring of the following tensor category $\mathcal{P}^l_{aff}$:

(i) Objects are all $\mathcal{L}$-modules in $\mathcal{O}^l_{int}$.
(ii) Morphisms are the usual morphisms of $\mathcal{L}$-modules.
(iii) The tensor product is the non-classical tensor product $\otimes^{\cdot}$.

Indeed in the setting of modular representations of the Chevalley group G we can define a tensor category $\mathcal{P}_{mod}$ by the following data :

(i) Objects are all tilting modules.
(ii) Morphisms are given by some functor T (not the usual one).
(iii) The tensor product is the usual tensor product.

Roughly speaking, our main result states that the tensor product multiplicities in the two categories are the same. In order to give a more precise statement, let us first introduce an auxiliary construction. Let J be a Dynkin diagram and let $\mathfrak{g}$ be the corresponding complex simple Lie algebra. Given an automorphism θ of a Dynkin diagram J we can associate an automorphism Θ of $\mathfrak{g}$. Then we define J/θ as the Dynkin diagram of the Lie algebra $\mathfrak{g}^{\Theta}$. Recall that $(J, \theta) \mapsto J/\theta$ is a bijective correspondence between

1) Pairs (J, θ), where J is a simply laced connected Dynkin diagram and θ is an automorphism with at least one fixed point,

2) connected Dynkin diagrams (J/θ).

For our (connected) Dynkin diagram I, let (J, θ) be the unique pair such that $I = J/\theta$ and let $\mathcal{L}$ be the complex twisted affine Lie algebra associated with the pair (J, θ). Also set $l = p - h^v$ where h^v is the dual Coxeter number . Assume $p > n(I)$ and $p > h^v$. Then the simple objects in the categories $\mathcal{P}^l_{aff}$ and $\mathcal{P}_{mod}$ are classified by the same alcove and the main result precisely states that the tensor product multiplicities are the same (see **[V]**, **[F]**, **[K]** for calculation of the multiplicities in $\mathcal{P}^l_{aff}$).

It is expected that this coincidence can be explained through a lifting to quantum groups and an equivalence of the category of tilting modules for quantum groups (with modified morphisms) at a p-root of unity and the category of integrable modules of the corresponding complex affine Lie algebras at a level $p - h^v$ (see **[GK]**, **[F]**).

The result about multiplicities $V^{\nu}_{\lambda,\mu}$ can be stated in a different way. Denote by $K(G)$ the Grothendieck ring of rational (finite-dimensional) G - modules. In a natural way (see section 4), $K(\mathcal{P}_{mod})$ is a quotient of $K(G)$. Recall that $Spec(\mathbf{C} \otimes K(G))$ is the set of semi-simple conjugacy classes in $G(\mathbf{C})$ (see **[S]**). We prove that the spectrum of $\mathbf{C} \otimes K(\mathcal{P}_{mod})$ is the set of (semi-simple) regular conjugacy classes $[g]$ in $G(\mathbf{C})$, such that g^p is central (Theorem 4.8).

Remarks. A very different construction of a tensor category is given in **[GK]**. It follows easily from **[GK]** and the present paper that these two categories are the same, at least when $p > n(I)$. A similar category has been considered for quantum groups in **[A2]**. Results of this paper have been announced in the conference and in a short note **[GM]**.

Acknowledgements. The authors thank M. Duflo, J. Lepowsky and J. Jones for helpful discussions. G.G. is supported by Rutgers University Graduate Excellence Fellowship. O.M. is partially supported by a NSF grant, a grant of the Sloan foundation and UA 748 du CNRS.

1. Preliminary results

(1.1) Let $A = (a_{i,j})_{i,j \in I}$ be an l by l indecomposable Cartan matrix. A real realization of A is the data of a l-dimensional real vector space $\mathfrak{h}_{\mathbf{R}}$, a scalar product $(,)$ on $\mathfrak{h}^*_{\mathbf{R}}$ and a set $\Pi = \{\alpha_i | i \in I\} \subset \mathfrak{h}^*_{\mathbf{R}}$ such that $2.(\alpha_i, \alpha_j)/(\alpha_i, \alpha_i) = a_{i,j}$. There exists such a realization and it is unique up to an isomorphism. Moreover the scalar product is definite and Π is a basis of $\mathfrak{h}^*_{\mathbf{R}}$. We will assume that $(,)$ is positive. Define $\rho \in \mathfrak{h}^*_{\mathbf{R}}$ by $2(\rho, \alpha_i) = (\alpha_i, \alpha_i)$, for any $i \in I$.

(1.2) Let W be the group generated by the hyperplane reflexions $s_i : \lambda \mapsto \lambda - 2.(\lambda, \alpha_i)/(\alpha_i, \alpha_i)\, \alpha_i$. Set $\Delta = W.\Pi$. Set $\Delta^+ = \{\alpha \in \Delta | (\alpha, \rho) \geq 0\}$. For any $\alpha \in \Delta$ let h_α be the unique element of $\mathfrak{h}_{\mathbf{R}}$ such that $1 - \alpha \otimes h_\alpha$ is an orthogonal hyperplane reflexion. There is a unique $\theta \in \Delta^+$ such that $\theta + \alpha_i \notin \Delta$, for any $i \in I$. Normalize $(,)$ by imposing $(\theta, \theta) = 2$.

By definition W is called Weyl group, Δ is the set of roots, Δ^+ is the set of positive roots, elements h_α are called coroots, θ is the highest root and h_θ is the highest short coroot. The number $h^v = \rho(h_\theta)$ is the dual Coxeter number. Let h_0 be the highest coroot and for $i \in I$ set $h_i = h_{\alpha_i}$. Also set $P = \{\lambda \in \mathfrak{h}^*_{\mathbf{R}} | \lambda(h_\alpha) \in \mathbf{Z}$ for any $i \in I\}$ and $P^+ = \{\lambda \in \mathfrak{h}^*_{\mathbf{R}} | \lambda(h_\alpha) \in \mathbf{N}$ for any $i \in I\}$. Elements of P (respectively P^+) are called integral weights (respectively dominant weights). Abusing the notations, we will denote by $I(A)$, or simply I, the Dynkin diagram, corresponding to A.

(1.3) Let n be a positive integer. Let $\mathcal{H}_n$ be the collection of all affine hyperplanes of $\mathfrak{h}^*_{\mathbf{R}}$ given by an equation $\lambda(h) = nj$, where h is a coroot and j is an integer. Let Ω_0 (respectively Ω_1) be the set of all $\lambda \in \mathfrak{h}^*_{\mathbf{R}}$ which belong to no hyperplanes from $\mathcal{H}_n$ (respectively which belong to exactly one hyperplane from $\mathcal{H}_n$). Elements of Ω_0 (respectively of $\mathfrak{h}^*_{\mathbf{R}} \setminus \Omega_0$) are called n-regular (respectively n-singular). The connected components of Ω_0 are called the alcoves and the connected components of Ω_1 are called the walls. Set $C_n = \{\lambda \in \mathfrak{h}^*_{\mathbf{R}} | \lambda(h_0) < n$ and

$\lambda(h_i) > 0$ for any $i \in I\}$. Then C_n is an alcove and it is called a fundamental alcove. Its walls are called fundamental walls. For any $H \in \mathcal{H}_n$ let s_H be the affine orthogonal hyperplane reflexion with respect to H. Similarily, for any wall M, set $s_M = s_{\overline{M}}$, where $\overline{M}$ is the affine hyperplane containing M. The group generated by all these hyperplane reflexions is called affine Weyl group. It is denoted by W_n.

LEMMA 1.4 *There is a constant $n(I)$ which depends only on the Dynkin diagramm I such that for $n > n(I)$ the following statement holds:*
Any fundamental wall M contains at least one integral weight λ_M such that $\lambda_M + \Delta \subset C_n \cup M \cup s_M C_n$.

Proof. For $n = 1$ define M_0 to be the unique fundamental wall in the hyperplane $\{\lambda | \lambda(h_0) = 1\}$ and for $i \in I$ define M_i to be the unique fundamental wall in the hyperplane $\{\lambda | \lambda(h_i) = 0\}$. Set $O_i = C_1 \cup M_i \cup s_{M_i} C_1$. Each O_i is an open set, thus for any $i \in I \cup \{0\}$ there is some $\lambda_i \in M_i$ and $\epsilon > 0$ such that $B(\lambda_i, \epsilon) \subset O_i$ (where $B(\lambda_i, \epsilon)$ is the open ball with center λ_i and radius ϵ). Set $D_i = B(\lambda_i, \epsilon/2) \cap M_i$. Clearly there is some constant n_i such that for any integer $n > n_i$, $n.D_i$ contains at least one integral weight. Choose $n(I)$ such that $n(I).\epsilon > 2.\sqrt{(\theta,\theta)}$ and $n(I) > n_i$ for any $i \in I \cup \{0\}$.

Let $n > n(I)$ be an integer. Then we have $C_n = n.C_1$ and the fundamental walls are $n.M_i$ where $i \in I \cup \{0\}$. Choose an integral weight $\lambda_i \in n.D_i$. Then we have $\lambda_i + \Delta \subset B(\lambda_i, \sqrt{(\theta,\theta)}) \subset B(\lambda_i, n.\epsilon/2) \subset n.O_i$. Since we have $C_n \cup M \cup s_M C_n = n.O_i$, the claim is proved.

LEMMA 1.5 *Assume that the integer n is greater than $n(I)$. Then for any wall M, there exists an integral weight $\lambda \in M$ such that $\lambda + \Delta \subset C \cup M \cup C'$ where C, C' are the two alcolves adjacent to M.*

Proof. Lemma 1.5 results from the previous lemma and the following facts:
(i) Δ is W-invariant,
(ii) P is W_n-invariant,
(iii) Any alcove is W_n-conjugated to the fundamental alcove. Q.E.D.

Set $P' = \{h \in \mathfrak{h} | \alpha(h) \in \mathbf{Z}\, for\, any\, \alpha \in \Delta\}$. Set $\alpha_0 = \theta$ (the highest root). Define the dual fundamental alcove as

$C'_n = \{h \in \mathfrak{h}_{\mathbf{R}} | \alpha_0(h) < n \text{ and } \alpha_i(h) > 0\, for\, any\, i \in I\}$.

LEMMA 1.6 *There is a linear isomorphism $\Phi : \mathfrak{h}_{\mathbf{R}} \to \mathfrak{h}^*_{\mathbf{R}}$ such that $\Phi(P') = P$ and $\Phi(C'_n) = C_n$.*

Proof. For $i \in I$ define $\omega_i \in \mathfrak{h}^*_{\mathbf{R}}$, $\omega'_i \in \mathfrak{h}_{\mathbf{R}}$ and $m_i, m'_i \in \mathbf{Z}$ by the requirements
(i) $\omega_i(h_j) = \delta_{i,j}$
(ii) $\alpha_i(\omega'_j) = \delta_{i,j}$
(iii) $\alpha_0 = \Sigma_{i \in I}\, m'_i \alpha_i$
(iv) $h_0 = \Sigma_{i \in I}\, m_i h_i$.

We claim that there is a bijection σ of the set I such that $m_{\sigma(i)} = m'_i$. Note that α_0 is the highest coroot and h_0 is the highest root of the root system of the dual Dynkin diagram I'. So when the Dynkin diagrams I and I' are isomorphic the claim is clear. When I is not isomorphic to I' then I is of type B_l or C_l. In this case all the integers m_i but one are equal to 2 and the last one equals 1. The same is true for the integers m'_i. Thus in all the cases we can find the desired set bijection σ.

Define Φ as the linear isomorphism sending ω_i' to $\omega_{\sigma(i)}$. Note that C_n (respectively C_n') is the interior of the simplex generated by 0 and $(n/m_i).\omega_i$ (respectively 0 and $(n/m_i').\omega_i'$). It is clear that Φ satisfies the requirement of the lemma. Q.E.D.

Denote by $\mathcal{X}_n$ the set of conjugacy classes $[g]$ of elements $g \in G(\mathbf{C})$ such that g is regular and g^n is central in $G(\mathbf{C})$.

LEMMA 1.7 *The map* $E : h \mapsto [e^{(i2\pi/n).h}]$ *defines a bijection from* $C_n' \cap P'$ *to* $\mathcal{X}_n$.

Proof. Any conjugacy class $[g] \in \mathcal{X}_n$ has a finite order. So $[g]$ contains an element in the compact form of $H(\mathbf{C})$, i.e. there is some $h \in \mathfrak{h}_{\mathbf{R}}$ such that $[g] = [e^{(i2\pi/n).h}]$. Since g^n is central and g is regular, h belongs to P' and is n-regular. Thus Lemma 1.7 results from the following well-known facts:

(i) Any n-regular element h in P' is conjugated by a unique element $\tau \in W_n$ to an element in C_n'.

(ii) The W_n-orbit of such an element h is $W.h + nQ'$, where $Q' = \oplus_{i \in I} \mathbf{Z}.h_i$.

(iii) Conjugacy classes of semi-simple elements in $G(\mathbf{C})$ are exactly W-orbits in $H(\mathbf{C})$. Q.E.D.

LEMMA 1.8 *The sets* $C_n \cap P$ *and* $\mathcal{X}_n$ *have the same number of elements.*

Proof. The composition of the maps Φ of Lemma 1.6 and E of Lemma 1.7 gives rise to a bijection between these two sets. Q.E.D.

2. The functor T. *In section 2, G stands for an arbitrary algebraic group over an algebraically closed field k of characteristic p.*

A representation M is called rational if it is finite dimensional and if its matrix coefficients are regular functions. Let M be a rational G-module. Denote by M^G the space of G-invariants and by M_G be the space of covariants (i.e. $M_G = ({M^*}^G)^*$). Then there are natural maps $M^G \to M \to M_G$. Define $T(M)$ as the image of the composite map $M^G \to M_G$ and by $U(M)$ its kernel.

Similarily, given two rational G-modules M and N of finite dimension, set $T(M, N) = T(Hom(M, N))$ and $U(M, N) = U(Hom(M, N))$.

There is a natural map $\pi : M^G \otimes N^G \to (M \otimes N)^G$. It is clear that we have $\pi(M^G \otimes U(N)) \subset U(M \otimes N)$ and $\pi(U(M) \otimes N^G) \subset U(M \otimes N)$. Thus π induces a map $T(M) \otimes T(N) \to T(M \otimes N)$.

Thus $T(M, M)$ is a k-algebra, wich is a quotient of $End_G(M)$. The algebra $T(M, M)$ has a unit, denoted by 1_M.

LEMMA 2.1. *(i) We have* $1_M = 0$ *if and only if* $T(M, M) = 0$.

(ii) Moreover if $T(M, M) = 0$, *then we have* $T(M, N) = T(N, M) = 0$ *and* $T(M \otimes N, M \otimes N) = 0$ *for any rational G-module N.*

Proof. Assertion (i) is clear. If $T(M, M) = 0$, then any (left, right or bi) $T(M, M)$-module is zero. Thus (ii) follows. Q.E.D.

LEMMA 2.2. *Let* M, N *be two indecomposable rational G-modules. If M and N are isomorphic of dimension not divisible by p, we have* $T(M, N) = k$. *Otherwise we have* $T(M, N) = 0$.

Proof. The dual of the G-module $Hom(M, N)$ is $Hom(N, M)$, and the duality pairing is given by $x, y \in Hom(M, N) \times Hom(N, M) \mapsto tr(y.x)$, where tr denotes the trace. Let $x \in Hom_G(M, N)$. Thus x belongs to $U(M, N)$ if and only if $tr(x.y) = 0$ for any $y \in Hom_G(N, M)$. Since M is indecomposable, any G-invariant endomorphism z of M has a unique eigenvalue, say $\lambda(z)$, with multiplicity $dim\, M$.

Thus we get $tr(x.y) = (dim\, M).\lambda(xy)$. Thus if M and N are not isomorphic or if the dimension of M is divisible by p, we have $T(M, N) = 0$.

Furthermore, if M and N are isomorphic of dimension not divisible by p, then the nilpotent radical of the k-algebra $End_G(M)$ is the kernel of λ, and we have $\lambda(x.y) = \lambda(x).\lambda(y)$. Thus $U(M, M) = ker\, \lambda$ and therefore $T(M, M) = End_G(M)/U(M, M) \simeq k$. Q.E.D.

LEMMA 2.3 *Let M be a rational G-module, and set $M = \oplus m_P P$ be a decomposition into indecomposable modules. Then $T(M, M)$ is a matrix algebra. More precisely we have $T(M, M) = \oplus_{P | dim\, P \neq 0\, modulo\, p}\, M_{m_P}(k)$, where $M_n(k)$ is the algebra of n by n matrices.*

Proof. Lemma 2.3 follows from Lemma 2.2. Q.E.D.

(2.4) *Grothendieck groups* Let $\mathcal{A}$ be an additive category. The group generated by all symbols $([M])_{M \in \mathcal{A}}$ and generated by relations $[M] + [N] - [L]$ whenever $L \simeq M \oplus N$ is denoted by $K'(\mathcal{A})$.

Similarily let $\mathcal{A}$ be an abelian category. The group generated by all symbols $([M])_{M \in \mathcal{A}}$ and generated by relations $[M] + [N] - [L]$ whenever there is an exact sequence $0 \to M \to L \to N \to 0$ is denoted by $K(\mathcal{A})$. When $\mathcal{A}$ is the category of all rational G-modules we set $K(G) = K(\mathcal{A})$.

There is a natural map $\kappa : K'(\mathcal{A}) \to K(\mathcal{A})$. When $\mathcal{A}$ is semi-simple κ is an isomorphism.

Now define *subtractive categories.* Let $\mathcal{C}$ be a full subcategory of an abelian category $\mathcal{A}$ (i.e. by definitions morphisms in category $\mathcal{C}$ are the same as in $\mathcal{A}$). Consider all triples L, M, N of objects of $\mathcal{C}$ such that $L \simeq M \oplus N$.

Recall that if for any such a triple we have

$$(M,\ N \in \mathcal{C}) \text{ implies } (L \in \mathcal{C}),$$

one says that $\mathcal{C}$ is additive. Similarily if for any such a triple we have

$$(L \in \mathcal{C}) \text{ implies } (M,\ N \in \mathcal{C}),$$

then we say that $\mathcal{C}$ is *subtractive.*

Obviously for an additive and subtractive subcategory $\mathcal{C}$ of an artinian abelian category $\mathcal{A}$, $K'(\mathcal{C})$ is the $\mathbf{Z}$- module freely generated by all isomorphisms classes of indecomposable objects in $\mathcal{C}$.

(2.5) *Categories $\mathcal{C}_T$.* Let $\mathcal{C}$ be a full category of rational G-modules. Define a new category $\mathcal{C}_T$ by the following requirements.

(i) The objects of $\mathcal{C}_T$ are the objects of $\mathcal{C}$.

(ii) The space of all morphisms between two objects M and N is $T(M, N)$.

LEMMA 2.5 *Let $\mathcal{C}$ be an additive and subtractive category of rational G-modules. Then $\mathcal{C}_T$ is abelian and semisimple. Moreover the simple objects in $\mathcal{C}_T$ are the indecomposable objects in $\mathcal{C}$ of dimension not divisible by p.*

Proof. It is clear that $\mathcal{C}_T$ is an additive category.

Let us prove the existence of cokernels. Let $\mathcal{S}$ the set of isomorphism classes of indecomposable modules of dimension not divible by p in $\mathcal{C}$. For $M \in \mathcal{C}_T$, let T_M be the functor $X \in \mathcal{C}_T \mapsto T(M, X)$.

Let Θ be an arbitrary additive functor from $\mathcal{C}_T$ to the category of finite-dimensional k- vector spaces, such that $\Theta(S) = 0$ for almost all $S \in \mathcal{S}$. Then $L = \oplus_{S \in \mathcal{S}}\, \Theta(S) \otimes_k S$ is a well defined object of $\mathcal{C}_T$. There is a natural functor transform $\mu : T_L \to \Theta$. Using Lemma 2.2 it is clear that $\mu : T_L(N) \to \Theta(N)$ is an isomorphism for any indecomposable N. Thus μ is a functor isomorphism i.e. Θ is representable by L.

Let $\nu : M \to N$ be an arbitrary morphism in $\mathcal{C}_T$. The kernel functor of $T_N \to T_M$ is representable by some object L, i.e. L is the cokernel of ν.

The proof of existence of kernels is similar. The assertions about the semisimplicity of $\mathcal{C}_T$ and about the simple objects of $\mathcal{C}_T$ follows from Lemma 2.3. Q.E.D.

Let $\mathcal{C}$ be an additive and subtractive category of rational G-modules. Assume that $\mathcal{C}$ is stable under tensor product. Then $\mathcal{C}_T$ is a tensor category. Thus $K'(\mathcal{C})$ and $K(\mathcal{C}_T)$ are associative commutative rings. The identity functor $\mathcal{C} \to \mathcal{C}_T$ induces a natural ring morphism $K'(\mathcal{C}) \to K'(\mathcal{C}_T) \simeq K(\mathcal{C}_T)$.

LEMMA 2.6 *(i) The ring morphism $K'(\mathcal{C}) \to K(\mathcal{C}_T)$ is surjective and its kernel is generated by $[S]$, where S runs over all indecomposable modules in $\mathcal{C}$ of dimension divisible by p.*

(ii) The ring $K(\mathcal{C}_T)$ is reduced.

Proof. Assertion (i) is obvious.

In order to prove (ii) we can assume that $\mathcal{C}$ is stable with respect to dualizing (indeed we can assume that $\mathcal{C}$ is the category of all rational G-modules). Any $x \in K(\mathcal{C}_T)$ can be uniquely written as $\Sigma_{S \in \mathcal{S}}\, m_S[S]$. Denote by k the trivial representation of G and set $I(x) = m_k$, $x^* = \Sigma_{S \in \mathcal{S}}\, m_S[S^*]$. It is clear that we have $I([M]) = dim\, T(M)$ for any $M \in \mathcal{C}_T$.

We have $I(x.x^*) = \Sigma_{S \in \mathcal{S}}\, m_S^2$. Thus we have $I(x.x^*) > 0$ unless $x = 0$.

Let x be a non-zero element in $K(\mathcal{C}_T)$. Set $y = x.x^*$. We have $I(y) \neq 0$, thus $y \neq 0$ and $I(y.y^*) \neq 0$. We have $y.y^* = x^2.(x^*)^2$. Hence we get $x \neq 0$ implies $x^2 \neq 0$. Thus the ring $K(\mathcal{C}_T)$ is reduced. Q.E.D.

Proof of Lemma 2.7, stated in the introduction. By Lemma 2.2 all indecomposable direct summands of $M \otimes N$ have dimension divisible by p if and only if $T(M \otimes N, M \otimes N) = 0$. Moreover $T(M, M) = 0$. Thus Lemma 2.7 follows from Lemma 2.1. Q.E.D.

3. Tilting modules and Ringel's theorem. *From now on we denote by G a simply connected Chevalley group over an algebraically closed field of characteristic p. We will use freely the notations, conventions and definitions from Section 1.*

For $\lambda \in P^+$ set $F(\lambda) = \{f \in k[G] | f(gb) = \lambda(b^{-1})f(g)\, for\, any\, b \in B\}$. Its dual $L(\lambda)$ is called the Weyl module with highest weight λ (because it satifies Weyl character formula). A good filtration of a G-module is a filtration whose successive quotients are isomorphic to $F(\lambda)$, for various λ.

A rational G-module is called a tilting module if M and M^* have a good filtration. Denote by $\mathcal{P}$ the category of all tilting modules. Actually $\mathcal{P}$ is additive, subtractive and closed under tensor products (this follows from a similar statement for the category of modules with a good filtration, see**[D1]**,**[M]**). Set $\mathcal{P}_{mod} = \mathcal{P}_T$.

THEOREM 3.1 (Ringel **[R]**).*For any $\lambda \in P^+$ there is a unique indecomposable tilting module $P(\lambda)$, such that*

(i) $P(\lambda)$ has a unique highest weight λ and $dim\, P(\lambda)_\lambda = 1$

(ii) any weight of $P(\lambda)$ is in the convex hull of $W.\lambda$.

Moreover any indecomposable tilting module is isomorphic to some $P(\lambda)$.

Proof. See **[R]** and **[D2]**.

COROLLARY 3.2 *We have* $K'(\mathcal{P}) \simeq K(G)$.

Proof. By Theorem 3.1 the set $\{[P(\lambda)] | \lambda \in P^+\}$ is a basis of $K'(\mathcal{P})$. An induction on the standard ordering in P^+ and the theorem easily imply that it is also a basis of $K(G)$. Thus the lemma follows. Q.E.D.

4. The fusion ring. *From now on we assume that* $p > h^v$, $p > n(I)$, *where* $n(I)$ *is the positive integer defined in Lemma 1.4.*

For $\lambda \in P^+$, let $\chi(\lambda)$ be the character of the Weyl module $L(\lambda)$. For a general $\lambda \in P$ it will be convenient to set $\chi(\lambda) = \epsilon(w).\chi(w(\lambda + \rho) - \rho)$ if $w \in W$ and $w.\lambda \in P^+$ (this implies $\chi(\lambda) = 0$ for singular $\lambda + \rho$). We will identify $\chi(\lambda)$ with the corresponding element in $K(G)$.

Let λ be a weight in the fundamental alcove. Define a linear map $t_\lambda : K(G) \to K(G)$ by $t_\lambda(\chi(\mu)) = \chi(\mu)$ if $\mu + \rho \in W_p.(\lambda + \rho)$ and $t_\lambda(\chi(\mu)) = 0$ otherwise. Actually t_λ is induced by an exact functor T_λ (called projection functor) on the category of rational G-modules (see **[A1]**, **[J]**). The linear maps t_λ will be called the projection functor maps.

By Lemmas 2.6 and 3.2 there is a natural surjective ring morphism $K(G) \to K(\mathcal{P}_{mod})$. The kernel will be denoted by I_{mod}. Let J_0 be the smallest ideal of $K(G)$ satisfying the following two conditions:

(i) J_0 contains $\chi(\lambda)$ for any weight λ such that $\lambda + \rho$ is p-singular

(ii) J_0 is stable under all projection functor maps.

Let J_1 be the ideal of all $f \in K(G)$ such that $f([h]) = 0$ for any $[h] \in \mathcal{X}_p$.

For any weight $\lambda \in C_p$ and any $f = \Sigma_{\mu \in P}\, a_\mu.\chi(\mu)$ set

$V_\lambda(f) = \Sigma_{w \in W_p}\, \epsilon(w) a_{w.(\lambda+\rho)-\rho}$. Put $J_2 = \cap_{\lambda \in P \cap C_p} ker(V_\lambda)$.

Also define S to be the span of $\chi(\lambda)$ where λ runs over $C_p \cap P$.

Let λ, μ be two regular weights, and let C_μ, C_λ be their respective alcoves. Say that λ and μ are adjacent if the alcoves C_λ and C_μ are distinct and adjacent and we have $\lambda = s_M \mu$ where M is the common wall of the two alcoves C_λ and C_μ.

LEMMA 4.1 *The group* J_2 *is spanned by the following elements:*

(i) the character $\chi(\lambda)$, *where* $\lambda + \rho$ *is p-singular,* $\lambda \in P^+$,

(ii) the elements $\chi(\lambda) + \chi(\mu)$ *where* $\lambda + \rho$ *and* $\mu + \rho$ *are p-regular and adjacent,* $\lambda, \mu \in P^+$.

Moreover we have $K(G) = J_2 \oplus S$.

Proof. If ξ and η are two regular weights which are conjugated under W_p, then we can find a integer n and a sequence $\nu_1, \nu_2, ..., \nu_n$ of weights such that $\nu_1 = \xi$, $\nu_n = \eta$ and ν_m and ν_{m+1} are adjacent for any $1 \leq m < n$. Thus the lemma is clear. Q.E.D.

LEMMA 4.2 *We have* $J_2 \subset J_0$.

Proof. By lemma 4.1 we have to prove the following assertion:

$$\chi(\lambda) + \chi(\mu) \text{ belongs to } J_0$$

for any weights λ and μ such that $\lambda + \rho$ and $\mu + \rho$ are p-regular and adjacent.

Let C, C' be the alcoves containing $\lambda + \rho$ and $\mu + \rho$, let M be their common wall and let $\pm\alpha$ be the two roots orthogonal to M. By lemma 1.5 we can find an integral weight τ such that $\tau + \rho \in M$ and $\tau + \rho + \Delta \subset C \cup M \cup C'$. Let Ad be the character of the adjoint representation. We have

$$(Ad.\chi)(\tau) = l.\chi(\tau) + \Sigma_{\beta \in \Delta}\, \chi(\tau + \beta).$$

Let $\mathcal{O}$ be the W_p-orbit of $\tau + \alpha + \rho$. By hypothesis we have

$$\mathcal{O} \cap (C \cup M \cup C') = \{\tau + \rho \pm \alpha\}.$$

As J_0 is stable under the projection functor maps, the element $\chi(\tau+\alpha)+\chi(\tau-\alpha)$ belongs to J_0. Using the maps, corresponding to the translation functors (see **[J]**), we get that $\chi(\lambda)+\chi(\mu)$ also belongs to J_0. Q.E.D.

LEMMA 4.3 *We have $J_0 \subset I_{mod}$.*

Proof. Any direct summand of a tilting module is a tilting module. Therefore, from Lemma 2.2 follows that I_{mod} is stable under the projection functor maps. So to prove Lemma 4.3, it suffices to prove that I_{mod} contains the characters $\chi(\lambda)$ for any weight λ such that $\lambda+\rho$ is p-singular.

Let λ be a dominant weight such that $\lambda+\rho$ is p-singular. By the Weyl dimension formula we have

$$dimL(\lambda) = \Pi_{\alpha\in\Delta^+}\ (\lambda+\rho)(h_\alpha)/\rho(h_\alpha).$$

We have $\rho(h_\alpha) \leq \rho(h_\theta) = h^v < p$. Thus the denominator in the Weyl dimension formula is not divisible by p. As one of the terms in the numerator is divisible by p, we get that the dimension of $L(\lambda)$ is divisible by p. The tilting module $P(\lambda)$ has a filtration whose successive quotients are some Weyl modules $L(\mu)$. By indecomposability of $P(\lambda)$ and by Andersen's linkage principle (see **[A1]**) for any such an highest weight μ the weight $\mu+\rho$ is p-singular. Therefore the dimension of $P(\lambda)$ is divisible by p and the character of $P(\lambda)$ belongs to I_{mod}.

Moreover the character of $P(\lambda)$ can be written as $chP(\lambda) = \chi(\lambda)+\Sigma\ a_{\lambda,\mu}\chi(\mu)$, where μ runs over the set of weights of norms $< ||\lambda||$ such that $\mu+\rho$ is p-singular. The matrix $1+(a_{\lambda,\mu})$ is invertible and we can write $chL(\lambda) = ch(P(\lambda))+\Sigma\ b_{\lambda,\mu}chP(\mu)$, where μ runs over the set of p-singular weights of norms $< ||\lambda||$ such that $\mu+\rho$ is p-singular. Hence $\chi(\lambda)$ belongs to I_{mod}. The Lemma is proved.

LEMMA 4.4 *For any dominant weight λ such that $\lambda+\rho$ is in the fundamental alcove the module $L(\lambda)$ is tilting and its dimension is not divisible by p.*

Proof. It follows from the linkage principle that $L(\lambda)$ is simple. Thus this module is tilting. Moreover its dimension is $\Pi_{\alpha\in\Delta^+}\ (\lambda+\rho)(h_\alpha)/\rho(h_\alpha)$. Each factor in the numerator and denominator is an integer $< p$. So its dimension is not divisible by p. Q.E.D.

LEMMA 4.5 *We have $I_{mod} = J_0 = J_2$ and $K(G) = I_{mod} \oplus S$.*

Proof. It follows from Lemmas 2.6 and 4.4 that the map $K(G) \to K(\mathcal{P}_{mod})$ is one to one when restricted to S. Thus we get $I_{mod} \cap S = 0$. Since $J_2 \subset J_0 \subset I_{mod}$ and $J_2 \oplus S = K(G)$ (Lemmas 4.1, 4.2 and 4.3) Lemma 4.5 follows. Q.E.D.

LEMMA 4.6 *We have $J_2 \subset J_1$.*

Proof. Let $h \in P'$ be a integral and p-regular element of $\mathfrak{h}_{\mathbf{R}}$. Let s be an orthogonal reflection in W_p and denote by $\bar{s}$ be its image in W. There is a root α and an integer m such that $s.\mu = \mu - (mp+1)\mu(h_\alpha)\alpha$ and $\bar{s}\mu = \mu - \mu(h_\alpha).\alpha$ for any weight μ. In particular we have $s.\mu(h) = \bar{s}.\mu(h)\ modulo\, p$, for any weight μ. Thus for any $w \in W$ we get

$$ws.\mu(h) = (wsw^{-1})w.\mu(h) = (w\bar{s}w^{-1})w.\mu(h) = w\bar{s}.\mu(h)\ modulo\, p.$$

For any weight μ set $\chi_\mu = \Sigma_{w\in W}\ \epsilon(w)e^{w.\mu}$. Note that we have $\chi_{w.\mu} = \epsilon(w).\chi_\mu$ for any $w \in W$. Set $g = e^{(i2\pi/p)h}$. We get

$$\chi_{s.\mu}(g) = \chi_{\bar{s}.\mu}(g) = -\chi_\mu(g).$$

Since g is regular, we have $\chi_\rho(g) \neq 0$ (by the classical denominator formula). Let λ be an arbitrary weight. Recall Weyl character formula: $\chi(\lambda) = \chi_{\lambda+\rho}/\chi_\rho$. Thus we get

$$\chi(\lambda)(g) + \chi(s.(\lambda+\rho)-\rho)(g) = 0.$$

Moreover when $\lambda+\rho$ is p singular, we can find a reflection $s \in W_p$ such that $s.(\lambda+\rho) = \lambda+\rho$, therefore

$$\chi(\lambda)(g) = 0.$$

Thus we have $f(g) = 0$ for any f in the spanning set of J_2 of Lemma 4.1 and any $g \in \mathcal{X}_p$ (by Lemma 1.7) i.e. $J_2 \subset J_1$.Q.E.D.

LEMMA 4.7 *We have* $J_1 = I_{mod}$.

Proof. Any conjugacy class [g] in $\mathcal{X}_p$ is of finite order. Thus they define points of $SpecK(G)$ with value in a ring of algebraic integers. Moreover $\mathcal{X}_p$ is stable under $Gal(\mathbf{C}/\mathbf{Q})$. Thus $K(G)/J_1$ is a free $\mathbf{Z}$ - module of rank $Card(\mathcal{X}_p)$. By Lemma 1.7 $K(G)/I_{mod}$ is a free $\mathbf{Z}$-module of rank $Card(C_p \cap P)$. By Lemmas 1.8 and 4.5 the $\mathbf{Z}$-modules $K(G)/J_1$ and $K(G)/I_{mod}$ have the same rank. Moreover by Lemmas 4.5 and 4.6 we have $I_{mod} \subset J_1$. Hence we have $I_T = J_1$.Q.E.D.

THEOREM 4.8 *The ring* $K(\mathcal{P}_{mod})$ *is a reduced quotient ring of* $K(G)$ *and the spectrum of* $\mathbf{C} \otimes K(\mathcal{P}_{mod})$ *is the space of regular conjugacy classes* $[g]$ *in* $G(\mathbf{C})$ *such that* g^p *is central.*

Proof Theorem 4.8 follows from Lemma 4.7.Q.E.D.

THEOREM 4.9 *For any dominant weight* λ *the tilting module* $P(\lambda)$ *satisfies one of the following assertions:*

(i) The dimension of $P(\lambda)$ *is divisible by* p, *or*

(ii) λ *is in the fundamental alcove and* $P(\lambda)$ *is simple of dimension not divisible by* p.

Proof. Theorem 4.9 follows from Lemmas 2.6, 4.4 and 4.5. Q.E.D.

For $\lambda, \mu, \nu \in P^+$ define the positive integers $K^{\nu}_{\lambda,\mu}$ by the following equality:

$$\chi(\lambda).\chi(\mu) = \Sigma_{\nu \in P^+} K^{\nu}_{\lambda,\mu}.\chi(\nu).$$

In other words, $K^{\nu}_{\lambda,\mu}$ are the usual tensor product multiplicities and they are given by the Kostant formula. Define the positive integers $V^{\nu}_{\lambda,\mu}$ by the following requirement :

$$P(\lambda) \otimes P(\mu) \simeq \oplus_{\nu \in P^+} V^{\nu}_{\lambda,\mu}.P(\nu).$$

It would be interesting to compute all these multiplicities $V^{\nu}_{\lambda,\mu}$. We can compute them only when $\nu+\rho$ is in the fundamental alcove.

THEOREM 4.10 *Let* $\lambda, \mu, \nu \in P^+$. *Assume that* $\nu+\rho$ *is in the fundamental alcove* C_p.

(i) If $\lambda+\rho$ *or* $\mu+\rho$ *is not in the fundamental alcove we have* $V^{\nu}_{\lambda,\mu} = 0$.

(ii) If $\lambda+\rho$ *and* $\mu+\rho$ *are in the fundamental alcove then we have*

$$V^{\nu}_{\lambda,\mu} = \sum_{\substack{w \in W_p \\ w^{-1}.(\nu+\rho)-\rho \in P^+}} \epsilon(w) K^{w^{-1}.(\nu+\rho)-\rho}_{\lambda,\mu}.$$

Proof. Assertion (i) follows from Lemmas 2.7 and 4.4. Let us prove Assertion (ii) (see **[K]**,Exercise 13.35, where an analogous formula is stated in completely different setting). Define an additive map $\Theta : K(G) \to S$ as follows. For any $\xi \in P^+$ set $\Theta(\chi(\xi)) = 0$ if $\xi+\rho$ is p-singular and $\Theta(\chi(\xi)) = \epsilon(w)\chi(\eta)$ if $\xi+\rho = w(\eta+\rho)$ with $(\eta+\rho) \in C_p$ and $w \in W_p$. Clearly Θ is a retraction of $K(G)$ on S and its kernel is J_2. By Lemma 4.5 its kernel is I_T. Let $\lambda, \mu \in P^+$ such that $\lambda+\rho$ and $\mu+\rho$ are in the fundamental alcove. Note that we have $P(\xi) = L(\xi)$ whenever $\xi+\rho$ belongs to $C_p \cap P$ (lemma 4.4). Thus we get

$$\Sigma_{(\nu+\rho)\in C_p\cap P} V^{\nu}_{\lambda,\mu}[P(\nu)] =$$

$$= \Theta([P(\lambda)].[P(\mu)]) =$$

$$= \Theta(\sum_{\nu\in P^+} K^{\nu}_{\lambda,\mu}.\chi(\nu)) =$$

$$= \sum_{\nu\in P^+} \sum_{\substack{w\in W_p \\ w^{-1}.(\nu+\rho)-\rho\in P^+}} \epsilon(w) K_{\lambda,\mu}^{w^{-1}.(\nu+\rho)-\rho}.\chi(\nu) =$$

$$= \sum_{\nu\in P^+} \sum_{\substack{w\in W_p \\ w^{-1}.(\nu+\rho)-\rho\in P^+}} \epsilon(w) K_{\lambda,\mu}^{w^{-1}.(\nu+\rho)-\rho}.[P(\nu)].$$

Thus Assertion (ii) is proved.

Bibliography

[A1] H.H.Andersen, *The strong linkage principle,* J.Reine Angew Math. 315 (1980), 53-55.

[A2] H.H. Andersen, *Tensor products of quantized tilting modules,* Commun. Math. Phys. 149 (1992), 149-159.

[D1] S.Donkin, *Rational representations of algebraic groups*, Springer Verlag, Lect. Notes in Math. 1140, 1985.

[D2] S.Donkin, *On tilting modules for algebraic groups,* preprint QMW, 1991.

[F] M.Finkelberg, *Fusion categories*, Ph.D. thesis, Harvard University, 1993.

[GK] S. Gelfand and D. Kazhdan, *Example of tensor categories*, Inv.Math. 109 (1992).

[GM] G. Georgiev and O. Mathieu, *Categorie de fusion pour les groupes de Chevalley*, Comptes Rendus Acad. Sc. Paris 315 (1992), 659-662.

[HL] Y.Z. Huang and J. Lepowsky, *Toward a theory of tensor products for representations of a vertex operator algebra*, Proc. 20th International Conference on Differential Geometry Methods in Theoretical Physics, New York 1991, ed. S. Catto and A. Rocha Carridi, World Sc., Singapore 1 (1992), 344-354.

[H] J.Humphreys, *Linear algebraic groups*, GTM 21, Springer-Verlag, Berlin, 1975.

[J] J.Jantzen, *Representations of algebraic groups*, Academic Press, Orlando, 1987.

[K] V.Kac, *Infinite dimensional Lie algebras*, 3rd edition, Cambridge University Press, Cambridge, 1990.

[KL0] D. Kazhdan and G. Lusztig, *Affine Lie algebras and quantum groups,* Int. Math. Research Notices 2, Duke Math. J. 62 (1991), 21-29.

[KL1÷5] D.Kazhdan and G.Lusztig, *Tensor structures, arising from affine Lie algebras*, to appear.

[M] O. Mathieu, *Filtrations of G-modules*, Ann. Ecole Norm. Sup. 23 (1990), 625-644.

[MS] G. Moore and N. Seiberg, *Classical and quantum conformal field theory*, Comm. Math. Phys. 123 (1988), 177-254.

[R] C.M. Ringel, *The category of modules with good filtrations over a quasi-hereditary algebra has almost split sequences*, Math.Zeitschrift 208 (1991), 209-225.

[S] R.Steinberg, *Regular elements of semisimple algebraic groups*, Publ. Math. IHES 25 (1965), 49-80.

[V] E. Verlinde, *Fusion rules and modular transformations in 2D conformal field theory*, Nucl. Phys. B 300 (1988), 360-375.

Authors' addresses:

Department of Mathematics, Rutgers University, New Brunswick NJ 08903, U.S.A.

E-mail address: georgiev@math.rutgers.edu

U.A. 748 du CNRS, 15 place Souham, 75013 Paris, FRANCE (current address) and

Department of Mathematics, Rutgers University, New Brunswick NJ 08903, U.S.A.

E-mail address: mathieu@mathp7.jussieu.fr

Contemporary Mathematics
Volume 175, 1994

QUANTUM GROUPS AND FLAG VARIETIES.

VICTOR GINZBURG , NICOLAI RESHETIKHIN , ERIC VASSEROT

1. Introduction

By classical Schur-Weyl duality there is a one-to-one correspondence between irreducible representations of the Symmetric group S_d and certain finite dimensional irreducible representations of the group $GL_n(\mathbb{C})$. The correspondence is obtained by decomposing the tensor space $(\mathbb{C}^n)^{\otimes d}$ with respect to the natural (commuting) actions of the two groups. The main goal of the present paper is to extend this construction to the *quantum affine* setup and to provide its geometric interpretation.

By "quantizing" we mean replacing objects by their q-analogues. The natural q-analogue of the Symmetric group S_d, or rather of its group algebra, is the Hecke algebra of type A_d (cf. e.g. [KL 1]). The natural q-analogue of $GL_n(\mathbb{C})$, or rather of the enveloping algebra of its Lie algebra, is the quantum group $\mathbf{U}_q(\mathfrak{gl}_n)$ introduced by Drinfeld and Jimbo. A q-analogue of the Weyl correspondence was found by Jimbo [J] about the same time as the definition of the quantum group itself. A very interesting geometric construction of that correspondence was given later by Grojnowski-Lusztig [GL] following the earlier work [BLM] where a "Hecke-like" interpretation of $\mathbf{U}_q(\mathfrak{gl}_n)$ was discovered.

The next step in that direction consists of replacing 'finite dimensional' quantum objects by their 'affine' counterparts, which amounts heuristically to replacing the complex field $\mathbb{C}$ by the p-adic field. Representation theory of *affine Hecke algebras* was studied in [Gi] and [KL 2] (see [CG] for details) and turned out to be quite an intersting problem in itself. A similar finite dimensional Representaion theory of $\mathbf{U}_q(\widehat{\mathfrak{gl}_n})$ was worked out in [GV] (some special cases were treated earlier in [CP] and [Ch 1]). In the present paper we 'match' the geometric approach to affine Hecke algebras used in [Gi] and [KL 2] with the corresponding geometric approach to $\mathbf{U}_q(\widehat{\mathfrak{gl}_n})$

used in [GV], thus providing a geometric approach to the 'quantum affine' Weyl reciprocity and 'explaining' the results of [Dr 3] for Yangians and [Ch 2] for quantum algebras.

The role of the tensor representation $(\mathbb{C}^n)^{\otimes d}$ above is played, in the quantum affine setup, by a *Polynomial Tensor Representation* introduced in section 4 below. In the subsequent sections we produce three different geometric constructions of the Polynomial Tensor Representation. The first one is based on equivariant K-theory, and the other two, which are "dual" to the first in the sense of Langlands, involve affine flag varieties over a finite field. Thus, this paper may be viewed as a continuation of [GV] with a new geometric ingredient being borrowed from [GL].

There are several reasons to suggest that it is the interplay between the three constructions of the paper that seems to be of most importance. We would like to mention some of them. First, the Polynomial Tensor Representation was constructed originally via an *R-matrix realization* of the algebra $\mathbf{U}_q(\widehat{\mathfrak{gl}_n})$ discovered in [FRT] and [RS 1]. On the other hand, the formulas for the representation of $\mathbf{U}_q(\widehat{\mathfrak{gl}_n})$ given in [GV] can be modified to produce the Polynomial Tensor Representation in Drinfeld's *loop-like realization* [Dr 2]. Comparison of those two approaches yields an independent proof of the main result of [DF]. Second, the Polynomial Tensor Representation arises naturally in Physics in connection with quantum integrable systems and Bethe Ansatz (see [KR]). It is a somewhat miraculous coincidence that the $d \to \infty$ limit of the construction, which comes out quite naturally from the point of view of [BLM] and [GV], is also of great importance in Physics, where it is known as the 'Thermodinamics Limit'. This subject is closely related to the so-called "Zamolodchikov conjecture" on deformations of conformal field theories. It was approached quite differently in [KKM] via the techniques of the crystal bases. Finally, the last two of geometric constructions presented below provide distinguished *Intersection homology bases* in the Polynomial Tensor Representation. Those bases are likely to enjoy various nice properties similar (and possibly related) to Lusztig-Kashiwara canonical bases.

2. The matrix loop realization of $\mathbf{U}_q(\widehat{\mathfrak{gl}_n})$

Let $\mathfrak{g} = \operatorname{End}(\mathbb{C}^n)$ be the endomorphism (associative) algebra of the vector space $\mathbb{C}^n$ with the standard basis formed by the matrix units E_{ij} , the $n\times n$-matrices with the only non-zero entry 1 at the $i\times j$-th place. Let $q \in \mathbb{C}^*$ be a complex number and z a formal variable. The Drinfeld-Jimbo R-*matrix* in the fundamental representation $\mathbb{C}^n$ of the affine algebra $\mathbf{U}_q(\widehat{\mathfrak{gl}_n})$ has the following form:

$$R(z) = \frac{z-1}{qz-q^{-1}} \sum_{1\leq i\neq j\leq n} E_{ii}\otimes E_{jj} + \sum_{1\leq i\leq n} E_{ii}\otimes E_{ii} + \tag{2.1}$$

$$+ \frac{z(q-q^{-1})}{qz-q^{-1}} \sum_{i<j} E_{ij}\otimes E_{ji} + \frac{q-q^{-1}}{qz-q^{-1}} \sum_{i>j} E_{ij}\otimes E_{ji}$$

Thus, $R(z)$ is a $\mathfrak{g}\otimes\mathfrak{g}$-valued rational function in z.

A somewhat mysterious role throughout the theory of affine quantum groups is played by a function θ that depends on a complex parameter $q \in \mathbb{C}^*$ and an integral parameter m, and which is given by the formula:

$$\theta_m(z) := \frac{q^m\, z - 1}{z - q^m} \tag{2.2}$$

The R-matrix (2.1) can be written in terms of the function θ_{-1} as follows

$$R(z) = \sum_{1\leq i,j\leq n} E_{ij}\otimes E_{ji} + \theta_{-1}(qz) \sum_{1\leq i\neq j\leq n} (E_{ii}\otimes E_{jj} - q^{\mathrm{sgn}(i-j)} E_{ij}\otimes E_{ji})$$

where $sgn(k)$ equals $+1$ if $k>0$ and -1 if $k<0$.

It can be verified that the R-matrix satisfies the following *Yang-Baxter identity* in $\mathfrak{g}\otimes\mathfrak{g}\otimes\mathfrak{g}$:

$$R_{12}(\tfrac{z_1}{z_2})\cdot R_{13}(\tfrac{z_1}{z_3})\cdot R_{23}(\tfrac{z_2}{z_3}) = R_{23}(\tfrac{z_2}{z_3})\cdot R_{13}(\tfrac{z_1}{z_3})\cdot R_{12}(\tfrac{z_1}{z_2}) \tag{2.3}$$

along with the *unitarity condition*: $\quad R_{21}(z)^{-1} = R(z^{-1})$.

Write $\mathfrak{g} = \mathfrak{n}^+ \oplus \mathfrak{h} \oplus \mathfrak{n}^-$ for the standard triangular decomposition, where $\mathfrak{h}$ and $\mathfrak{n}^\pm$ stand for the subalgebras of diagonal and strictly upper (resp. lower) triangular matrices. Let $\mathfrak{g}[t,t^{-1}]$ be the algebra of $\mathfrak{g}$-valued Laurent polynomials; it contains $\mathfrak{g}$ as the subalgebra of constant polynomials. We introduce the following subalgebras:

$$\mathbf{L}^\pm\mathfrak{g} = \mathfrak{n}^\pm \oplus t^{\pm1}\,\mathfrak{g}[t^{\pm1}] \;\subset\; \mathfrak{g}[t,t^{-1}]$$

Let $\mathbf{K} = \mathbb{C}[E_{11}, E_{11}^{-1}, \ldots, E_{nn}, E_{nn}^{-1}]$ be the polynomial algebra. Form the vector space:

$$\mathbf{L}\mathfrak{g} = \mathbf{L}^+\mathfrak{g} \oplus \mathbf{K} \oplus \mathbf{L}^-\mathfrak{g}$$

There are two imbeddings of the space $\mathfrak{h}$ of diagonal matrices into $\mathbf{L}\mathfrak{g}$ that arise from two different imbeddings $\mathfrak{h} \hookrightarrow \mathbf{K}$ given by the formulas:

$$\mathfrak{h} \ni \sum a_i E_{ii} = A \mapsto A^{\pm} = \sum a_i E_{ii}^{\pm 1} \in \mathbf{K}$$

For $A \in \mathfrak{h}$, introduce two generating functions in the formal variable z:

$$A^{\pm}(z) = A^{\pm} \oplus \sum_{i=1}^{\infty} (A\, t^{\pm i})\, z^{\mp i} \quad \in \mathbf{K} \oplus \mathbf{L}^{\pm}\mathfrak{g}[[z^{\mp 1}]]$$

Similarly, for any $A \in \mathfrak{n}^{\pm}$ form the generating function:

$$A(z) = \sum_{i=-\infty}^{\infty} (A\, t^i)\, z^{-i} \quad \in (\mathfrak{g}[t, t^{-1}])[[z, z^{-1}]]$$

Observe that this power series has a unique decomposition:

$$A(z) = A^+(z) + A^-(z) \quad \text{so that} \quad A^{\pm}(z) \in \mathbf{L}^{\pm}\mathfrak{g}[[z^{\mp 1}]]$$

Thus, to each element A of $\mathfrak{h}$, $\mathfrak{n}^+$ or $\mathfrak{n}^-$, we assigned two generating functions $A^+(z)$ and $A^-(z)$.

Let $T(\mathbf{L}\mathfrak{g})[C, C^{-1}]$ be the Laurent polynomial ring in the variables $C^{\pm 1}$ with coefficients in the tensor algebra of the $\mathbb{C}$-vector space $\mathbf{L}\mathfrak{g}$. It is convenient to introduce the following 'universal' generating functions, usually referred to as 'L-operators'

$$L^{\pm}(z) := \sum_{\alpha,\beta=1}^{n} E_{\alpha\beta} \otimes E^{\pm}_{\alpha\beta}(z) \in \mathfrak{g} \otimes \mathbf{L}^{\pm}\mathfrak{g}[[z^{\mp 1}]] \tag{2.4}$$

and set

$$L_1^{\pm}(z) = \sum_{\alpha,\beta=1}^{n} E_{\alpha\beta} \otimes Id \otimes E^{\pm}_{\alpha\beta}(z) \in \mathfrak{g} \otimes \mathfrak{g} \otimes T(\mathbf{L}\mathfrak{g})[[z^{\mp 1}]]$$

$$L_2^{\pm}(z) = \sum_{\alpha,\beta=1}^{n} Id \otimes E_{\alpha\beta} \otimes E^{\pm}_{\alpha\beta}(z) \in \mathfrak{g} \otimes \mathfrak{g} \otimes T(\mathbf{L}\mathfrak{g})[[z^{\mp 1}]]$$

Following [FRT] and [RS 1], define an associative $\mathbb{C}$-algebra $\mathbf{U}(R)$ to be the quotient of $T(\mathbf{L}\mathfrak{g})[C, C^{-1}]$ modulo the following relations involving the R-matrix (2.1) and the generating functions introduced above:

$$A \cdot B = A\, B, \qquad \forall\, A, B \in \mathbf{K}$$

$$R(\tfrac{z}{w}) \cdot L_1^{\pm}(z) \cdot L_2^{\pm}(w) \;=\; L_2^{\pm}(w) \cdot L_1^{\pm}(z) \cdot R(\tfrac{z}{w})$$

$$R(C^2 \tfrac{z}{w}) \cdot L_1^{+}(z) \cdot L_2^{-}(w) \;=\; L_2^{-}(w) \cdot L_1^{+}(z) \cdot R(C^{-2} \tfrac{z}{w})$$

Thus, for all $A, B \in \mathfrak{h}$, $\mathfrak{n}^{\pm}$, one gets an infinite number of quadratic relations on generators from $\mathbf{L}\mathfrak{g}$. The dot-product denotes either the product in $T(\mathbf{L}\mathfrak{g})[C, C^{-1}]$ or the product in $\mathfrak{g}$ and in the above formulas we identify $R(\frac{z}{w})$ with $R(\frac{z}{w}) \otimes 1 \in \mathfrak{g} \otimes \mathfrak{g} \otimes T(\mathbf{L}\mathfrak{g})[C, C^{-1}]$.

We record the following result that will be used later.

Theorem 2.5 [DF]. *There exists a factorization*

$$L^{\pm}(z) = \left(Id \otimes 1 + \sum_{\alpha > \beta} E_{\alpha\beta} \otimes A^{\pm}_{\alpha\beta}(z) \right) \cdot$$
$$\cdot \left(\sum_{\alpha} E_{\alpha\alpha} \otimes A^{\pm}_{\alpha\alpha}(z) \right) \cdot \left(Id \otimes 1 + \sum_{\alpha < \beta} E_{\alpha\beta} \otimes A^{\pm}_{\alpha\beta}(z) \right)$$

where $A^{\pm}_{\alpha\beta}(z) \in \mathbf{U}(R)[[z^{\mp 1}]]\,, 1 \leq \alpha, \beta \leq n$ *are certain uniquely determined elements.* □

Next recall Drinfeld's "loop-like" presentation [Dr 2] of $\mathbf{U}_q(\widehat{\mathfrak{gl}_n})$ as the associative $\mathbb{C}$-algebra with unit 1 and generators

$$E_{\alpha k}\,,\; F_{\alpha k}\,,\; K^{\pm}_{\beta l} \quad,\quad \alpha \in [1, n{-}1]\,,\; \beta \in [1, n]\,,\; k \in \mathbb{Z}\,,\; l \in \pm\mathbb{N} \tag{2.6}$$

The relations among the generators are conveniently expressed in terms of the following generating functions in the formal variable z :

$$E_{\alpha}(z) = \sum_{k=-\infty}^{\infty} E_{\alpha k} z^{-k} \quad F_{\alpha}(z) = \sum_{k=-\infty}^{\infty} F_{\alpha k} z^{-k} \quad K^{\pm}_{\beta}(z) = \sum_{k \in \pm\mathbb{N}} K^{\pm}_{\beta k} z^{-k}.$$

Write also

$$\delta(z) := \sum_{n \in \mathbb{Z}} z^n \quad,\quad \epsilon_{\alpha\beta} := \delta_{\alpha+1\,,\beta} - \delta_{\alpha,\,\beta}. \tag{2.7}$$

Using the Cartan matrix $\mathbf{m}_{\alpha\beta}$ of the Lie algebra $\mathfrak{sl}_n$ the relations among the generating functions are written as follows

$$K^{+}_{\beta 0}K^{-}_{\beta 0} = K^{-}_{\beta 0}K^{+}_{\beta 0} = 1 \tag{2.8.1}$$

$$[K^{\pm}_{\alpha}(z), K^{\pm}_{\beta}(w)] = 0 = [K^{\pm}_{\alpha}(z), K^{\mp}_{\alpha}(w)] \tag{2.8.2}$$

$$\theta_1(q^{-1}C^{\pm 2}\tfrac{z}{w})K^{\pm}_{\alpha}(z)K^{\mp}_{\beta}(w) = \theta_1(q^{-1}C^{\mp 2}\tfrac{z}{w})K^{\mp}_{\beta}(w)K^{\pm}_{\alpha}(z) \quad \text{if } \alpha < \beta \tag{2.8.3}$$

$$K^{\pm}_{\beta}(z)E_{\alpha}(w) = \theta_{\epsilon_{\alpha\beta}}(q^{\epsilon_{\alpha\beta}}C^{\pm 1}\tfrac{z}{w})\, E_{\alpha}(w)K^{\pm}_{\beta}(z) \tag{2.8.4}$$

$$K^{\pm}_{\beta}(z)F_{\alpha}(w) = \theta_{-\epsilon_{\alpha\beta}}(q^{\epsilon_{\alpha\beta}}C^{\mp 1}\tfrac{z}{w})\, F_{\alpha}(w)K^{\pm}_{\beta}(z) \tag{2.8.5}$$

$$E_{\alpha}(z)E_{\beta}(w) = \theta_{\mathbf{m}_{\alpha\beta}}(q^{\alpha-\beta}\tfrac{z}{w})\, E_{\beta}(w)E_{\alpha}(z) \tag{2.8.6}$$

$$F_{\alpha}(z)F_{\beta}(w) = \theta_{-\mathbf{m}_{\alpha\beta}}(q^{\alpha-\beta}\tfrac{z}{w})\, F_{\beta}(w)F_{\alpha}(z) \tag{2.8.7}$$

$$[E_{\alpha}(z), F_{\beta}(w)] = (q-q^{-1})\cdot\delta_{\alpha\beta}\cdot \tag{2.8.8}$$

$$\Big(\delta(C^{-2}\tfrac{z}{w})K^{+}_{\alpha+1}(Cw)/K^{+}_{\alpha}(Cw) - \delta(C^{2}\tfrac{z}{w})K^{-}_{\alpha+1}(Cz)/K^{-}_{\alpha}(Cz)\Big)$$

$$\{E_{\alpha}(z_1)E_{\alpha}(z_2)E_{\beta}(w) - (q+q^{-1})E_{\alpha}(z_1)E_{\beta}(w)E_{\alpha}(z_2) + \tag{2.8.9}$$

$$+ E_{\beta}(w)E_{\alpha}(z_1)E_{\alpha}(z_2)\} + \{z_1 \leftrightarrow z_2\} = 0 \ \text{ if } \ |\alpha-\beta| = 1$$

$$\{F_{\alpha}(z_1)F_{\alpha}(z_2)F_{\beta}(w) - (q+q^{-1})F_{\alpha}(z_1)F_{\beta}(w)F_{\alpha}(z_2) + \tag{2.8.10}$$

$$+ F_{\beta}(w)F_{\alpha}(z_1)F_{\alpha}(z_2)\} + \{z_1 \leftrightarrow z_2\} = 0 \ \text{ if } \ |\alpha-\beta| = 1.$$

Remark 2.9. Drinfeld's realization of the subalgebra $U_q(\widehat{\mathfrak{sl}_n}) \subset \mathbf{U}_q(\widehat{\mathfrak{gl}_n})$ is given, in the notation of [Dr 2], by the formulas (see [DF], [Dr 2]):

$$\xi^{+}_{\alpha}(z) = (q-q^{-1})^{-1}E_{\alpha}(zq^{-\alpha}) \ , \qquad \xi^{-}_{\alpha}(z) = (q-q^{-1})^{-1}F_{\alpha}(zq^{-\alpha})$$

$$\psi_{\alpha}(z) = K^{+}_{\alpha+1}(zq^{-\alpha})/K^{+}_{\alpha}(zq^{-\alpha}),\ \varphi_{\alpha}(z) = K^{-}_{\alpha+1}(zq^{-\alpha})/K^{-}_{\alpha}(zq^{-\alpha}). \ \square$$

Thus we have defined the algebras $\mathbf{U}(R)$ and $\mathbf{U}_q(\widehat{\mathfrak{gl}_n})$. The connection between the two was studied in detail in [DF], and the result is

Theorem 2.10 [DF]. *The assignment* (see the notation of theorem 2.5)

$$F_\alpha(z) \mapsto A^+_{\alpha,\alpha+1}(C^{-1}z^{-1}) - A^-_{\alpha,\alpha+1}(C\,z^{-1}),$$

$$E_\alpha(z) \mapsto A^+_{\alpha+1,\alpha}(C\,z^{-1}) - A^-_{\alpha+1,\alpha}(C^{-1}z^{-1}),$$

$$K^\pm_\beta(z) \mapsto A^\mp_{\beta\beta}(z^{-1}),$$

extends uniquely to an algebra isomorphism $\mathbf{U}_q(\widehat{\mathfrak{gl}_n}) \xrightarrow{\sim} \mathbf{U}(R)$. □

Let $\mathbf{U}$ be the quotient of the algebra $\mathbf{U}_q(\widehat{\mathfrak{gl}_n}) \simeq \mathbf{U}(R)$ modulo relations $C^{\pm 1} = 1$.

Remark 2.11. The assignment

$$K^\pm_\beta(z) \mapsto K^\pm_\beta(z), \qquad E_\alpha(z) \rightleftarrows F_\alpha(z), \qquad C \rightleftarrows C^{-1},$$

gives rise to an anti-involution on the algebra $\mathbf{U}_q(\widehat{\mathfrak{gl}_n})$ preserving both the subalgebra $\mathbf{U}_q(\widehat{\mathfrak{sl}_n})$ and the quotient algebra $\mathbf{U}$. □

3. Quantum groups and Hecke algebras via K-theory

This section is to a large extent a review of the constructions of [Gi], [GV] and [KL 2]; we refer the reader to [CG] for further details.

Given a complex linear algebraic group A and an algebraic A-variety Z, let $K^A(Z)$ stand for the *complexified Grothendieck group* of A-equivariant coherent sheaves on Z. Observe that $K^A(pt) = \mathbf{R}(A)$ is the complexified Representation ring of A, and for any A-variety Z the group $K^A(Z)$ has a natural $\mathbf{R}(A)$-module structure.

Let M_1, M_2, M_3 be smooth quasi-projective A-varieties. Let $p_{ij} : M_1 \times M_2 \times M_3 \to M_i \times M_j$ denote the projection along the factor not named. The projections p_{ij} commute with the A-actions.

Let $Z_{ij} \subset M_i \times M_j$, $(i,j) = (1,2)$ or $(2,3)$, be A-stable closed subvarieties. Assume, in addition, that the map

$$p_{13}: \; p_{12}^{-1}(Z_{12}) \cap p_{23}^{-1}(Z_{23}) \to M_1 \times M_3 \qquad \textbf{is proper.} \tag{3.1}$$

Then we let $Z_{12} \circ Z_{23}$ denote its image, a closed A-stable subvariety of $M_1 \times M_3$. In that case there is a *convolution* on the equivariant K-groups

$$\star : \; K^A(Z_{12}) \otimes K^A(Z_{23}) \;\; \to \;\; K^A(Z_{12} \circ Z_{23}) \tag{3.2}$$

defined as follows. Let $[\mathcal{F}_{ij}] \in K^A(Z_{ij})$ be the classes of certain sheaves $\mathcal{F}_{ij}$ on Z_{ij}. Set (cf. [CG])

$$\mathcal{F}_{12} \star \mathcal{F}_{23} = (Rp_{13})_* \left(p_{12}^* \mathcal{F}_{12} \overset{L}{\underset{\mathcal{O}_{M_1 \times M_2 \times M_3}}{\bigotimes}} p_{23}^* \mathcal{F}_{23} \right)$$

In this formula the upper star stands for the pull-back morphism, well-defined for smooth maps (e.g., $p_{12}^*\mathcal{F}_{12} = \mathcal{F}_{12} \boxtimes \mathcal{O}_{M_3}$), and $\overset{L}{\otimes}$ for the derived tensor product.

Given an integer $d \geq 1$, let $\mathcal{F}$ denote (cf. [BLM]) the set of all n-step partial flags in $\mathbb{C}^d$ of the form $F = (0 = F_0 \subseteq F_1 \subseteq \ldots \subseteq F_n = \mathbb{C}^d)$. The set $\mathcal{F}$ is a smooth projective variety whose connected components are parametrized by partitions $\mathbf{d} = (0 = i_0 \leq i_1 \leq \ldots \leq i_n = d)$, of the segment $[1, d]$ into n possibly empty segments $]i_0, i_1]$, $]i_1, i_2], \ldots$, $]i_{n-1}, i_n]$. The component $\mathcal{F}_{\mathbf{d}}$ associated to the partition $\mathbf{d}$ consists of all flags $F = (F_0 \subseteq F_1 \subseteq \ldots \subseteq F_n)$ such that $\dim F_k = i_k$, $k = 0, 1, \ldots, n$. The group $G := GL(\mathbb{C}^d)$ acts naturally on each component $\mathcal{F}_{\mathbf{d}}$. The cotangent bundle on $\mathcal{F}$ can be identified in a natural way with the set $M = \{(F, x) \in \mathcal{F} \times \mathfrak{gl}_d \mid x(F_i) \subset F_{i-1}, \forall i = 1, 2, \ldots, n\}$.

Let N be the variety of all linear maps $x : \mathbb{C}^d \to \mathbb{C}^d$ such that $x^n = 0$. Define a morphism $\pi : M \to N$ to be the second projection $\pi : (F, x) \mapsto x$. Put $A_d := \mathbb{C}^* \times G$. Observe that M and N are A_d-varieties (the group G acts by conjugation and $z \in \mathbb{C}^*$ acts by multiplication by z^{-2} on $x \in N$). Furthermore, the variety M is smooth and π is a proper A_d-equivariant morphism.

Put $Z_d = M \times_N M \subset M \times M$. If there is no ambiguity we will simply write Z and A instead of Z_d and A_d. There are natural isomorphisms: $M \times M \simeq T^*\mathcal{F} \times T^*\mathcal{F} \simeq T^*(\mathcal{F} \times \mathcal{F})$ where the first isomorphism acts as multiplication by (-1) on the second factor (the sign of the standard symplectic 2-form on the second factor $T^*\mathcal{F}$ is also changed). The variety Z, viewed as a subvariety of $T^*(\mathcal{F} \times \mathcal{F})$, can be shown (see [CG]) to be the union of conormal bundles on all the G-orbits in $\mathcal{F} \times \mathcal{F}$. Furthermore, we have

$$Z \circ Z = (M \times_N M) \circ (M \times_N M) = M \times_N M = Z$$

It is clear also that Z is stable with respect to the diagonal A-action on $M \times M$. Thus the convolution makes $K^A(Z)$ an associative (non-commutative) algebra with unit. We recall one of the main results of [GV].

Theorem 3.3 [GV]. *There is a surjective algebra homomorphism:*

$$\mathbf{U} \twoheadrightarrow K^A(Z). \qquad \square$$

Given any semisimple element $a = (t, s) \in A$, let $\epsilon : \mathbf{R}(A) \to \mathbb{C}$ be the algebra homomorphism given by evaluation of characters of the group A at the point a, and let $\mathbb{C}_a$ denote the 1-dimensional $\mathbf{R}(A)$-module with $\chi \in \mathbf{R}(A)$ acting via the multiplication by $\epsilon(\chi)$. Further, let A^a denote the centralizer of a in A. We shall often view $\mathbb{C}_a$ as an $\mathbf{R}(A^a)$-module, by restriction. Given an A-variety Y, let Y^a denote the a-fixed point subvariety. In particular Y^a is an A^a-stable subvariety in Y.

Observe that, for $Z = M \times_N M$ and any semisimple element $a \in A$, we have $Z^a \circ Z^a = Z^a$. Thus, the K-group $K^{A^a}(Z^a)$ has again a natural associative algebra structure via convolution. Furthermore, the Bivariant Localization theorem (cf. [CG, ch.4]) yields a natural algebra homomorphism

$$r_a : \mathbb{C}_a \otimes_{\mathbf{R}(A^a)} K^A(Z) \to \mathbb{C}_a \otimes_{\mathbf{R}(A^a)} K^{A^a}(Z^a) \tag{3.4}$$

Fix a semisimple (diagonal) matrix $s \in G = GL_d$ with k distinct eigenvalues with multiplicities $d_1, d_2, \dots, d_k$, and write $a = (1, s) \in \mathbb{C}^* \times G$. Then there are natural isomorphisms

$$G^s \simeq GL_{d_1} \times \ldots \times GL_{d_k} \quad , \quad Z^a \simeq Z_{d_1} \times \ldots \times Z_{d_k}$$

The algebra $\mathbf{U}(R)$ (see sect.2) has a *Hopf algebra* structure with *coproduct* morphism Δ given by the formula

$$\Delta : E^{\pm}_{\alpha\gamma}(z) \mapsto \sum_{\beta=1}^{n} E^{\pm}_{\alpha\beta}(C^{\pm 1}z) \otimes E^{\pm}_{\beta\gamma}(C^{\mp 1}z).$$

The result below provides a geometric interpretation of the coproduct in the quotient $\mathbf{U} = \mathbf{U}(R)/(C^{\pm 1} = 1)$; cf. [Gr] for a closely related result.

Theorem 3.5. (i) *There is a natural Kunneth isomorphism*

$$K^{A^a}(Z^a) \simeq \Big(\bigotimes_{\mathbf{R}(A^a)}\Big)_{i=1}^{k} K^{A_{d_i}}(Z_{d_i})$$

(ii) *Moreover, the following natural diagram of algebra homomorphisms commutes*

$$\begin{array}{ccccc}
\mathbf{U} & & \xrightarrow{\quad\Delta^k\quad} & & \mathbf{U}\otimes\ldots\otimes\mathbf{U} \\
\big\downarrow{\scriptstyle thm.\ 3.3} & & & & \big\downarrow{\scriptstyle thm.\ 3.3} \\
\mathbb{C}_a\otimes K^{A}(Z) & \to & \mathbb{C}_a\otimes K^{A^a}(Z^a) & \underset{part\ (i)}{=\!=\!=} & \mathbb{C}_a\otimes K^{A_{d_1}}(Z_{d_1})\otimes\ldots\otimes K^{A_{d_k}}(Z_{d_k})
\end{array}$$

Proof of the first claim can be deduced from a general Kunneth formula in equivariant algebraic K-theory of cellular fibrations (cf. [CG]). The second claim is verified by a direct computation based on explicit formulas for the vertical maps in the diagram, given in [GV]. $\square$

Theorem 3.3 is in fact an analogue of a geometric construction discovered earlier for *affine Hecke algebras.* Recall that the affine Hecke algebra $\mathbf{H} = \mathbf{H}_d$ of type GL_d is defined algebraically as the $\mathbb{C}[q, q^{-1}]$-algebra on generators $T_1, T_2, \ldots, T_{d-1}$, and $X_1^{\pm 1}, \ldots, X_d^{\pm 1}$ subject to the relations

$$\begin{gathered}
(T_a - q^2)(T_a + 1) = 0, \\
T_a T_{a+1} T_a = T_{a+1} T_a T_{a+1} \quad , \quad [T_a, T_b] = 0 \quad \text{if} \quad |a-b| \geq 2, \qquad (3.6) \\
T_a X_a T_a = q^2 X_{a+1} \quad , \quad [X_a, T_b] = 0 \quad \text{if} \quad a \neq b,\, b+1.
\end{gathered}$$

Now let $\dot{\mathcal{F}}$ be the variety of all complete flags in $\mathbb{C}^d$ and $\dot{N}$ the variety of all nilpotent $d \times d$-matrices. The cotangent bundle on $\dot{\mathcal{F}}$ can be identified naturally (cf. [CG]) with the set $\dot{M} = \{(F, x) \in \dot{\mathcal{F}} \times \dot{N} \mid x(F_i) \subset F_{i-1}, \forall i = 1, 2, \ldots, d\}$. The second projection gives a proper morphism $\dot{M} \to \dot{N}$. Mimicing the construction at the beginning of the section, we put $\dot{Z} = \dot{M} \times_{\dot{N}} \dot{M} \subset \dot{M} \times \dot{M}$. The variety $\dot{Z}$, the so-called Steinberg variety, is clearly stable with respect to the diagonal action of A in $\dot{M} \times \dot{M}$. Furthermore, we have $\dot{Z} \circ \dot{Z} = \dot{Z}$. Thus the convolution makes $K^A(\dot{Z})$ an associative algebra. Here is an analogue of theorem 3.3:

Theorem 3.7 (cf. [Gi], [KL 2], [CG]). *There is an algebra isomorphism*

$$\mathbf{H} \simeq K^A(\dot{Z}). \qquad \square$$

4. The Polynomial Tensor Representation

For a finite dimensional vector space V, there is a canonical isomorphism $\mathrm{End}(V \otimes V) \simeq V^* \otimes V^* \otimes V \otimes V \simeq \mathrm{Hom}(\mathrm{End}V, \mathrm{End}V)$. Using this isomorphism, one may view the R-matrix (2.1) as a linear map

$$\hat{R} : \mathfrak{g} \to \mathbb{C}(X) \otimes \mathfrak{g}, \quad \text{i.e.,} \quad E_{ij} \mapsto \hat{R}(E_{ij})(X) = \sum_{k,l} R^{ijkl}(X)\, E_{kl}$$

where $R^{ijkl}(X)$ are rational $\mathbb{C}$-valued functions given by (2.1):

$$R(X) = \sum_{i,j,k,l} R^{ijkl}(X)\, E_{ij} \otimes E_{kl}$$

Define a linear map $\pi : \mathbf{U} \to \mathfrak{g}[X^{\pm 1}, q^{\pm 1}] := \mathfrak{g} \otimes_{\mathbb{C}} \mathbb{C}[X, X^{-1}, q, q^{-1}]$ by the following assignment of generating functions:

$$A^{\pm}(z) \mapsto \hat{R}(A)(z\,X), \qquad A \in \mathfrak{h}\,,\, \mathfrak{n}^+\,,\, \mathfrak{n}^- \tag{4.1}$$

where the right hand side is understood as a power series expansion in z^{-1} in the "$A^+(z)$" case, and in z in the "$A^-(z)$" case, respectively. In particular

$$(Id \otimes \pi)(L^{\pm}(z)) = R(z\,X). \tag{4.2}$$

Using the Yang-Baxter equation (2.3) and the unitarity condition, one deduces the following result.

Proposition 4.3. *The map π defined by (4.1) is an algebra homomorphism.*□

The standard $\mathfrak{g}$-action on the vector space $\mathbb{C}^n$ gives rise to a $\mathbb{C}[X^{\pm 1}, q^{\pm 1}]$-linear action of the algebra $\mathfrak{g}[X^{\pm 1}, q^{\pm 1}]$ on the vector space $\mathbf{P} := \mathbb{C}^n[X^{\pm 1}, q^{\pm 1}]$. This makes $\mathbf{P}$ an $\mathbf{U}(R)$-module via the homomorphism π, cf. (4.1). In a similar way, there is an algebra homomorphism $\rho : \mathbf{U}_q(\widehat{\mathfrak{gl}_n}) \to \mathfrak{g}[X^{\pm 1}, q^{\pm 1}]$, which is a special case of the general construction [GV, (4.1)] in the particular case $d = 1$ (in [GV] the notation $K = \mathbb{C}^n[X^{\pm 1}, q^{\pm 1}]$ was used). The homomorphism ρ is given by the following formulas

$$E_\alpha(z) \mapsto (q - q^{-1})\,\delta(\tfrac{X}{z})\,E_{\alpha\,\alpha+1} \tag{4.4.1}$$

$$F_\alpha(z) \mapsto (q - q^{-1})\,\delta(\tfrac{X}{z})\,E_{\alpha+1\,\alpha} \tag{4.4.2}$$

$$K^{\pm}_\beta(z) \mapsto \sum_{\gamma<\beta} \theta^{\pm}_{-1}(q^{-1}\tfrac{X}{z})\,E_{\gamma\gamma} + E_{\beta\beta} + \sum_{\gamma>\beta} \theta^{\pm}_{-1}(q\tfrac{X}{z})\,E_{\gamma\gamma} \tag{4.4.3}$$

where $\theta^{\pm}_m(z)$ is the power series expansion in $z^{\mp 1}$ of the rational function (2.2).

Observe that the homomorphism ρ makes $\mathbf{P}$ an $\mathbf{U}_q(\widehat{\mathfrak{gl}_n})$-module via the pull-back of the natural $\mathfrak{g}[X^{\pm 1}, q^{\pm 1}]$-module structure. A straightforward computation based on formulas (2.10) and (4.4.1-3) yields the following result.

Proposition 4.5. *The above defined $\mathbf{U}(R)$- and $\mathbf{U}_q(\widehat{\mathfrak{gl}_n})$-module structures on $\mathbf{P} = \mathbb{C}[X^{\pm 1}, q^{\pm 1}]$ are compatible with the isomorphism of theorem 2.10, i.e. the following algebra homomorphisms form a com-*

mutative triangle

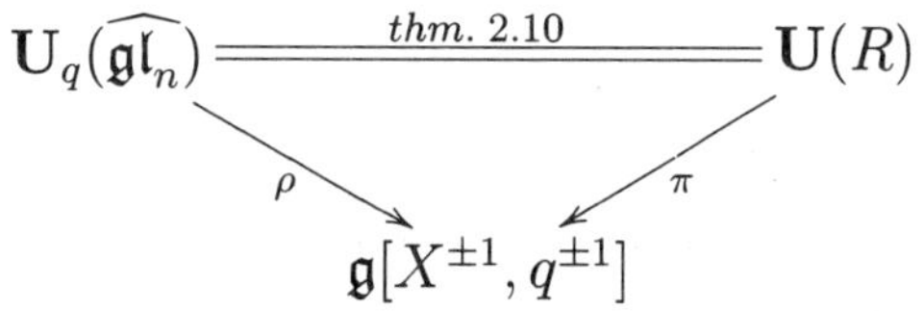

The coproduct on the Hopf algebra $\mathbf{U}$ provides, for each integer $d \geq 1$, the composition of algebra homomorphisms:

$$\pi^d \; : \; \mathbf{U} \to \mathbf{U}^{\otimes d} \to (\mathfrak{g}[X^{\pm1}, q^{\pm1}])^{\otimes d} \simeq \mathfrak{g}^{\otimes d}[X_1^{\pm1}, \dots, X_d^{\pm1}, q^{\pm1}] \quad (4.6)$$

Here the first arrow is given by the iterated coproduct map (tensor product is taken over $\mathbb{C}[q, q^{-1}]$), and the second arrow is a tensor power of the homomorphism π introduced above. Formula (4.2) yields the following property of the map (4.6)

$$(Id \otimes \pi^d)(L^{\pm}(z)) = R_{0,d}(z\, X_d)\, R_{0,d-1}(z\, X_{d-1}) \dots R_{0,1}(z\, X_1). \quad (4.7)$$

Set

$$\mathbf{P}^{\otimes d} = \mathbf{P} \otimes_{\mathbb{C}[q,q^{-1}]} \dots \otimes_{\mathbb{C}[q,q^{-1}]} \mathbf{P} \simeq (\mathbb{C}^n)^{\otimes d}[X_1^{\pm1}, \dots, X_d^{\pm1}, q^{\pm1}]$$

Observe that the rightmost vector space above has a natural action of the associative algebra $\mathfrak{g}^{\otimes d}[X_1^{\pm1}, \dots, X_d^{\pm1}, q^{\pm1}] \simeq (\mathfrak{g}[X^{\pm1}, q^{\pm1}])^{\otimes d}$. That makes $\mathbf{P}^{\otimes d}$ an $\mathbf{U}$-module via the homomorphism (4.6). It will be referred to as the *Polynomial Tensor Representation.*

Let $e_1, \dots, e_n$ be the standard base of $\mathbb{C}^n$ and $[1, n]^d$ the set of d-tuples of integers from the segment $[1, n]$. Then the vectors

$$e_{\mathbf{j}} = e_{j_1} \otimes \dots \otimes e_{j_d} \quad , \quad \mathbf{j} = (j_1, j_2, \dots, j_d) \in [1, n]^d$$

form the standard base of $(\mathbb{C}^n)^{\otimes d}$. The Symmetric group S_d acts naturally on the set $[1, n]^d$ by permutations of elements of a d-tuple $(j_1, j_2, \dots, j_d)$. That induces the standard S_d-action $\sigma : e_{\mathbf{j}} \mapsto e_{\sigma(\mathbf{j})}$ on $(\mathbb{C}^n)^{\otimes d}$. Let further $P \mapsto P^\sigma$ be the natural S_d-action on the polynomial ring $\mathbb{C}[X_1^{\pm1}, \dots, X_d^{\pm1}, q^{\pm1}]$ by permutation of the variables $X_1, \dots, X_d$. The simultaneous actions on $(\mathbb{C}^n)^{\otimes d}$ and on the polynomial ring give rise to an S_d-module structure on the tensor product $\mathbf{P}^{\otimes d} = (\mathbb{C}^n)^{\otimes d}[X_1^{\pm1}, \dots, X_d^{\pm1}, q^{\pm1}]$. There is also a natural action of the additive group $\mathbb{Z}^d$ on the same space $\mathbf{P}^{\otimes d}$ by multiplication $\mathbb{Z}^d \ni (m_1, \dots, m_d) : P \mapsto P \cdot X_1^{m_1} \cdot \dots \cdot X_d^{m_d}$. The S_d- and $\mathbb{Z}^d$-module structures fit together making $\mathbf{P}^{\otimes d}$ a module over the semi-direct product $W_{af} := S_d \ltimes \mathbb{Z}^d$, the affine Weyl group.

We now 'quantize' the above defined W_{af}-action by introducing a right $\mathbf{H}$-module structure on $\mathbf{P}^{\otimes d}$, where $\mathbf{H}$ is the affine Hecke algebra. For $\alpha = 1, 2, \dots, n-1$, let $s_\alpha \in S_d$ be the transposition $(\alpha, \alpha+1)$ and T_α the corresponding generator of the Hecke algebra (cf. (3.6)). The vector space $\mathbf{P}^{\otimes d}$ is spanned over $\mathbb{C}$ by elements of the form $e_{\mathbf{j}} \otimes P$, $P \in \mathbb{C}[X_1^{\pm 1}, \dots, X_d^{\pm 1}, q^{\pm 1}]$. For each $\alpha = 1, 2, \dots, n-1$ and $P \in \mathbb{C}[X_1^{\pm 1}, \dots, X_d^{\pm 1}, q^{\pm 1}]$, we put

$$(e_{\mathbf{j}} \otimes P) \cdot X_\alpha = e_{\mathbf{j}} \otimes (P \cdot X_\alpha) \tag{4.8}$$

$$(e_{\mathbf{j}} \otimes P) \cdot T_\alpha = \begin{cases} (1-q^2) e_{\mathbf{j}} \otimes \frac{X_{\alpha+1}(P^{s_\alpha} - P)}{X_{\alpha+1} - X_\alpha} + e_{s_\alpha(\mathbf{j})} \otimes P^{s_\alpha} & \text{if } j_\alpha < j_{\alpha+1}, \\ (1-q^2) e_{\mathbf{j}} \otimes \frac{X_{\alpha+1}(P^{s_\alpha} - P)}{X_{\alpha+1} - X_\alpha} + q^2 e_{\mathbf{j}} \otimes P^{s_\alpha} & \text{if } j_\alpha = j_{\alpha+1}, \\ (1-q^2) e_{\mathbf{j}} \otimes \frac{X_\alpha P^{s_\alpha} - X_{\alpha+1} P}{X_{\alpha+1} - X_\alpha} + q^2 e_{s_\alpha(\mathbf{j})} \otimes P^{s_\alpha} & \text{if } j_\alpha > j_{\alpha+1}, \end{cases}$$

In the next section we will prove the following result.

Theorem 4.9. *The above formulas give rise to a well-defined right $\mathbf{H}_d$-module structure on $\mathbf{P}^{\otimes d}$.*

Recall that the Representation ring $\mathbf{R}(A)$ is isomorphic to

$$\mathbf{R}(A) = \mathbf{R}(\mathbb{C}^* \times GL_d) = \mathbb{C}[X_1^{\pm 1}, \dots, X_d^{\pm 1}, q^{\pm 1}]^{S_d},$$

the algebra of symmetric polynomials. Formulas (4.8) yield

Corollary 4.10. (i) *The $\mathbf{H}$-action on $\mathbf{P}^{\otimes d}$ is $\mathbf{R}(A)$-linear.*

(ii) *Let $\mathbf{j} \in [1, n]^d$ be a non-decreasing d-tuple, and $\mathbf{j}'$ a d-tuple obtained from it by permutation. Let $w \in S_d$ be a permutation of minimal length such that $\mathbf{j}' = w(\mathbf{j})$. Then we have* $(e_{\mathbf{j}} \otimes 1) \cdot T_w = e_{\mathbf{j}'} \otimes 1$. □

The right $\mathbf{H}$-module structure on $\mathbf{P}^{\otimes d}$ enables us to describe the left $\mathbf{U}$-module structure on $\mathbf{P}^{\otimes d}$ given by the Polynomial Tensor Representation in a rather explicit way. We have the following result.

Theorem 4.11. *The* $\mathbf{U}$*-action arising from homomorphism (4.6) is the unique* $\mathbf{R}(A)$*-linear left* $\mathbf{U}$*-module structure on* $\mathbf{P}^{\otimes d}$ *that commutes with the right* $\mathbf{H}$*-action and such that, for any non-decreasing* d*-tuple* $\mathbf{j}$*, we have*

$$E_\alpha(z)\cdot e_{\mathbf{j}} = (q^2-1)q^{-\#\mathbf{j}^{-1}(\alpha+1)} e_{\mathbf{j}^-_{s+1}} \cdot \delta(\tfrac{X_{s+1}}{z})(1+\sum_{m=s+1}^{t-1} T_{s+1}T_{s+2}\dots T_m) \tag{4.12.1}$$

$$F_\alpha(z)\cdot e_{\mathbf{j}} = (q^2-1)\, q^{-\#\mathbf{j}^{-1}(\alpha)} e_{\mathbf{j}^+_s} \cdot \delta(\tfrac{X_s}{z})(1+\sum_{m=r+1}^{s-1} T_{s-1}T_{s-2}\dots T_m) \tag{4.12.2}$$

$$K^\pm_\alpha(z)\cdot e_{\mathbf{j}} = \prod_{l_m<\alpha} \theta^\pm_{-1}\left(q^{-1}\tfrac{X_m}{z}\right) \prod_{l_m>\alpha} \theta^\pm_{-1}\left(q\tfrac{X_m}{z}\right) e_{\mathbf{j}}, \tag{4.12.3}$$

where $]r,s] := \mathbf{j}^{-1}(\alpha)$, $]s,t] := \mathbf{j}^{-1}(\alpha+1)$, and one puts

$$\begin{cases} e_{\mathbf{j}^-_{s+1}} = 0 & \text{if } \mathbf{j}^{-1}(\alpha+1)=\emptyset, \\ \mathbf{j}^-_{s+1} = (j_1,\dots,j_s,\alpha,j_{s+2},\dots,j_d) & \text{else,} \end{cases}$$

$$\begin{cases} e_{\mathbf{j}^+_s} = 0 & \text{if } \mathbf{j}^{-1}(\alpha)=\emptyset, \\ \mathbf{j}^+_s = (j_1,\dots,j_{s-1},\alpha+1,j_{s+1},\dots,j_d) & \text{else.} \end{cases}$$

5. The Polynomial Tensor Representation via K-theory

In this section and in section 7 we will give three geometric constructions of the Polynomial Tensor Representation. The first construction, presented here, is based on the equivariant K-theory. The other two, of section 7, involve affine flag varieties over a finite field and are dual to the first one in the sense of Langlands. Those constructions may be viewed as an extension to the affine case of the Grojnowski-Lusztig construction [GL] in the finite case.

Recall the n-step flag variety $\mathcal{F}$ and the complete flag variety $\dot{\mathcal{F}}$ introduced in section 3. The corresponding cotangent bundles $M = T^*\mathcal{F}$ and $\dot{M} = T^*\dot{\mathcal{F}}$ have natural $A = \mathbb{C}^* \times GL_d$-actions. Following [GL], we introduce the A-stable subvariety $W = M \times_N \dot{M} \subset M \times \dot{M}$. The variety W 'links' Z with $\dot{Z}$ and it will play a major role in the rest of the paper. Observe that we have

$$Z \circ W = W\,, \quad W \circ \dot{Z} = W$$

Hence there are natural convolution morphisms

$$K^A(Z) \otimes K^A(W) \to K^A(W)\,, \quad K^A(W) \otimes K^A(\dot{Z}) \to K^A(W) \quad (5.1)$$

Thus the group $K^A(W)$ has natural *left* $K^A(Z)$-module and *right* $K^A(\dot{Z})$-module structures respectively. Here is one of the main results of this section.

Theorem 5.2. *There is an* $\mathbf{R}(A)$*-linear isomorphism* $\psi : \mathbf{P}^{\otimes d} \simeq K^A(W)$ *that intertwines both the left* $\mathbf{U}$*- and the right* $\mathbf{H}$*-module structures, i.e., makes the following diagram commute*

$$\begin{array}{ccccc} \mathbf{U} & \xrightarrow{(4.6)} & \operatorname{End}\mathbf{P}^{\otimes d} & \xleftarrow{thm.\ 4.8} & \mathbf{H} \\ \downarrow{\scriptstyle thm.\ 3.3} & & \|{\scriptstyle \psi} & & \|{\scriptstyle thm.\ 3.7} \\ K^A(Z) & \xrightarrow[(5.1)]{} & \operatorname{End} K^A(W) & \xleftarrow[(5.1)]{} & K^A(\dot{Z}) \end{array}$$

Sketch of Proof. Following [GL] we observe that the G-orbits in $\mathcal{F}\times\dot{\mathcal{F}}$ are parametrized by the set $[1,n]^d$. To a d-tuple $\mathbf{j} = (j_1, j_2, \dots, j_d)$ one associates the G-orbit $\mathcal{W}_{\mathbf{j}}$ of the pair $(F, \dot{F})$ of flags

$$F = (\{0\} \subseteq \bigoplus_{k_i \leq 1} \mathbb{C}\cdot e_i \subseteq \bigoplus_{k_i \leq 2} \mathbb{C}\cdot e_i \subseteq \ldots \subseteq \mathbb{C}^d)$$

$$\dot{F} = (\{0\} \subset \mathbb{C}\cdot e_1 \subset \mathbb{C}\cdot e_1 \oplus \mathbb{C}\cdot e_2 \subset \ldots \subset \mathbb{C}^d)$$

where $e_1, \dots, e_d$ is the standard base of $\mathbb{C}^d$. Further, set

$$\mathbf{a}(\mathbf{j}) = \sum_{1\leq i<k\leq n} \#\mathbf{j}^{-1}(i)\cdot \#\mathbf{j}^{-1}(k) \quad (5.3)$$

The integer $\mathbf{a}(\mathbf{j})$ is shown to be the dimension of the fiber of the first projection $\mathcal{F} \times \dot{\mathcal{F}} \supset \mathcal{W}_{\mathbf{j}} \to \mathcal{F}$.

Now let $\mathbf{j} \in [1, n]^d$ be a *nondecreasing* d-tuple. We show that the corresponding G-orbit $\mathcal{W}_\mathbf{j}$ is closed. Let $\Omega_\mathbf{j}$ of the sheaf of maximal rank relative forms along the first projection $\mathcal{W}_\mathbf{j} \to \mathcal{F}$ (see [V, §2]). Let $[\pi^*\Omega_\mathbf{j}] \in K^A(W)$ be the class of the pull-back of $\Omega_\mathbf{j}$ via the natural conormal bundle projection $\pi : T^*_{\mathcal{W}_\mathbf{j}}(\mathcal{F} \times \dot{\mathcal{F}}) \to \mathcal{W}_\mathbf{j}$. Thus, the sheaf $\pi^*\Omega_\mathbf{j}$ is supported on $T^*_{\mathcal{W}_\mathbf{j}}(\mathcal{F} \times \dot{\mathcal{F}})$, and we set

$$\mathcal{E}_\mathbf{j} = q^{\mathbf{a}(\mathbf{j})} \cdot [\pi^*\Omega_\mathbf{j}] \in K^A(W)$$

For a general $\mathbf{j} \in [1, n]^d$, which is not necessarily non-decreasing, we define a class $\mathcal{E}_\mathbf{j} \in K^A(W)$ by means of the isomorphism of theorem 3.7 as follows. First, find a *non-decreasing* d-tuple $\mathbf{j}_+$ obtained from $\mathbf{j}$ by a permutation. Let $w \in S_d$ be a permutation of minimal length such that $\mathbf{j} = w(\mathbf{j}_+)$ and $T_w \in \mathbf{H}$ the corresponding element of the Hecke algebra. Let $[T_w] \in K^A(\dot{Z})$ be the image of T_w under the isomorphism of theorem 3.7. The K-group $K^A(\dot{Z})$ acts on $K^A(W)$ on the right via convolution, and we put by definition $\mathcal{E}_\mathbf{j} := \mathcal{E}_{\mathbf{j}_+} \star [T_w]$.

Next let $\Delta \subset \dot{\mathcal{F}} \times \dot{\mathcal{F}}$ be the diagonal and $\dot{Z}_\Delta \subset \dot{Z}$ its conormal bundle. The group $K^A(\dot{Z}_\Delta)$ is a subalgebra of $K^A(\dot{Z})$. Moreover, this subalgebra is isomorphic to $\mathbb{C}[X_1^{\pm 1}, \ldots, X_d^{\pm 1}, q^{\pm 1}]$ in a canonical way (cf. [CG]). Hence, the right $K^A(\dot{Z}_\Delta)$-action on $K^A(W)$ gives an $\mathbb{C}[X_1^{\pm 1}, \ldots, X_d^{\pm 1}, q^{\pm 1}]$-module structure on $K^A(W)$. We define an $\mathbf{R}(A)$-linear map $\psi : \mathbf{P}^{\otimes d} \to K^A(W)$ by the following assingment

$$(\mathbb{C})^{\otimes d}[X_1^{\pm 1}, \ldots, X_d^{\pm 1}, q^{\pm 1}] \ni e_\mathbf{j} \otimes P \;\mapsto\; \mathcal{E}_\mathbf{j} \star P$$

This map can be shown to be bijective. Furthermore, one verifies by a lengthy but straightforward computation that the $K^A(Z)$- and $K^A(\dot{Z})$-actions on $K^A(W)$ correspond via the bijection ψ to the action given by the formulas (4.8) and (4.12) respectively. That completes the proof of the theorem. □

Remark 5.4. The commutativity of the right square in the diagram of theorem 5.2, combined with theorem 3.7, implies theorem 4.9. □

Next fix a diagonal matrix $s \in GL_d$ with k distinct eigenvalues with multiplicities $d_1, d_2, \ldots, d_k$. Put $a = (s, 1) \in GL_d \times \mathbb{C}^*$. Then $Z^a \circ W^a = W^a$, so that the group $K^{A^a}(W^a)$ acquires a $K^{A^a}(Z^a)$-module structure. Moreover, we have natural isomorphisms

$$Z^a \simeq Z_{d_1} \times \ldots \times Z_{d_k} \qquad\qquad W^a \simeq W_{d_1} \times \ldots \times W_{d_k}.$$

Proposition 5.5. (i) *There is a Künneth isomorphism*

$$K^{A^a}(W^a) \simeq \bigotimes_{i=1}^{k} K^{A_{d_i}}(W_{d_i})$$

here the tensor product is taken over $\mathbb{C}[q,q^{-1}]$.

(ii) *The following diagram of algebra homomorphisms commutes:*

$$\begin{array}{ccccc}
K^A(Z)_a & \longrightarrow & K^{A^a}(Z^a)_a & \overset{3.5(i)}{=\!=\!=} & \Big(\bigotimes_{\mathbf{R}(A^a)}\Big)_{i=1}^{k} K^{A_{d_i}}(Z_{d_i})_a \\
\big\downarrow{\scriptstyle (5.1)} & & \big\downarrow{\scriptstyle (5.1)} & & \big\downarrow{\scriptstyle (5.1)} \\
\operatorname{End}_{\mathbf{R}(A)} K^A(W)_a & \longrightarrow & \operatorname{End}_{\mathbf{R}(A^a)} K^{A^a}(W^a)_a & = & \Big(\bigotimes_{\mathbf{R}(A^a)}\Big)_{i=1}^{k} \operatorname{End}_{\mathbf{R}(A_{d_i})} K^{A_{d_i}}(W_{d_i})_a \\
\Big\| & & & & \Big\| \\
\operatorname{End}_{\mathbf{R}(A)} \mathbf{P}_a^{\otimes d} & \longrightarrow & \operatorname{End}_{\mathbf{R}(A^a)} \mathbf{P}_a^{\otimes(d_1+\ldots+d_k)} & = & \Big(\bigotimes_{\mathbf{R}(A^a)}\Big)_{i=1}^{k} \operatorname{End}_{\mathbf{R}(A_{d_i})} \mathbf{P}_a^{\otimes d_i}
\end{array}$$

Where the subscript 'a' indicates specialization at the point $a \in \operatorname{Specm} \mathbf{R}(A)$, e.g., $K^A(Z)_a := \mathbb{C}_a \otimes_{\mathbf{R}(A)} K^A(Z)$.

6. Geometric Weyl correspondence

In this section we establish geometrically an equivalence between the category of finite dimensional $\mathbf{H}$-module and a certain category of finite dimensional $\mathbf{U}$-modules (cf. [Ch 2], [Dr 3]).

We first remind the results of [Gi],[GV] (see also [KL 2] and [CG]) about simple finite dimensional modules over the algebras $\mathbf{U}$ and $\mathbf{H}$. They have a similar geometric description. Observe that in the setup of (4.1), (4.2) the K-theoretic convolution has a counterpart in *Borel-Moore homology* with complex coefficients defined e.g. in [CG, ch. 2] :

$$\star : \quad H_k(Z_{12}) \times H_l(Z_{23}) \longrightarrow H_{l+k-\dim M_2}(Z_{12} \circ Z_{23})$$

and there is a *bivariant Riemann-Roch* map (see [CG]):

$$c : \quad K(Z_{ij}) \longrightarrow H_*(Z_{ij}) \tag{6.1}$$

which commutes with convolution. Moreover, for varieties like Z and $\dot{Z}$, which are built out of complex cells, the Riemann-Roch map (6.1) is known (see [CG, Cellular Fibration Lemma]) to be an isomorphism.

Fix a semisimple element $a = (t, s) \in A$ and the corresponding 1-dimensional $\mathbf{R}(A)$-module $\mathbb{C}_a$ given by evaluation at a. We have a chain of natural morphisms :

$$\mathbb{C}_a \otimes_{\mathbf{R}(A)} K^A(Z) \xrightarrow{(3.4)} \mathbb{C}_a \otimes_{\mathbf{R}(A)} K^{A^a}(Z^a) \xrightarrow{forget} K(Z^a) \xrightarrow{(6.1)} H_*(Z^a) \tag{6.2}$$

all of which commute with convolution. Furthermore, for varieties built out of complex cells, each of the morphisms above is actually an isomorphism (cf. [CG] for details). It follows that the composite map yields *algebra isomorphisms*

$$\mathbb{C}_a \otimes_{\mathbf{R}(A)} K^A(Z) \xrightarrow{\sim} H_*(Z^a)$$

$$\mathbb{C}_a \otimes_{\mathbf{R}(A)} K^A(\dot{Z}) \xrightarrow{\sim} H_*(\dot{Z}^a).$$

Thus from Theorem 4.3 and (6.1) we have *surjective* algebra homomorphisms

$$\mathbf{U} \twoheadrightarrow H_*(Z^a) \quad , \quad \mathbf{H} \twoheadrightarrow H_*(\dot{Z}^a) \tag{6.3}$$

Next, let $\dot{N} \subset \mathrm{End}\,(\mathbb{C}^d)$ denote the variety of all nilpotent endomorphisms, and $\dot{N}^a$ the fixed point subvariety. Explicitly, we have $\dot{N}^a = \{x \in \dot{N} \mid s \cdot x \cdot s^{-1} = t^2 \cdot x\}$ and $N^a = \{x \in \dot{N}^a \mid x^n = 0\}$. The centralizer G^s of s in G acts on M^a, N^a (resp. on $\dot{M}^a$, $\dot{N}^a$) in such a way that the projection $\pi : M^a \longrightarrow N^a$ (resp. $\dot{\pi} : \dot{M}^a \longrightarrow \dot{N}^a$) is G^s-equivariant. We form the following G^s-equivariant constructible complexes

$$\mathcal{L} = \pi_* \mathbb{C}_{M^a}[\dim M^a] \in D^b(N^a), \qquad \dot{\mathcal{L}} = \dot{\pi}_* \mathbb{C}_{\dot{M}^a}[\dim \dot{M}^a] \in D^b(\dot{N}^a)$$

The varieties N^a and $\dot{N}^a$ are known to consist of finite union of G^s-orbits each. Moreover, any G^s-equivariant local system on any G^s-orbit $\mathbf{O}$ is constant, for its isotropy group is connected. Thus, $\mathcal{L}$ and $\dot{\mathcal{L}}$ decompose in the following way (Decomposition theorem [BBD]):

$$\mathcal{L} = \bigoplus_{\mathbf{O}\subset N^a,\, i\in\mathbb{Z}} L_{\mathbf{O}}[i] \otimes IC_{\mathbf{O}}[i] \tag{6.4.1}$$

$$\dot{\mathcal{L}} = \bigoplus_{\mathbf{O}\subset \dot{N}^a,\, i\in\mathbb{Z}} \dot{L}_{\mathbf{O}}[i] \otimes IC_{\mathbf{O}}[i] \tag{6.4.2}$$

where $IC_{\mathbf{O}}$ is the *Intersection cohomology complex* associated with the constant sheaf on $\mathbf{O}$ and $L_{\mathbf{O}}[i]$, $\dot{L}_{\mathbf{O}}[i]$, are certain finite dimensional vector spaces. Put $L_{\mathbf{O}} := \oplus_i L_{\mathbf{O}}[i]$ and $\dot{L}_{\mathbf{O}} := \oplus_i \dot{L}_{\mathbf{O}}[i]$. Then (6.4.1-2) yield

$$\begin{aligned}\mathrm{Ext}(\mathcal{L},\mathcal{L}) \simeq &\bigoplus_{\mathbf{O}\subset N^a} \mathrm{Hom}(L_{\mathbf{O}}, L_{\mathbf{O}}) \bigoplus \\ &\bigoplus_{\mathbf{O},\mathbf{O}'\subset N^a,\, k>0} \mathrm{Hom}(L_{\mathbf{O}}, L_{\mathbf{O}'}) \otimes \mathrm{Ext}^k(IC_{\mathbf{O}}, IC_{\mathbf{O}'})\end{aligned} \tag{6.5.1}$$

$$\begin{aligned}\mathrm{Ext}(\dot{\mathcal{L}},\dot{\mathcal{L}}) \simeq &\bigoplus_{\mathbf{O}\subset \dot{N}^a} \mathrm{Hom}(\dot{L}_{\mathbf{O}}, \dot{L}_{\mathbf{O}}) \bigoplus \\ &\bigoplus_{\mathbf{O},\mathbf{O}'\subset \dot{N}^a,\, k>0} \mathrm{Hom}(\dot{L}_{\mathbf{O}}, \dot{L}_{\mathbf{O}'}) \otimes \mathrm{Ext}^k(IC_{\mathbf{O}}, IC_{\mathbf{O}'}).\end{aligned} \tag{6.5.2}$$

We now introduce the fixed point subvarieties Z^a and $\dot{Z}^a$. Observe that we have

$$Z^a = M^a \times_{N^a} M^a \quad , \quad \dot{Z}^a = \dot{M}^a \times_{\dot{N}^a} \dot{M}^a$$

It follows (see [Gi],[CG]) that there are natural algebra isomorphisms (not grading-preserving):

$$H_*(Z^a) \simeq \mathrm{Ext}(\mathcal{L},\mathcal{L}) \quad , \quad H_*(\dot{Z}^a) \simeq \mathrm{Ext}(\dot{\mathcal{L}},\dot{\mathcal{L}}) \tag{6.6}$$

We see that the second sums in both (6.5.1) and (6.5.2) are the *radicals* $\mathrm{Rad}(H_*(Z^a))$, $\mathrm{Rad}(H_*(\dot{Z}^a))$ of the algebras $H_*(Z^a)$ and $H_*(\dot{Z}^a)$ respectively, while the first sums are, respectively, their maximal semisimple quotients. Thus, formulas (6.3) and (6.6) yield the following result, cf. [CG].

Theorem 6.7 [Gi], [GV]. (i) *Each vector space $L_{\mathbf{O}}$ (resp. $\dot{L}_{\mathbf{O}}$) has a natural $\mathbf{U}$-module (resp. $\mathbf{H}$-module) structure.*

(ii) *Fix $t \in \mathbb{C}^*$ which is not a root of unity. Then the set $\{L_{\mathbf{O}}\}_{a,\mathbf{O}}$ is a collection of non-isomorphic simple finite dimensional*

$\mathbf{U}_{|q=t}$*-modules, and the set* $\{\dot{L}_{\mathrm{O}}\}_{a,\mathrm{O}}$ *is a complete collection of isomorphism classes of simple* $\mathbf{H}_{|q=t}$*-modules.*

We now state the main result of the section, which is a quantum affine counterpart of the classical Weyl correspondence between simple GL_n- and S_d-modules in the tensor representation.

Theorem 6.8. *Fix a positive integer* $d \leq n$*, and* $t \in \mathbb{C}^*$ *which is not a root of unity. Then the functor*

$$V \mapsto \mathbf{P}^{\otimes d} \otimes_{\mathbf{H}_d} V$$

provides an equivalence between the category of finite dimensional $\mathbf{H}_d$*-modules and the category of finite dimensional* $\mathbf{U}_q(\widehat{\mathfrak{gl}_n})$*-modules whose simple components are among the* $\{L_{\mathrm{O}}\}_{a,\mathrm{O}}$ *with fixed d.*

Sketch of proof: The arguments are similar to those leading to theorem 6.7 with the variety Z being replaced by W. First, given any semisimple element $a \in A$, one has

$$W^a = M^a \times_{N^a} \dot{M}^a. \tag{6.9}$$

It follows from (6.3) and from theorem 4.3 that both the K-group $K^A(W)$ and the homology group $H_*(W^a)$ have left $\mathbf{U}$-module structures and right $\mathbf{H}$-module structures each. Moreover, it was shown in [CG] that those structures commute with the localization isomorphism, cf. [CG] and (6.2) :

$$c \circ r_a : \mathbb{C}_a \otimes_{\mathbf{R}(A)} K^A(W) \xrightarrow{\sim} H_*(W^a).$$

Assume from now on that $d \leq n$. Then, for any nilpotent operator x in $\mathbb{C}^d$ we have $x^n = 0$, hence $x^d = 0$. It follows that $N = \dot{N}$. Formula (6.9) and the general results of [CG, ch. 7] yield a natural isomorphism

$$H_*(W^a) \simeq \operatorname{Ext}(\dot{\mathcal{L}}, \mathcal{L}).$$

Thus, using the decompositions (6.4.1-2) one finds

$$H_*(W^a) \simeq \bigoplus_{\mathbf{O}\subset N^a} \mathrm{Hom}(\dot{L}_{\mathbf{O}}, L_{\mathbf{O}}) \ \bigoplus \tag{6.10}$$

$$\bigoplus_{\mathbf{O},\mathbf{O}'\subset N^a,\, k>0} \mathrm{Hom}(\dot{L}_{\mathbf{O}'}, L_{\mathbf{O}}) \otimes \mathrm{Ext}^k(IC_{\mathbf{O}'}, IC_{\mathbf{O}}).$$

Similarly, as we have observed earlier, formulas (6.5.2) and (6.6) yield an algebra isomorphism

$$H_*(\dot{Z}^a) \simeq \Big(\bigoplus_{\mathbf{O}\subset N^a} \mathrm{End}\,\dot{L}_{\mathbf{O}}\Big) \bigoplus \Big(\bigoplus_{\mathbf{O}.\mathbf{O}'\subset N^a,\, k>0} \mathrm{Hom}(\dot{L}_{\mathbf{O}}, \dot{L}_{\mathbf{O}'}) \otimes \mathrm{Ext}^k(IC_{\mathbf{O}}, IC_{\mathbf{O}'})\Big)$$

Let $p_{\mathbf{O}}$ denote the identity element of the simple algebra $\mathrm{End}\,\dot{L}_{\mathbf{O}}$, viewed as the projector to the corresponding simple direct summand in the maximal semisimple subalgebra

$$\mathcal{A} := \bigoplus_{\mathbf{O}\subset N^a} \mathrm{End}\,\dot{L}_{\mathbf{O}} \ \subset \ H_*(\dot{Z}^a)$$

The projectors form the canonical central decomposition $1 = \sum_{\mathbf{O}} p_{\mathbf{O}} \in \mathcal{A}$. It follows from the formulas above that, for any $\mathbf{O} \subset N^a$, the right ideal $p_{\mathbf{O}} \cdot H_*(\dot{Z}^a) \subset H_*(\dot{Z}^a)$ has the form

$$p_{\mathbf{O}} H_*(\dot{Z}^a) = \mathrm{End}\,\dot{L}_{\mathbf{O}} \ \bigoplus \ \Big(\bigoplus_{\mathbf{O}'\subset N^a,k>0} \mathrm{Hom}(\dot{L}_{\mathbf{O}'}, \dot{L}_{\mathbf{O}}) \otimes \mathrm{Ext}^k(IC_{\mathbf{O}'}, IC_{\mathbf{O}})\Big) \tag{6.11}$$

Hence, we find

$$\mathrm{Hom}(\dot{L}_{\mathbf{O}}, L_{\mathbf{O}}) \ \bigoplus \ \Big(\bigoplus_{\mathbf{O}'\subset N^a,\, k>0} \mathrm{Hom}(\dot{L}_{\mathbf{O}'}, L_{\mathbf{O}}) \otimes \mathrm{Ext}^k(IC_{\mathbf{O}'}, IC_{\mathbf{O}})\Big) =$$

$$= \mathrm{Hom}(\dot{L}_{\mathbf{O}}, L_{\mathbf{O}}) \bigotimes_{\mathcal{A}} \Big(\mathrm{End}\,\dot{L}_{\mathbf{O}} \ \bigoplus_{\mathbf{O}'\subset N^a,\, k>0} \mathrm{Hom}(\dot{L}_{\mathbf{O}'}, \dot{L}_{\mathbf{O}}) \otimes \mathrm{Ext}^k(IC_{\mathbf{O}'}, IC_{\mathbf{O}})\Big)$$

The last expression can be rewritten, by (6.11), as

$$\mathrm{Hom}(\dot{L}_{\mathbf{O}}, L_{\mathbf{O}}) \bigotimes_{\mathcal{A}} p_{\mathbf{O}} \cdot H_*(\dot{Z}^a) \quad ,$$

which implies that it is a projective right $H_*(\dot{Z}^a)$-module, for $p_{\mathbf{O}} \cdot H_*(\dot{Z}^a)$ is a direct summand of the free module $H_*(\dot{Z}^a)$. Observe now that the RHS of (6.10) is a direct sum of right $H_*(\dot{Z}^a)$-modules of that form. Thus, we conclude that

$$H_*(W^a) \simeq \bigoplus_{\mathbf{O}\subset N^a} \Big(\mathrm{Hom}(\dot{L}_{\mathbf{O}}, L_{\mathbf{O}}) \bigotimes_{\mathcal{A}} p_{\mathbf{O}} \cdot H_*(\dot{Z}^a)\Big) \tag{6.12}$$

is a projective right $H_*(\dot{Z}^a)$-module. Hence, the functor $H_*(W^a)\otimes_{H_*(\dot{Z}^a)}(\bullet)$ is exact.

Finally, for any orbit $\mathbf{O}'$, we calculate using (6.12)

$$H_*(W^a)\otimes_{H_*(\dot{Z}^a)} \dot{L}_{\mathbf{O}'}$$

$$= \Big(\bigoplus_{\mathbf{O}\subset N^a} \mathrm{Hom}(\dot{L}_{\mathbf{O}}, L_{\mathbf{O}}) \bigotimes_{\mathcal{A}} p_{\mathbf{O}} H_*(\dot{Z}^a)\Big)\otimes_{H_*(\dot{Z}^a)} \dot{L}_{\mathbf{O}'} =$$

$$= \Big(\bigoplus_{\mathbf{O}\subset N^a} \mathrm{Hom}(\dot{L}_{\mathbf{O}}, L_{\mathbf{O}})\Big) \bigotimes_{\mathcal{A}} \dot{L}_{\mathbf{O}'} = L_{\mathbf{O}'}$$

Thus the functor is exact and takes simple $H_*(\dot{Z}^a)$-modules into the corresponding simple $H_*(Z^a)$-modules. The result now follows easily from theorem 5.2. □

Remark 6.13. Similar arguments yield the duality theorem of [Dr 3] between the *Yangian* $Y(\widehat{\mathfrak{gl}_n})$ and the *degenerate affine Hecke algebra* D_d of type GL_d. To that end one uses, instead of (5.1) and Theorem 4.3, analogously constructed algebra morphisms (see [Lu 2], [GV, Remark 8.7]) :

$$Y(\widehat{\mathfrak{gl}_n}) \twoheadrightarrow H^A(Z) \qquad\qquad D_d \simeq H^A(\dot{Z})$$

where H^A stands for the *equivariant Borel-Moore homology* with complex coefficients. □

7. Dual approach to the Polynomial Tensor Representation

Fix a finite field $\mathbb{F}$ with p elements and let $\mathbb{K} = \mathbb{F}((z))$ be the field of Laurent power series. By a *lattice* we mean a free $\mathbb{F}[[z]]$-submodule in $\mathbb{K}^d$ of rank d. Let $\mathcal{B}$ be the flag variety of all *n-periodic* sequences of lattices $F_i \subset \mathbb{K}^d$, $i \in \mathbb{Z}$, of the form

$$F = (\cdots \subseteq F_{i-1} \subseteq F_i \subseteq F_{i+1} \subseteq \cdots), \qquad \text{where} \quad F_{i+n} = z^{-1} \cdot F_i, \ \forall i \in \mathbb{Z}.$$

Similarly, let $\dot{\mathcal{B}}$ be the flag variety of all *complete* sequences of lattices

$F_i \subset \mathbb{K}^d$, $i \in \mathbb{Z}$, of the form

$$F = (\cdots \subset F_{i-1} \subset F_i \subset F_{i+1} \subset \cdots),$$

where $F_{i+d} = z^{-1} \cdot F_i$, and $\dim(F_{i+1}/F_i) = 1$, $\forall i \in \mathbb{Z}$.

Let $G(\mathbb{K}) = GL_d(\mathbb{K})$ be the group of invertible $\mathbb{K}$-valued $d \times d$-matrices. The group $G(\mathbb{K})$ acts naturally on the set of lattices, inducing a $G(\mathbb{K})$-action on $\mathcal{B}$ and on $\dot{\mathcal{B}}$. Let $\mathbb{C}_{G(\mathbb{K})}[\mathcal{B} \times \mathcal{B}]$ be the complex vector space formed by all $G(\mathbb{K})$-invariant functions on $\mathcal{B} \times \mathcal{B}$ whose support is a finite union of $G(\mathbb{K})$-orbits. For any two orbits Z, $Z' \subset \mathcal{B} \times \mathcal{B}$ the map (3.1) has finite fibers (where $M_1 = M_2 = M_3 = \mathcal{B}$). Hence, the convolution makes $\mathbb{C}_{G(\mathbb{K})}[\mathcal{B} \times \mathcal{B}]$ an associative algebra.

Let $\mathbf{d} = (0 = d_0 \leq d_1 \leq \ldots \leq d_n = d)$ denote a partition of the set $\{1, \ldots, d\}$ into n (possibly empty) segments of lengths $d_1 - d_0, \ldots, d_n - d_{n-1}$, respectively. Introduce the set $\mathcal{P} = \{\textit{partitions of } d \textit{ into } n \textit{ segments}\} \cup \{\nabla\}$ where ∇ is a formal element, the *ghost-partition*. For each $\alpha = 1, \ldots, n$, define two transformations $\mathbf{d} \mapsto \mathbf{d}^{\pm}_{\alpha}$ of the set $\mathcal{P}$ by the following rules:

(i) The element ∇ is kept fixed by both transformations;

(ii) Given a partition $\mathbf{d}$ and $\alpha \neq n$, set $\mathbf{d}^{\pm}_{\alpha} = (0 = d_0 \leq \ldots \leq d_{\alpha-1} \leq d_\alpha \pm 1 \leq d_{\alpha+1} \leq \ldots \leq d_n = d)$ unless $d_\alpha = d_{\alpha \pm 1}$, in which case we set $\mathbf{d}^{\pm}_{\alpha} = \nabla$;

(iii) Set $\mathbf{d}^{\pm}_{n} = (d_0 \leq d_1 \mp 1 \leq \cdots \leq d_{n-1} \mp 1 \leq d_n = d)$ unless $d_0 = d_1$ or $d_{n-1} = d_n$, in which case we let $\mathbf{d}^{\pm}_{n}$ to be ∇.

Given a partition $\mathbf{d} \in \mathcal{P}$, let $\mathcal{B}_{\mathbf{d}}$ be the subset of $\mathcal{B}$ formed by all flags F such that $\dim(F_k/F_0) = d_k$, for any $k = 1, \ldots, n$. Put $\mathcal{B}_\nabla = \emptyset$. For each $\alpha = 1, \ldots, n$ and each $\mathbf{d} \in \mathcal{P}$ such that $\mathbf{d}^{+}_{\alpha} \neq \nabla$, resp. $\mathbf{d}^{-}_{\alpha} \neq \nabla$, set:

$$\hat{Y}_{d^+_\alpha, d} = \left\{(F, F') \in \mathcal{B}_{d^+_\alpha} \times \mathcal{B}_{\mathbf{d}} \;\middle|\; \begin{array}{ll} F_i = F'_i & \forall i \in \{1, \ldots, n\} \setminus \{\alpha\} \\ F'_\alpha \subset F_\alpha & \& \quad \dim(F_\alpha/F'_\alpha) = 1 \end{array} \right\}$$

$$\hat{Y}_{d_\alpha^-,d} = \{(F,F') \in \mathcal{B}_{d_\alpha^-} \times \mathcal{B}_{\mathbf{d}} \mid \begin{matrix} F_i = F'_i & \forall i \in \{1,\dots,n\} \setminus \{\alpha\} \\ F_\alpha \subset F'_\alpha & \& \quad \dim(F'_\alpha/F_\alpha) = 1 \end{matrix}\}$$

Each of the sets $\hat{Y}_{\mathbf{d}_\alpha^\pm,\mathbf{d}}$ is a single $G(\mathbb{K})$-orbit. Let $\mathbf{1}_{\mathbf{d}_\alpha^\pm,\mathbf{d}} \in \mathbb{C}_{G(\mathbb{K})}[\mathcal{B}\times\mathcal{B}]$ denote the characteristic function of the corresponding orbit. Further, let $\mathbf{1}_{\mathbf{d},\mathbf{d}}$ denote the characteristic function of the diagonal in $\mathcal{B}_{\mathbf{d}} \times \mathcal{B}_{\mathbf{d}}$. Analogous to the construction of [BLM] in the finite case, for $\alpha = 1,\dots,n$, define the following elements of the algebra $\mathbb{C}_{G(\mathbb{K})}[\mathcal{B} \times \mathcal{B}]$:

$$\mathbf{e}_\alpha = \sum_{\mathbf{d}\in\mathcal{P}} p^{(d_{\alpha-1}-d_\alpha)/2}\, \mathbf{1}_{\mathbf{d}_\alpha^+,\mathbf{d}} \quad , \quad \mathbf{f}_\alpha = \sum_{\mathbf{d}\in\mathcal{P}} p^{(d_\alpha-d_{\alpha+1})/2}\, \mathbf{1}_{\mathbf{d}_\alpha^-,\mathbf{d}}$$

(where $d_{n+1} := d + d_1$) and also

$$\mathbf{k}_\alpha = \sum_{\mathbf{d}\in\mathcal{P}} p^{(d_\alpha-d_{\alpha-1})/2}\, \mathbf{1}_{\mathbf{d},\mathbf{d}}.$$

Now, let $K_\alpha^{\pm1}, E_\alpha, F_\alpha$, $\alpha = 1,\dots,n$, be the images in $\mathbf{U}$ of the standard Kac-Moody generators of the algebra $\mathbf{U}_q(\widehat{\mathfrak{gl}_n})$.

Theorem 7.1.[GV] *The assignment:*

$$q \mapsto p^{1/2} \quad , \quad K_\alpha \mapsto \mathbf{k}_\alpha \quad , \quad E_\alpha \mapsto \mathbf{e}_\alpha \quad , \quad F_\alpha \mapsto \mathbf{f}_\alpha$$

can be extended (uniquely) to a surjective algebra homomorphism: $\mathbf{U} \twoheadrightarrow \mathbb{C}_{G(\mathbb{K})}[\mathcal{B} \times \mathcal{B}]$. $\square$

We turn now to the Polynomial Tensor Representation. Let $\mathbb{C}_{G(\mathbb{K})}[\mathcal{B} \times \dot{\mathcal{B}}]$ be the complex vector space formed by $G(\mathbb{K})$-invariant functions on $\mathcal{B} \times \dot{\mathcal{B}}$ whose support is a finite union of $G(\mathbb{K})$-orbits. The convolution product endows the space $\mathbb{C}_{G(\mathbb{K})}[\mathcal{B} \times \dot{\mathcal{B}}]$ with a left $\mathbb{C}_{G(\mathbb{K})}[\mathcal{B} \times \mathcal{B}]$-module structure. Hence, by Theorem 7.1, the vector space $\mathbb{C}_{G(\mathbb{K})}[\mathcal{B} \times \dot{\mathcal{B}}]$ acquires a left $\mathbf{U}$-module structure.

Theorem 7.2 *The* $\mathbf{U}$*-module* $\mathbb{C}_{G(\mathbb{K})}[\mathcal{B}\times\dot{\mathcal{B}}]$ *is isomorphic to the Polynomial Tensor Representation, i.e., there is an isomorphism* ψ : $\left(\mathbf{P}^{\otimes d}{}_{|q=p}\right) \xrightarrow{\sim} \mathbb{C}_{G(\mathbb{K})}[\mathcal{B}\times\dot{\mathcal{B}}]$ *making the following diagram commute*

$$\begin{array}{ccc} \mathbf{U} & \xrightarrow{(4.6)} & \mathrm{End}\left(\mathbf{P}^{\otimes d}{}_{|q=p}\right) \\ \Big\downarrow{\scriptstyle thm.\ 7.1} & & \Big\|{\scriptstyle \psi} \\ \mathbb{C}_{G(\mathbb{K})}[\mathcal{B}\times\mathcal{B}] & \longrightarrow & \mathrm{End}\,\mathbb{C}_{G(\mathbb{K})}[\mathcal{B}\times\dot{\mathcal{B}}] \end{array}$$

In order to construct the isomorphism ψ of the theorem (that will not be done here) one has do describe first the $\mathbf{U}$-action on $\mathbb{C}_{G(\mathbb{K})}[\mathcal{B}\times\dot{\mathcal{B}}]$ in an explicit way. To that end we introduce some combinatorial objects.

For any $s\in\mathbb{Z}$, write $s=\underline{s}\cdot d+\overline{s}$, where $\underline{s}$ is a certain integer and $\overline{s}\in\{1,\dots,d\}$, is the remainder. Let $\mathcal{J}$ be the set of all functions

$$\mathcal{J}=\{\mathbf{j}:\mathbb{Z}\to\mathbb{Z}\quad\text{such that}\quad \mathbf{j}(s+d)=\mathbf{j}(s)+n,\quad \forall s\in\mathbb{Z}\}$$

To such a function we assign the d-tuple $(\mathbf{j}(1),\mathbf{j}(2),\dots,\mathbf{j}(d))\in\mathbb{Z}^d$. This way one gets a bijection $\mathcal{J}\simeq\mathbb{Z}^d$. We shall often identify a function $\mathbf{j}$ with the corresponding d-tuple, an element of $\mathbb{Z}^d$. Further, given an integer s, we define two transformations $\mathbf{j}\mapsto\mathbf{j}_s^{\pm}$ of the set $\mathbb{Z}^d$ by the following formulas

$$\mathbf{j}_s^{\pm}=(\mathbf{j}(1),\dots,\mathbf{j}(\overline{s}-1),\mathbf{j}(\overline{s})\pm1,\mathbf{j}(\overline{s}+1),\dots,\mathbf{j}(d))$$

We claim next that there is a bijection between the set $\mathcal{J}$ and the set of $G(\mathbb{K})$-orbits in $\mathcal{B}\times\dot{\mathcal{B}}$. If $e_1,\dots,e_d$ denotes the standard base of $\mathbb{K}^d$, then to a function $\mathbf{j}\in\mathcal{J}$ we assign the $G(\mathbb{K})$-orbit that contains the pair $(F,\dot{F})\in\mathcal{B}\times\dot{\mathcal{B}}$ where $F=(\ldots\subset F_i\subset\ldots)$ and $\dot{F}=(\ldots\subset\dot{F}_i\subset\ldots)$ are given by

$$F_i=\bigoplus_{\mathbf{j}(k+jd)\le i} z^{-j}\,\mathbb{F}[[z]]\cdot e_k\quad,\quad \dot{F}_i=\bigoplus_{k+jd\le i} z^{-i}\,\mathbb{F}[[z]]\cdot e_k$$

Let $\mathbf{1}_{\mathbf{j}}$ denote the characteristic function of this $G(\mathbb{K})$-orbit. The family $\{\mathbf{1}_{\mathbf{j}}\}_{\mathbf{j}\in\mathcal{J}}$ forms a $\mathbb{C}$-basis of $\mathbb{C}_{G(\mathbb{K})}[\mathcal{B}\times\dot{\mathcal{B}}]$. The $\mathbf{U}$-action in that basis can be computed explicitly. It is given by the following formulas (compare with [GL] in the finite case).

$$E_\alpha \cdot \mathbf{1}_{\mathsf{j}} = q^{-\#\mathsf{j}^{-1}(\alpha)} \sum_{s\in \mathsf{j}^{-1}(\alpha+1)} q^{2\#\{z\in\mathsf{j}^{-1}(\alpha):z>s\}} \mathbf{1}_{\mathsf{j}_s^-},$$

$$F_\alpha \cdot \mathbf{1}_{\mathsf{j}} = q^{-\#\mathsf{j}^{-1}(\alpha+1)} \sum_{s\in \mathsf{j}^{-1}(\alpha)} q^{2\#\{z\in\mathsf{j}^{-1}(\alpha+1):z<s\}} \mathbf{1}_{\mathsf{j}_s^+},$$

$$K_\alpha \cdot \mathbf{1}_{\mathsf{j}} = q^{\#\mathsf{j}^{-1}(\alpha)} \mathbf{1}_{\mathsf{j}}.$$

Here $\alpha \in \{1, 2, \dots, n\}$. □

Finally, we introduce yet another realization of the Polynomial Tensor Representation in terms of functions on the *periodic flag manifold.* This manifold was independently introduced (over $\mathbb{C}$) by Feigin-Frenkel [FF] and less explicitly by Lusztig [Lu 3]. Mimicing their construction over the finite field, we define $\mathring{\mathcal{B}}$ to be the set of pairs:

$$\mathring{\mathcal{B}} = \{(\mathring{F}, L) \mid \mathring{F} = (\{0\} = \mathring{F}_0 \subset \mathring{F}_1 \subset \cdots \subset \mathring{F}_d = \mathbb{K}^d), \\ L = (L_1 \subset \mathring{F}_1/\mathring{F}_0, \dots, L_d \subset \mathring{F}_d/\mathring{F}_{d-1})\}$$

where $\mathring{F}_i$ is an i-dimensional $\mathbb{K}$-vector space and $L_i \subset \mathring{F}_i/\mathring{F}_{i-1}$ is a 1-dimensional $\mathbb{F}$-vector subspace, for any $i = 1, \dots, d$.

Replacing the set $\dot{\mathcal{B}}$ by $\mathring{\mathcal{B}}$ in the previous constructions, we obtain a new left $\mathbf{U}$-module $\mathbb{C}_{G(\mathbb{K})}[\mathcal{B} \times \mathring{\mathcal{B}}]$. Similarly, to any d-tuple $\mathbf{j} \in \mathcal{J}$, one associates the $G(\mathbb{K})$-orbit in $\mathcal{B} \times \mathring{\mathcal{B}}$ of the pair $\big(F, (\mathring{F}, L)\big)$ where $F = (\dots \subset F_i \subset \dots)$, $\mathring{F} = (\dots \subset \dot{F}_i \subset \dots)$ and $L = (\dots \subset L_i \subset \dots)$ are given by

$$F_i = \bigoplus_{\mathsf{j}(k+jd)\le i} z^{-j}\,\mathbb{F}[[z]] \cdot e_k, \quad \mathring{F}_i = \bigoplus_{k=1}^{i} \mathbb{K} \cdot e_k, \quad L_i = \mathbb{F} \cdot e_i \ (\mathrm{mod}\, \mathring{F}_{i-1})$$

Let ε_{j} be the characteristic function of this $G(\mathbb{K})$-orbit.

Recall the function $\mathbf{a}$ defined by (5.3) and other notation involved in the proof of theorem 5.2. Make the renormalization: $\hat{X}_s = q^{2s-d-1} X_s$, where $s = 1, \dots, d$, and , for any $\mathbf{j} \in \mathbb{Z}^d$, write

$$\mathsf{j} = \bar{\mathsf{j}} + n \cdot \underline{\mathsf{j}} \quad \text{where} \quad \bar{\mathsf{j}} \in [1, n]^d \ , \ \underline{\mathsf{j}} \in \mathbb{Z}^d$$

Theorem 7.3 *The assignment*

$$\phi : e_{\bar{\mathbf{j}}} \otimes \hat{X}_1^{-\mathbf{j}_1} \cdot \ldots \cdot \hat{X}_d^{-\mathbf{j}_d} \mapsto q^{\mathbf{a}(\bar{\mathbf{j}})} \varepsilon_{\mathbf{j}}$$

gives an isomorphism of $\mathbf{U}$*-modules* $\phi : \left(\mathbf{P}^{\otimes d}{}_{|q=p}\right) \xrightarrow{\sim} \mathbb{C}_{G(\mathbb{K})}[\mathcal{B} \times \mathring{\mathcal{B}}]$ *,*
i.e., the following diagram commutes

$$\begin{array}{ccc} \mathbf{U} & \xrightarrow{(4.6)} & \mathrm{End}\left(\mathbf{P}^{\otimes d}{}_{|q=p}\right) \\ \Big\downarrow{\scriptstyle thm.\ 7.1} & & \Big\Vert{\scriptstyle \phi} \\ \mathbb{C}_{G(\mathbb{K})}[\mathcal{B} \times \mathcal{B}] & \longrightarrow & \mathrm{End}\,\mathbb{C}_{G(\mathbb{K})}[\mathcal{B} \times \mathring{\mathcal{B}}] \end{array}$$

Proof of theorem 7.3 is entirely analogous to that of theorem 7.2. We observe first that the $G(\mathbb{K})$-orbits in $\mathcal{B} \times \mathring{\mathcal{B}}$ are parametrized by the set $\mathcal{J}$ so that the characteristic functions $\varepsilon_{\mathbf{j}}$ form a base of the $\mathbb{C}$-vector space $\mathbb{C}_{G(\mathbb{K})}[\mathcal{B} \times \mathring{\mathcal{B}}]$. In that basis, the $\mathbf{U}$-action can be computed explicitly, and is given by the following formulas:

$$E_\alpha \cdot \varepsilon_{\mathbf{j}} = q^{-\#\mathbf{j}^{-1}(\alpha)} \sum_{s \in \mathbf{j}^{-1}(\alpha+1)} q^{2\#\{z \in \mathbf{j}^{-1}(\alpha) : \bar{z} > \bar{s}\}} \varepsilon_{\mathbf{j}_s^-},$$

$$F_\alpha \cdot \varepsilon_{\mathbf{j}} = q^{-\#\mathbf{j}^{-1}(\alpha+1)} \sum_{s \in \mathbf{j}^{-1}(\alpha)} q^{2\#\{z \in \mathbf{j}^{-1}(\alpha+1) : \bar{z} < \bar{s}\}} \varepsilon_{\mathbf{j}_s^+},$$

$$K_\alpha \cdot \varepsilon_{\mathbf{j}} = q^{\#\mathbf{j}^{-1}(\alpha)} \varepsilon_{\mathbf{j}}.$$

The rest of the proof is a straightforward exercise. □

References

[BBD] A. Beilinson, J. Bernstein, P. Deligne. Faisceaux Pervers. *Astérisque* **100** (1981).

[BLM] A. Beilinson, G. Lusztig, R. MacPherson. A geometric setting for quantum groups. *Duke Math. J.* **61** (1990), 655-675.

[CP] V. Chari, A. Pressley. Quantum affine algebras. *Commun. Math. Phys.* **142** (1991), 261-283.

[Ch 1] I. Cherednik. Quantum groups as hidden symmetries of the classical Representation theory. *Proceedings of the 17-th. Intern. conf. of Differential Geometry methods in Theoretical Physics* (1989).

[Ch 2] I. Cherednik. A new interpretation of Gelfand-Tzetlin bases. *Duke Math. J.* **54** (1987), 563-578.

[CG] N. Chriss, V. Ginzburg. Representation theory and Complex Geometry (Geometric technique in Representation theory of Reductive groups). Progress in Mathem., Birkhäuser (1994), to appear.

[DF] J. Ding, I. Frenkel. Isomorphism of two realizations of quantum affine algebra $\mathbf{U}_q(\widehat{\mathfrak{gl}_n})$. *Commun. Mathem. Phys.* (1993).

[Dr 1] V. Drinfeld. Quantum Groups. *Proceedings of the ICM* , Berkeley 1986.

[Dr 2] V. Drinfeld. A new realization of Yangians and Quantum affine algebras. *Soviet Math. Dokl.* **36** (1988), 212 - 216.

[Dr 3] V. Drinfeld. Degenerate affine Hecke algebras and Yangians. *Funct. Anal. and Appl.* **20:1** (1986), 69-70.

[FRT] L. Faddeev, N. Reshetikhin, L. Takhtajan. Quantization of Lie groups and Lie algebras, Yang-Baxter equation in Integrable Systems. *Advanced Series in Mathem. Physics.* **10** (1989), 299-309. World Scientific.

[FF] B. Feigin, E. Frenkel. Affine Lie algebras and semi-infinite flag manifold. *Commun. Math. Phys.* **128** (1990), 16.

[Gi] V. Ginzburg. Deligne - Langlands conjecture and Representations of affine Hecke algebras. *Preprint*, Moscow 1985.

[GV] V. Ginzburg, E. Vasserot. Langlands Reciprocity for Affine Quantum groups of type A_n . *Intern. Mathem. Research Notices. (Duke Math. J.)* **3** (1993), 67-85.

[Gr] I. Grojnowski. The coproduct for quantum GL_n. *Preprint* 1992.

[GL] I. Grojnowski, G. Lusztig. On bases of irreducible representations of quantum GL_n. *Contemp. Mathem.* (1992).

[J] M. Jimbo. A q-Analogue of $U(\mathfrak{gl}(n+1))$, Hecke Algebra, and the Yang-Baxter Equation. *Lett. in Mathem. Phys.* **11** (1986), 247-252.

[KKM] S.-J. Kang, et al. Perfect crystals of quantum affine algebra. *Duke Math. Journ.* **68** (1992), 499-607.

[KL 1] D. Kazhdan, G. Lusztig. Representations of Coxeter groups and Hecke algebras. *Invent. Math.* **53** (1979), 165-184.

[KL 2] D. Kazhdan, G. Lusztig. Proof of the Deligne - Langlands conjecture for affine Hecke algebras. *Invent. Math.* **87** (1987), 153-215.

[KR] A. Kirillov, N. Reshetikhin. The Yangian, Bethe ansatz and Combinatorics. *Lett. in Mathem. Phys.* **12** (1986), 199-208.

[Lu 1] G. Lusztig. Canonical bases arising from Quantized enveloping algebras. *Journ. A.M.S.* **3** (1990), 447-498.

[Lu 2] G. Lusztig. Cuspidal local systems and graded Hecke algebras. *Publ. Mathem. I.H.E.S.* **67** (1988), 145-212.

[Lu 3] G. Lusztig. Hecke algebras and Jantzen generic decomposition pattern. *Advances Math.* (1981).

[RS 1] N. Reshetikhin, M. Semenov-Tian-Shansky. Central extensions of quantum current groups. *Lett. Mathem. Phys.* **19** (1990), 133-142.

[V] E. Vasserot. Représentations de groupes quantiques et permutations. *Annales Sci. ENS*, **26**, 747–773, (1993).

V.G.: The University of Chicago, Dept. Mathematics,
Chicago, IL 60637, USA; ginzburg@math.uchicago.edu

N.R.: University of California at Berkeley, Dept. of Mathematics,
Berkeley, CA 94720, USA ; reshetik@math.berkeley.edu

E.V.: Ecole Normale Supérieure, 45 rue d'Ulm, 75005 Paris, FRANCE
vasserot@dmi.ens.fr

Contemporary Mathematics
Volume **175**, 1994

Operadic formulation of the notion of vertex operator algebra

YI-ZHI HUANG AND JAMES LEPOWSKY

Dedicated to the memory of Daniel Gorenstein,
January 1, 1923 – August 26, 1992.

ABSTRACT. A reformulation of the notion of vertex operator algebra in terms of operads is presented. This reformulation shows that the rich geometric structure revealed in the study of conformal field theory and the rich algebraic structure of the theory of vertex operator algebras share a precise common foundation in basic operations associated with a certain kind of (two-dimensional) "complex" geometric object, in the sense in which classical algebraic structures (groups, algebras, Lie algebras and the like) are always implicitly based on (one-dimensional) "real" geometric objects. In effect, the standard analogy between point-particle theory and string theory is being shown to manifest itself at a more fundamental mathematical level.

Operads arose in the homotopy-theoretic characterization of iterated loop spaces. The first important example of operad-like structures already occurred in Stasheff's notion of A_∞-space ([**St1**], [**St2**]), before May's notion of operad [**M1**] was introduced. In this paper, we use the language of operads to reformulate the notion of vertex operator algebra, as defined in [**FLM**] or [**FHL**] (rather than the notion of vertex algebra introduced by Borcherds in [**B**]). More precisely, it has already been established that the category of vertex operator algebras is isomorphic to a certain category – the category of geometric vertex operator algebras – defined in terms of a certain moduli space of spheres with punctures and local coordinates ([**H1**], [**H2**]), and here we reformulate the category of geometric vertex operator algebras using operads. First, this moduli space naturally induces a certain (partial) operad, which admits "$\mathbb{C}$-extensions" constructed from the determinant line bundle. Naturally defined "meromorphic

1991 *Mathematics Subject Classification.* Primary 08C99; Secondary 17B68, 32G15, 81T40.
This paper is in final form and no version of it will be submitted for publication elsewhere.

associative algebras" for these extension-operads can then be thought of as "vertex associative algebras." The main theorem of [**H1**] and the announcement [**H2**] then in fact says that the category of vertex operator algebras with central charge say c is isomorphic to the category of vertex associative algebras with central charge c.

This reformulation shows that the rich geometric structure revealed in the study of conformal field theory (see especially [**BPZ**], [**FS**], [**Se**] and [**V**]) and the rich algebraic structure of the theory of vertex operator algebras share a precise common foundation in basic operations associated with a certain kind of two-dimensional object, in a sense in which classical algebraic structures (groups, algebras, Lie algebras and the like) are always implicitly based on one-dimensional geometric objects such as punctured circles and binary trees. In fact, the present paper provides the foundation for developing the whole theory of vertex operator algebras and related structures and concepts in parallel with such traditional theory. In particular, the theory of tensor products of modules for a general class of vertex operator algebras, as developed beginning in [**HL1**], can be expressed using the foundation presented in this paper. And it now appears that this view of the theory of vertex operator algebras can help us gain a new level of insight into the phenomena associated with monstrous moonshine (cf. [**FLM**]); see [**HL3**], which, along with the present paper, elaborates the authors' talks at this Conference.

The viewpoint and results in this paper are briefly summarized in [**HL2**].

We are very grateful to Todd Trimble and to Jim Stasheff for their illuminating comments on [**H1**] and [**H3**], respectively, which led to the reformulation using operads in this paper. Mikhail Kapranov and Vadim Schechtman also suggested to us that operads should be worthwhile to consider. We thank Peter May for comments on the original version of this paper (see Sections 1 and 2 below). Y.-Z. H. is supported in part by NSF grant DMS-9104519. J. L. is supported in part by NSF grant DMS-9111945 and the Rutgers University Faculty Academic Study Program. J. L. would also like to thank the Institute for Advanced Study for its hospitality.

Notations:

$\mathbb{C}$: the (structured set of) complex numbers.

$\mathbb{C}^{\times}$: the nonzero complex numbers.

$\mathbb{Z}$: the integers.

$\mathbb{N}$: the nonnegative integers.

1. Operads

DEFINITION 1. (cf. [**M1**]) An *operad* $\mathcal{C}$ consists of a family of sets $\mathcal{C}(j)$, $j \in \mathbb{N}$, together with *(abstract) substitution maps* γ, one for each $k \in \mathbb{N}$, $j_1, \dots, j_k \in \mathbb{N}$,

$$\begin{array}{rrcl} \gamma: & \mathcal{C}(k) \times \mathcal{C}(j_1) \times \cdots \times \mathcal{C}(j_k) & \longrightarrow & \mathcal{C}(j_1 + \cdots + j_k) \\ & (c; d_1, \dots, d_k) & \longmapsto & \gamma(c; d_1, \dots, d_k), \end{array} \tag{1}$$

an *identity element* $I \in \mathcal{C}(1)$ and a (left) action of the symmetric group S_j on $\mathcal{C}(j)$, $j \in \mathbb{N}$ (where S_0 is understood to be the trivial group), satisfying the following axioms:

(i) *Operad-associativity*: For any $k \in \mathbb{N}$, $j_s \in \mathbb{N}$ $(s = 1, \dots, k)$, $i_t \in \mathbb{N}$ $(t = 1, \dots, j_1 + \cdots + j_k)$, $c \in \mathcal{C}(k)$, $d_s \in \mathcal{C}(j_s)$ $(s = 1, \dots, k)$ and $e_t \in \mathcal{C}(i_t)$ $(t = 1, \dots, j_1 + \cdots + j_k)$,

$$\gamma(\gamma(c; d_1, \dots, d_k); e_1, \dots, e_{j_1+\cdots+j_k}) = \gamma(c; f_1, \dots, f_k), \tag{2}$$

where

$$f_s = \gamma(d_s; e_{j_1+\cdots+j_{s-1}+1}, \dots, e_{j_1+\cdots+j_s}). \tag{3}$$

(ii) For any $j, k \in \mathbb{N}$, $d \in \mathcal{C}(j)$ and $c \in \mathcal{C}(k)$,

$$\gamma(I; d) = d, \tag{4}$$

$$\gamma(c; I, \dots, I) = c. \tag{5}$$

(In particular, for $k = 0$, $\gamma : \mathcal{C}(0) \longrightarrow \mathcal{C}(0)$ is the identity map.)

(iii) For any $k \in \mathbb{N}$, $j_s \in \mathbb{N}$ $(s = 1, \dots, k)$, $c \in \mathcal{C}(k)$, $d_s \in \mathcal{C}(j_s)$ $(s = 1, \dots, k)$, $\sigma \in S_k$ and $\tau_s \in S_{j_s}$ $(s = 1, \dots, k)$,

$$\gamma(\sigma(c); d_1, \dots, d_k) = \sigma(j_1, \dots, j_k)(\gamma(c; d_{\sigma(1)}, \dots, d_{\sigma(k)})), \tag{6}$$

$$\gamma(c; \tau_1(d_1), \dots, \tau_k(d_k)) = (\tau_1 \oplus \cdots \oplus \tau_k)(\gamma(c; d_1, \dots, d_k)), \tag{7}$$

where $\sigma(j_1, \dots, j_k)$ denotes the permutation of $j = \sum_{s=1}^{k} j_s$ letters which permutes the k blocks of letters determined by the given partition of j as σ permutes k letters, and $\tau_1 \oplus \cdots \oplus \tau_k$ denotes the image of $(\tau_1, \dots, \tau_k)$ under the obvious inclusion of $S_{j_1} \times \cdots \times S_{j_k}$ in S_j; that is,

$$\sigma(j_1, \dots, j_k)(j_{\sigma(1)} + \cdots + j_{\sigma(i-1)} + l) = j_1 + \cdots + j_{\sigma(i)-1} + l \tag{8}$$

for $l = 1, \dots, j_{\sigma(i)}$, $i = 1, \dots, k$ and

$$(\tau_1 \oplus \cdots \oplus \tau_k)(j_1 + \cdots + j_{i-1} + l) = j_1 + \cdots + j_{i-1} + \tau_i(l) \tag{9}$$

for $l = 1, \dots, j_i$, $i = 1, \dots, k$. (May [**M1**] uses right actions, so that the description of the permutation $\sigma(j_1, \dots, j_k)$ might have a different interpretation.)

EXAMPLE 2. The moduli space B of binary trees discussed in [**H1**], with a unique degenerate tree with one external node adjoined, can be naturally given the structure of an operad B for which $B(0) = \emptyset$.

Remark 3. The definition above is the same as that in [**M1**] with the following exceptions:

(i) $\mathcal{C}(0)$ need not consist of exactly one element.

(ii) The sets in the definition need not be (certain kinds of) topological spaces (and correspondingly the maps need not be continuous).

Below it will be convenient to have the following notions available:

DEFINITION 4. Suppose that in the definition above, the substitution maps γ are only partially defined, that is, each map γ in formula (1) takes a subset of $\mathcal{C}(k) \times \mathcal{C}(j_1) \times \cdots \times \mathcal{C}(j_k)$ to $\mathcal{C}(j_1 + \cdots + j_k)$; all the other data remain the same; each of the formulas (2) and (4) – (7) holds whenever *both* sides exist; and the left-hand sides of (4) and (5) always exist. Then we call such a family of sets $\mathcal{C}(j)$ together with the partial maps γ, the identity I and the actions of S_j on $\mathcal{C}(j)$ a *partial operad.* In addition, we define a *partial pseudo-operad* to be a family of sets $\mathcal{C}(j)$, $j \in \mathbb{N}$, together with partially defined substitution maps γ, an identity I and actions of S_j on $\mathcal{C}(j)$, $j \in \mathbb{N}$, satisfying all the axioms for partial operads except the operad-associativity. (Later, we shall typically denote partial operads by the symbol $\mathcal{P}$ rather than $\mathcal{C}$.)

A concept of "partial operad," together with examples, has already occurred informally in the work of May [**M2**]. In fact, in the topological situation which May was considering, the use of the partial operad could be avoided since a genuine operad could be constructed and used in place of the original partial operad [**Ste**]. But for vertex operator algebras, since the vertex operators are not elements of the endomorphism operad of the infinite-dimensional vector space V, we have to consider a certain "endomorphism partial pseudo-operad" associated with V (see Sections 4-6 below), and we are in fact required to consider the notions of partial operad and of partial pseudo-operad. We thank Peter May for drawing our attention to [**M2**] and [**Ste**].

Remark 5. If in Definitions 1 and 4 the sets $\mathcal{C}(j)$, $j \in \mathbb{N}$, are assumed to be objects in given categories (e.g., have certain kinds of topological, smooth or analytic structure) and the maps γ and the actions of S_j, $j \in \mathbb{N}$, are morphisms in these categories (e.g., are continuous or smooth or analytic), we have the notions of operads in these categories, and we use the names of these categories plus the word "operads" to designate them (e.g., *topological operads, smooth operads* or *analytic operads*). In the case of partial operads, we also require that the domains of the substitution maps are in the category we are considering.

A *morphism* $\psi : \mathcal{C} \longrightarrow \mathcal{C}'$ of operads $\mathcal{C}$ and $\mathcal{C}'$ is a sequence of S_j-equivariant maps $\psi_j : \mathcal{C}(j) \longrightarrow \mathcal{C}'(j)$ such that $\psi_1(I) = I'$ and the following diagram commutes:

$$\begin{array}{ccc} \mathcal{C}(k) \times \mathcal{C}(j_1) \times \cdots \times \mathcal{C}(j_k) & \xrightarrow{\gamma} & \mathcal{C}(j_1 + \cdots + j_k) \\ \downarrow & & \downarrow \\ \mathcal{C}'(k) \times \mathcal{C}'(j_1) \times \cdots \times \mathcal{C}'(j_k) & \xrightarrow{\gamma'} & \mathcal{C}'(j_1 + \cdots + j_k). \end{array} \tag{10}$$

For partial operads we also require that the domains of the substitution maps for $\mathcal{C}$ are mapped into the domains of the substitution maps for $\mathcal{C}'$; the diagram (10) is interpreted in the obvious way. Morphisms for partial pseudo-operads are defined in the same way as morphisms for partial operads.

2. $\mathcal{C}$-spaces and $\mathcal{C}$-algebras

Here we discuss the sense in which operads describe "operations." In this and the next section, all the operads will be ordinary operads (i.e., in the category of sets and not partial).

DEFINITION 6. Let X be a set and Y a subset of X. We define the *endomorphism operad* $\mathcal{E}_{X,Y}$ (cf. [**M1**]) as follows: Let $*$ be a one-element set (a terminal object in the category of sets) and take $X^0 = Y^0 = *$. Let $\mathcal{E}_{X,Y}(j)$, $j \in \mathbb{N}$, be the set of maps from X^j to X which map Y^j to Y; then $\mathcal{E}_{X,Y}(0) = Y$. The substitution maps are defined by

$$\gamma(f; g_1, \dots, g_k) = f \circ (g_1 \times \cdots \times g_k) \tag{11}$$

for $f \in \mathcal{E}_{X,Y}(k)$, $k \in \mathbb{N}$ and $g_s \in \mathcal{E}_{X,Y}(j_s)$, $s = 1, \dots, k$. The identity $I_{X,Y}$ is the identity map of X. For $f \in \mathcal{E}_{X,Y}(j)$, $\sigma \in S_j$, $\mathbf{x} = (x_1, \dots, x_j) \in X^j$,

$$(\sigma(f))(\mathbf{x}) = f(\sigma^{-1}(\mathbf{x})) \tag{12}$$

where

$$\sigma(\mathbf{x}) = \sigma(x_1, \dots, x_j) = (x_{\sigma^{-1}(1)}, \dots, x_{\sigma^{-1}(j)}). \tag{13}$$

It is easy to see that $\mathcal{E}_{X,Y}$ is an operad. The corresponding definition in [**M1**] amounts to the case in which Y has one element. Note the special cases $Y = \emptyset$ and $Y = X$. Observe that operad-associativity (the associativity of substitution) is unrelated to any associativity properties that the j-ary operations might or might not have.

DEFINITION 7. (cf. [**M1**]) Let $\mathcal{C}$ be an operad. A *$\mathcal{C}$-space* consists of a set X, a subset $Y \subset X$ and a morphism ψ of operads from $\mathcal{C}$ to $\mathcal{E}_{X,Y}$ such that $\psi_0(\mathcal{C}(0)) = \mathcal{E}_{X,Y}(0)$ $(= Y)$. It is denoted (X, Y, ψ). An element of $\psi_0(\mathcal{C}(0)) = Y$ is called a *quasi-identity element of X (for $\mathcal{C}$ and ψ).* Note that each element of $\mathcal{C}(j)$, $j \in \mathbb{N}$, defines a j-ary operation on X. A *morphism* from a $\mathcal{C}$-space (X, Y, ψ) to a $\mathcal{C}$-space (X', Y', ψ') is a map $\eta : X \longrightarrow X'$ such that $\eta(Y) \subset Y'$ and $\eta \circ \psi_j(c) = \psi'_j(c) \circ \eta^j$ for $j \in \mathbb{N}$ and $c \in \mathcal{C}(j)$; it follows that $\eta(Y) = Y'$. An *isomorphism* of $\mathcal{C}$-spaces is defined in the obvious way. The definition of "$\mathcal{C}$-space" in [**M1**] amounts to the case in which $\mathcal{C}(0)$ and Y consist of one element.

In order to define the notion of "$\mathcal{C}$-algebra," we need:

DEFINITION 8. Let V be a vector space and W a subspace of V. We define the corresponding *(multilinear) endomorphism operad* $\mathcal{M}_{V,W}$ as follows: Let $\mathcal{M}_{V,W}(j)$, $j \in \mathbb{N}$, be the set of multilinear maps from V^j to V which map W^j to W; it is understood that $V^0 = W^0$ is the one-element set as above, and that a "multilinear map" (a "zero-linear map") from this set to V (or W) is a map of sets, i.e., an element of the target set. In particular, $\mathcal{M}_{V,W}(0) = W$. The substitution maps, the identity and the actions of the symmetric groups are defined just as in the definition of endomorphism operads for sets. Then $\mathcal{M}_{V,W}$ is an operad. Note the special cases $W = 0$ and $W = V$.

DEFINITION 9. A *$\mathcal{C}$-algebra* (V, W, ν) consists of a vector space V, a subspace W and a morphism ν from $\mathcal{C}$ to $\mathcal{M}_{V,W}$ such that the subspace of V spanned by $\nu_0(\mathcal{C}(0))$ is $\mathcal{M}_{V,W}(0)$ $(= W)$. We call $\nu_0(\mathcal{C}(0))$ the set of *quasi-identity elements.* Each element of $\mathcal{C}(j)$, $j \in \mathbb{N}$, defines a multilinear j-ary operation on V. Morphisms and isomorphisms of $\mathcal{C}$-algebras are defined in the obvious ways.

EXAMPLE 10. Let B be the moduli space of binary trees. The notion of B-space is equivalent to the notion of set with a binary operation and the notion of B-algebra is equivalent to the notion of nonassociative algebra. (Note that here we must have $Y = \emptyset$ and $W = 0$, respectively.)

Remark 11. The notions of endomorphism operad for a set and for a vector space and the notions of $\mathcal{C}$-space and $\mathcal{C}$-algebra can be defined in unified ways as follows: Let $(\mathcal{S}, \times, \mathbf{1})$ be a symmetric strict monoidal category, X, Y two objects in $\mathcal{S}$ and $\phi : Y \longrightarrow X$ a monic morphism. Let $\mathcal{E}_{X,Y,\phi}(j)$ be the set of morphisms $X^j \longrightarrow X$ such that the compositions of ϕ^j with these morphisms agree with the compositions of (necessarily unique) morphisms $Y^j \longrightarrow Y$ with ϕ; here $X^0 = Y^0 = \mathbf{1}$ and $\phi^0 = 1$. Note that $\mathcal{E}_{X,Y,\phi}(0)$ may be identified with $\mathrm{Hom}(\mathbf{1}, Y)$. The substitution maps are defined in the obvious way, the identity $I_{X,Y,\phi}$ is the identity morphism $X \longrightarrow X$ and the actions of the symmetric groups are also defined in the obvious ways. It is easy to verify that $\mathcal{E}_{X,Y,\phi}$ is an operad. We shall call it the *endomorphism operad* of (X, Y, ϕ). Note the special case $Y = X$, $\phi = 1$. A *$\mathcal{C}$-space in the category* $(\mathcal{S}, \times, \mathbf{1})$ is a morphism ψ of operads from $\mathcal{C}$ to $\mathcal{E}_{X,Y,\phi}$ such that $\psi_0(\mathcal{C}(0))$, which is a certain set of morphisms from $\mathbf{1}$ to Y, is epic in the obvious sense (i.e., we are using the term "epic" for a set of morphisms rather than a single morphism). We denote such a $\mathcal{C}$-space by (X, Y, ϕ, ψ) and we call $\psi_0(\mathcal{C}(0))$ the set of *quasi-identity elements.* A *morphism* from a $\mathcal{C}$-space (X, Y, ϕ, ψ) to a $\mathcal{C}$-space (X', Y', ϕ', ψ') is defined in the expected way; some diagram-chasing shows that the induced morphism $Y \longrightarrow Y'$ is epic. Isomorphisms of $\mathcal{C}$-spaces are also defined in the obvious way. In the case in which $\mathcal{S}$ is the category of sets, $\times$ is Cartesian product, $\mathbf{1}$ is the one-element set and ϕ is the inclusion, we have $\mathcal{E}_{X,Y,\phi} = \mathcal{E}_{X,Y}$ and a $\mathcal{C}$-space in this category is a $\mathcal{C}$-space as defined in Definition 7. When $\mathcal{S}$ is the category of vector spaces, $\times$ is tensor product, $\mathbf{1}$ is the coefficient field and ϕ is the inclusion, $\mathcal{E}_{X,Y,\phi} = \mathcal{M}_{X,Y}$ and a $\mathcal{C}$-space in this category is a $\mathcal{C}$-algebra as defined in Definition 9.

Remark 12. In fact, in view of the coherence theorem, the definitions and properties in Remark 11 all generalize naturally to an arbitrary symmetric monoidal category, not necessarily strict (cf. [**ML**]). Indeed, the three operad axioms in this case come from the three coherence properties of a symmetric monoidal category: coherence of associativity, of the identity and of symmetry.

One can think of a $\mathcal{C}$-algebra as a kind of analogue of a *representation* $G \longrightarrow \text{Aut } V$ of an algebraic structure such as a group G on a vector space V, rather than as an analogue of a *module* structure $G \times V \longrightarrow V$. In [**HL2**], we used the term "$\mathcal{C}$-modules" for the stuctures designated "$\mathcal{C}$-algebras" in the present paper. (We thank Peter May for mentioning to us that the term "algebra" is standard for this concept.)

3. Associative operads and associated algebraic structures

The notion of *suboperad* of an operad is defined in the obvious way. An intersection of suboperads is again a suboperad. Let $\mathcal{C}$ be an operad and U a subset of the disjoint union of the sets $\mathcal{C}(j)$, $j \in \mathbb{N}$. The *suboperad of* $\mathcal{C}$ *generated by* U is the smallest suboperad of $\mathcal{C}$ such that the disjoint union of the family of sets in the suboperad contains U. The operad $\mathcal{C}$ is said to be *generated by* U if the suboperad generated by U is $\mathcal{C}$ itself.

DEFINITION 13. Let $\mathcal{C}$ be an operad. We call an element a of $\mathcal{C}(2)$ *associative* if

$$\gamma(a; a, I) = \gamma(a; I, a). \tag{14}$$

DEFINITION 14. We say that an operad $\mathcal{C}$ is *associative* if $\mathcal{C}$ is generated by $\mathcal{C}(0)$ and an associative element $a \in \mathcal{C}(2)$.

EXAMPLE 15. We define an operad C, the *circle operad*, which will be associative. Let $C(j)$, $j \in \mathbb{N}$, be the moduli space of circles (i.e., compact connected smooth one-dimensional manifolds) with $j+1$ ordered points (called *punctures*), the zeroth negatively oriented, the others positively oriented, and with local coordinates vanishing at these punctures. Then it is easy to see that $C(j)$ can be identified naturally with the set of permutations $(\sigma(1), \ldots, \sigma(j))$ of $(1, \ldots, j)$. Since this set can in turn be identified in the obvious way with the symmetric group S_j, the moduli space $C(j)$ can also be naturally identified with S_j, with the group S_j acting on $C(j)$ according to the left multiplication action on S_j. That is, S_j permutes the orderings of the positively oriented punctures. Given any two circles with punctures and local coordinates vanishing at the punctures, we can define the notion of sewing them together at any positively oriented puncture on the first circle and the negatively oriented puncture on the second circle by cutting out an open interval of length $2r$ centered at the positively oriented puncture on the first circle (using the local coordinate) and cutting out an open interval of length $2/r$ around the negatively oriented puncture on the second circle in the same way, and then by identifying the boundaries of the remaining parts using the two local coordinates and the map $t \longmapsto 1/t$; we assume that the corresponding closed intervals contain no other punctures. The ordering of the punctures of the sewn circle is given by "inserting" the ordering for the second circle into that for the first. Note that in general not every two circles with punctures and local coordinates can be sewn together at a given positively oriented puncture on the first circle. But it is clear that such a pair of circles with punctures and local coordinates is equivalent to a pair which can be sewn together. Also, the equivalence class of the sewn circle with punctures and local coordinates depends only on the two equivalence classes. Thus we obtain a sewing operation on the moduli space of circles with punctures and local coordinates. Given an element of $C(k)$ and an element of $C(j_s)$ for each j_s, $s = 1, \dots, k$, we define an element of $C(j_1 + \cdots + j_k)$ by sewing the element of $C(k)$ at its s-th positively oriented puncture with the element of $C(j_s)$ at its negatively oriented puncture, for $s = 1, \dots, k$, and by "inserting" the orderings for the elements of the $C(j_s)$ into the ordering for the element of $C(k)$. The identity is the unique element of $C(1)$. If we identify $C(j)$ with the set S_j as above, the substitution maps are given (using the notation of (6) and (7)) by

$$\gamma(\sigma; \tau_1, \dots, \tau_k) = (\tau_1 \oplus \cdots \oplus \tau_k)\sigma(j_1, \dots, j_k) = \sigma(j_1, \dots, j_k)(\tau_{\sigma(1)} \oplus \cdots \oplus \tau_{\sigma(k)}) \tag{15}$$

where $\sigma \in S_k$, $\tau_s \in S_{j_s}$, $s = 1, \dots, k$. It is straightforward to verify the operad axioms.

It is easy to see that C is associative, with the associative element a either of the two elements of $C(2)$.

Since in one dimension smooth structures and conformal structures are the same, the moduli space $C(j)$ can also be thought of as the moduli space of circles with conformal structures, with $j+1$ ordered punctures with the zeroth negatively oriented and the others positively oriented, and with local conformal coordinates vanishing at these punctures. In fact, we have just defined three operads, namely, C, the corresponding conformal moduli space and $\{S_j\}_{j \in \mathbb{N}}$, and these three operads are isomorphic.

EXAMPLE 16. Let $T(j)$, $j \in \mathbb{N}$, be the moduli space of genus-zero compact connected smooth two-dimensional manifolds with one negatively oriented puncture and j ordered positively oriented punctures and with local coordinates vanishing at these punctures. Then it is easy to see that $T(j)$ consists of exactly one element for each $j \in \mathbb{N}$, and it is obvious that the $T(j)$, $j \in \mathbb{N}$, give an associative operad. We denote this operad by T.

EXAMPLE 17. Let M be a commutative monoid. We define an operad (which will still be denoted by M) as follows: Let $M(j) = M$, $j \in \mathbb{N}$. We define the substitution map

$$\gamma_M : M(k) \times M(j_1) \times \cdots \times M(j_k) \longrightarrow M(j_1 + \cdots + j_k) \tag{16}$$

by

$$\gamma_M(a; a_1, \dots, a_k) = aa_1 \cdots a_k. \tag{17}$$

The identity element of $M(1)$ is 1, the identity element of the monoid M. The permutation group S_j acts on $M(j)$ trivially. The collection $M(j)$, $j \in \mathbb{N}$, with the given substitution maps, identity element and actions of S_j, is an associative operad, with the associative element taken to be any element of $M = M(2)$. Note that for $j \neq 1$ the element $1 \in M(j)$ plays no role, so that strictly speaking, each such $M(j)$ is naturally just a commutative-monoid-without-identity.

DEFINITION 18. Let $\mathcal{C}$ be an associative operad with associative element $a \in \mathcal{C}(2)$. We call a $\mathcal{C}$-space a *$\mathcal{C}$-monoid* and a $\mathcal{C}$-algebra a *$\mathcal{C}$-associative algebra*, with *$\mathcal{C}$-associative binary product* given by the image of a and with *quasi-identity elements* given by the image of $\mathcal{C}(0)$.

EXAMPLE 19. For the operad C discussed in Example 15, a C-monoid is a monoid in the usual sense and a C-associative algebra is an associative algebra in the usual sense.

EXAMPLE 20. A T-monoid (recall Example 16) is a commutative monoid in the usual sense and a T-associative algebra is a commutative associative algebra in the usual sense.

4. Rescaling groups for partial operads, rescalable partial operads and associated algebraic structures

Let $\mathcal{P}$ be a partial operad. A subset G of $\mathcal{P}(1)$ is called a *rescaling group* for $\mathcal{P}$ if G contains I; the substitution maps γ from a subset of $\mathcal{P}(1) \times \mathcal{P}(k)$ to $\mathcal{P}(k)$ and from a subset of $\mathcal{P}(k) \times (\mathcal{P}(1))^k$ to $\mathcal{P}(k)$ are defined on $G \times \mathcal{P}(k)$ and on $\mathcal{P}(k) \times G^k$, respectively, for each $k \in \mathbb{N}$; both sides of (2) exist if $c \in G$ or $d_1, \ldots, d_k \in G$ or $e_1, \ldots, e_{j_1+\cdots+j_k} \in G$ and if either side of (2) exists; γ maps $G \times G$ into G; and inverses of the elements of G exist with respect to γ and I; then G is in fact a group. (Note that $G = \{I\}$ is always an example of a rescaling group for $\mathcal{P}$.) Given a rescaling group G for $\mathcal{P}$, we define a corresponding equivalence relation on $\mathcal{P}$: Two elements c_1 and c_2 of $\mathcal{P}(j)$, $j \in \mathbb{N}$, are said to be *G-equivalent* if there exists $d \in G$ such that

$$c_2 = \gamma(d; c_1); \tag{18}$$

our assumptions insure that this is an equivalence relation.

DEFINITION 21. A *(G-)rescalable partial operad* is a partial operad $\mathcal{P}$ together with a rescaling group G for $\mathcal{P}$ satisfying the following condition: For any $c \in \mathcal{P}(k)$, $k \in \mathbb{N}$, $d_1 \in \mathcal{P}(j_1), \ldots, d_k \in \mathcal{P}(j_k)$, $j_i \in \mathbb{N}$, there exist $d_1' \in \mathcal{P}(j_1), \ldots, d_k' \in \mathcal{P}(j_k)$ which are G-equivalent to $d_1, \ldots, d_k$, respectively, such that $\gamma(c; d_1', \ldots, d_k')$ exists.

Suppose that we have a set-theoretic category with a reasonable notion of "induced substructure," such as a topological, smooth or analytic category. A *partial operad with rescaling group* (or *rescalable partial operad*) *in this category* (e.g., a *topological*, *smooth* or *analytic rescalable partial operad*) is a partial operad $\mathcal{P}$ with rescaling group (or a rescalable partial operad) such that its underlying partial operad is in the category we are considering and the rescaling group is a group in this category, with the structure induced from that on $\mathcal{P}(1)$.

Morphisms and isomorphisms of partial operads with rescaling groups are defined in the obvious ways.

The definitions of $\mathcal{P}$-space and $\mathcal{P}$-algebra in Section 3 also make sense when $\mathcal{P}$ is a partial operad. But for a partial operad $\mathcal{P}$ with a rescaling group G, it is more relevant to look for some kind of "(multilinear) endomorphism partial operad" of a G-module, and then to define a "$\mathcal{P}$-algebra" to be a morphism from $\mathcal{P}$ to such a (multilinear) endomorphism partial operad. However, we must be content with only the following "(multilinear) endomorphism partial pseudo-operads" (recall Definition 4):

DEFINITION 22. Let G be a group, V a completely reducible G-module and W a G-submodule of V. Then $V = \coprod_{M\in A} V_{(M)}$, where A is the set of equivalence classes of irreducible G-modules and $V_{(M)}$ is the sum of the G-submodules of V in the class M, and similarly for W. Assume that dim $V_{(M)} < \infty$ for every $M \in A$. We define a *(multilinear) endomorphism partial pseudo-operad* $\mathcal{H}^G_{V,W}$ as follows: For any $j \in \mathbb{N}$ the set $\mathcal{H}^G_{V,W}(j)$ is the set of all multilinear maps from V^j to $\overline{V} = \prod_{M\in A} V_{(M)} = V'^*$ such that W^j is mapped to $\overline{W} = \prod_{M\in A} W_{(M)} = W'^*$, where $'$ denotes the graded dual of an A-graded vector space and * denotes the dual space of a vector space. As in Definition 8, it is understood that $V^0 = W^0$ is the one-element set, so that the set $\mathcal{H}^G_{V,W}(0)$ is equal to $\overline{W}$. The identity $I^G_{V,W}$ is the embedding map from V to $\overline{V}$. The symmetric group S_j acts on $\mathcal{H}^G_{V,W}(j)$ in the obvious way. To define the substitution maps, we first define a contraction operation on $\mathcal{H}^G_{V,W}$: Given $f \in \mathcal{H}^G_{V,W}(k)$ and $g \in \mathcal{H}^G_{V,W}(j)$ $(k, j \in \mathbb{N})$ and a positive integer $s \leq k$, we say that the *contraction of f at the s-th argument and g at the zeroth argument exists* if for any $v_1, \dots, v_{k+j-1} \in V$ and $v' \in V'$, the series

$$\sum_{M\in A} \langle v', f(v_1, \dots, v_{s-1}, P_M(g(v_s, \dots, v_{s+j-1})), v_{s+j}, \dots, v_{k+j-1})\rangle \tag{19}$$

converges absolutely, where $P_M : \overline{V} \longrightarrow V_{(M)}$ is the projection operator. In this case the (well-defined) limits for all $v_1, \dots, v_{k+j-1} \in V$, $v' \in V'$ define an element $f\ {}_s{*}_0\ g$ of $\mathcal{H}^G_{V,W}(k + j - 1)$, the *contraction.* More generally, given any subset of $\{1, \dots, k\}$ and any element of $\cup_{j\in\mathbb{N}}\mathcal{H}^G_{V,W}(j)$ for each element of the subset, we have the analogous *contraction*, defined using the appropriate multisums, when they are absolutely convergent. The substitution map

$$\begin{aligned} \gamma^G_{V,W} : \ \mathcal{H}^G_{V,W}(k) \times \mathcal{H}^G_{V,W}(j_1) \times \cdots \times \mathcal{H}^G_{V,W}(j_k) &\longrightarrow \mathcal{H}^G_{V,W}(j_1 + \cdots + j_k) \\ (f; g_1, \dots, g_k) &\longmapsto \gamma^G_{V,W}(f; g_1, \dots, g_k) \end{aligned} \tag{20}$$

is defined by this procedure, using the whole set $\{1, \dots, k\}$. Of course, the cases of proper subsets of $\{1, \dots, k\}$ are recovered by letting some of the g_l be $I^G_{V,W}$. The family $\mathcal{H}^G_{V,W}$ of sets $\mathcal{H}^G_{V,W}(j)$, $j \in \mathbb{N}$, equipped with the substitution maps

$\gamma^G_{V,W}$, the identity $I^G_{V,W}$ and the actions of S_j on $\mathcal{H}^G_{V,W}(j)$, $j \in \mathbb{N}$, satisfies all the axioms for a partial operad except the operad-associativity and therefore is a partial pseudo-operad. The operad-associativity fails because in general we cannot expect to have the absolute convergence of the multisums corresponding to a sequence of substitutions.

Using (multilinear) endomorphism partial pseudo-operads, we define the following notions of $\mathcal{P}$-pseudo-algebra and of $\mathcal{P}$-algebra:

DEFINITION 23. Let $\mathcal{P}$ be a partial operad with rescaling group G. A *$\mathcal{P}$-pseudo-algebra* (V, W, ν) is a completely reducible G-module $V = \coprod_{M \in A} V_{(M)}$ with dim $V_{(M)} < \infty$, together with a submodule W of V and a morphism ν from $\mathcal{P}$ (viewed as a partial pseudo-operad) to the partial pseudo-operad $\mathcal{H}^G_{V,W}$, such that the submodule of V generated by the homogeneous components of the elements of $\nu_0(\mathcal{P}(0))$ is W and the map from G to $\mathcal{H}^G_{V,W}(1)$ induced from ν_1 is the given representation of G on V. An element of $\nu_0(\mathcal{P}(0))$ is called a *quasi-identity element of $\overline{V}$ for $\mathcal{P}$*. A *morphism* from a $\mathcal{P}$-pseudo-algebra (V, W, ν) to a $\mathcal{P}$-pseudo-algebra $(\tilde{V}, \tilde{W}, \tilde{\nu})$ is a G-module morphism $\eta : V \longrightarrow \tilde{V}$ such that $\eta(W) \subset \tilde{W}$ and $\overline{\eta} \circ \nu_j(c) = \tilde{\nu}_j(c) \circ \eta^j$ for $j \in \mathbb{N}$ and $c \in \mathcal{P}(j)$, where η is extended naturally to $\overline{\eta} : \overline{V} \longrightarrow \overline{\tilde{V}}$; it follows that $\eta(W) = \tilde{W}$. Isomorphisms of $\mathcal{P}$-pseudo-algebras are defined in the obvious way. For a $\mathcal{P}$-pseudo-algebra (V, W, ν), the image of $\mathcal{P}$ under ν (where it is understood that the substitution maps are the substitution maps for $\mathcal{H}^G_{V,W}$ restricted to the images of the domains of the substitution maps for $\mathcal{P}$) is a partial pseudo-operad. We define a *$\mathcal{P}$-algebra* to be a $\mathcal{P}$-pseudo-algebra (V, W, ν) such that the image of $\mathcal{P}$ under ν is a partial operad, that is, such that operad-associativity holds for the image. Morphisms and isomorphisms of $\mathcal{P}$-algebras are defined to be morphisms and isomorphisms of the underlying $\mathcal{P}$-pseudo-algebras, respectively.

Remark 24. Let $\mathcal{P}$ be a partial operad with rescaling group G and (V, W, ν) a $\mathcal{P}$-algebra. Then $\nu_1(G)$ is a rescaling group of the image partial operad $\nu(\mathcal{P})$. In the case in which $\mathcal{P}$ is a G-rescalable partial operad, $\nu(\mathcal{P})$ is a $\nu_1(G)$-rescalable partial operad.

Though the definition of $\mathcal{P}$-algebra above is conceptually natural, it is in practice typically very difficult to determine whether a $\mathcal{P}$-pseudo-algebra is a $\mathcal{P}$-algebra. The issue is to insure that operad-associativity holds for certain families of multilinear maps. In the case of vertex operator algebras (see Section 6 below), the positive energy axiom and the analyticity axiom in [**H1**] and [**H2**] in fact insure that operad-associativity holds for corresponding families of multilinear maps.

The notion of *partial suboperad* of a partial operad is defined in the obvious way; we require that substitutions in a partial suboperad exist if and only if the corresponding substitutions in the original partial operad exist. An intersection of partial suboperads of a partial operad is a partial suboperad. We also have the notion of *partial suboperad generated by a subset*. If a partial operad is the partial suboperad generated by a given subset, we say that this partial operad is *generated by the subset.*

DEFINITION 25. Let $\mathcal{P}$ be a partial operad with rescaling group G. We call an element $a \in \mathcal{P}(2)$ *associative* if there exists $a' \in \mathcal{P}(2)$ which is G-equivalent to a (that is, there exists $b_0 \in G$ such that $a' = \gamma(b_0; a)$) and there exist unique $b_i \in G$, $i = 1, \dots, 5$, which depend on a', such that $\gamma(a; a', I)$ exists and

$$\gamma(a; a', I) = \gamma(d_1; I, d_2), \tag{21}$$

where

$$d_1 = \gamma(b_1; \gamma(a; b_2, b_3)), \quad d_2 = \gamma(a; b_4, b_5). \tag{22}$$

Remark 26. Using the operad-associativity and the definition of rescaling group, we see that formula (21) is equivalent to

$$\gamma(a; I, a'') = \gamma(d_1'; d_2', I), \tag{23}$$

where

$$a'' = \gamma(b_3; a), \tag{24}$$

$$d_1' = \gamma(b_1^{-1}; \gamma(a; b_0, b_5^{-1})), \quad d_2' = \gamma(a; b_2^{-1}, b_4^{-1}). \tag{25}$$

DEFINITION 27. We call a partial operad $\mathcal{P}$ with rescaling group G *associative* if $\mathcal{P}$ is generated by $\mathcal{P}(0)$, G and an associative element $a \in \mathcal{P}(2)$.

DEFINITION 28. Let $\mathcal{P}$ be an associative partial operad with rescaling group G and associative element $a \in \mathcal{P}(2)$. We call a $\mathcal{P}$-pseudo-algebra a *$\mathcal{P}$-associative pseudo-algebra* and a $\mathcal{P}$-algebra a *$\mathcal{P}$-associative algebra*, with *$\mathcal{P}$-associative binary product* given by the image of a and with *quasi-identity elements* given by the image of $\mathcal{P}(0)$.

5. The moduli space of spheres with tubes as a partial operad and its $\mathbb{C}$-extensions

In this and the next section, we reformulate the geometric interpretation of vertex operator algebras using the language of operads we have developed. For the details of the proofs of the results given in these two sections, see [**H4**].

We recall some structures from [**H1**] and [**H2**]. A *sphere with n tubes* ($n > 0$) is a sphere (a genus-zero compact connected one-dimensional complex manifold) S with n distinct, ordered points $p_0, \dots, p_{n-1}$ (called *punctures*) on S with p_0 negatively oriented and the other punctures positively oriented, and with local analytic coordinates $(U_0, \varphi_0), \dots, (U_{n-1}, \varphi_{n-1})$ vanishing at the punctures $p_0, \dots, p_{n-1}$, respectively, where for $i = 0, \dots, n-1$, U_i is a local coordinate neighborhood at p_i (i.e., an open set containing p_i) and $\varphi_i : U_i \longrightarrow \mathbb{C}$, satisfying $\varphi_i(p_i) = 0$, is a local analytic coordinate map vanishing at p_i. Let S_1 and S_2 be spheres with m and n tubes, respectively. Let $p_0, \dots, p_{m-1}$ be the punctures of

S_1, $q_0, \dots, q_{n-1}$ the punctures of S_2, (U_i, φ_i) the local coordinate at p_i for some fixed i, $0 < i \leq m-1$, and (V_0, ψ_0) the local coordinate at q_0. Assume that there exists a positive number r such that $\varphi_i(U_i)$ contains the closed disc $\bar{B}_0^r$ centered at 0 with radius r and $\psi_0(V_0)$ contains the closed disc $\bar{B}_0^{1/r}$ centered at 0 with radius $1/r$. Assume also that p_i and q_0 are the only punctures in $\varphi_i^{-1}(\bar{B}_0^r)$ and $\psi_0^{-1}(\bar{B}_0^{1/r})$, respectively. In this case we say that *the i-th puncture of the first sphere with tubes can be sewn with the zeroth puncture of the second sphere with tubes.* From these two spheres with tubes we can obtain a sphere with $m+n-2$ tubes by cutting $\varphi_i^{-1}(B_0^r)$ and $\psi_0^{-1}(B_0^{1/r})$ from S_1 and S_2, respectively, and then identifying the boundaries of the resulting surfaces using the map $\varphi_i^{-1} \circ J \circ \psi_0$ where J is the map from $\mathbb{C}^\times$ to itself given by $J(z) = 1/z$. The punctures (with ordering) of this sphere with tubes are $p_0, \dots, p_{i-1}, q_1, \dots, q_{n-1}$, $p_{i+1}, \dots, p_{m-1}$. The local coordinates vanishing at these punctures are given in the obvious way. Thus we have a partial operation. We define the notion of conformal equivalence between two spheres with tubes in the obvious way. The space of equivalence classes of spheres with tubes is called the *moduli space of spheres with tubes.* The moduli space of spheres with n tubes ($n \geq 2$) can be identified with $K_n = M^{n-2} \times H \times H_c^{n-1}$ where H is the set of all sequences A of complex numbers such that $\exp(\sum_{j>0} A_j x^{j+1} \frac{d}{dx}) \cdot x$ is a convergent power series in some neighborhood of 0, $H_c = \mathbb{C}^\times \times H$, and M^{n-2} is the set of elements of $\mathbb{C}^{n-2}$ with nonzero and distinct components. We think of each element of K_n, $n \geq 2$, as the sphere $\mathbb{C} \cup \{\infty\}$ equipped with ordered punctures ∞, $z_1, \dots, z_{n-2}$, 0, with an element of H specifying the local coordinate at ∞ and with $n-1$ elements of H_c specifying the local coordinates at the other punctures. Analogously, the moduli space of spheres with one tube can be identified with $K_1 = \{B \in H \mid B_1 = 0\}$. Then the moduli space of spheres with tubes can be identified with $K = \cup_{n>0} K_n$. From now on we will refer to K as the moduli space of spheres with tubes. The sewing operation for spheres with tubes induces a partial operation on K. It is still called the sewing operation and is denoted ${}_i\infty_0$.

We now give the moduli space K the structure of an associative $\mathbb{C}^\times$-rescalable partial operad. This operad is a natural two-dimensional analogue of the operad C discussed in Example 15, very different from the two-dimensional analogue in Example 16.

The sets of the partial operad K are

$$K(j) = K_{j+1}, \quad j \in \mathbb{N}. \tag{26}$$

The composition maps

$$\begin{array}{rrcl} \gamma_K : & K(k) \times K(j_1) \times \cdots \times K(j_k) & \longrightarrow & K(j_s + \cdots + j_k) \\ & (Q; Q_1, \dots, Q_k) & \longmapsto & \gamma_K(Q; Q_1, \dots, Q_k) \end{array} \tag{27}$$

are defined by

$$\gamma_K(Q; Q_1, \ldots, Q_k) = (\cdots((Q_1 {}_1\infty_0 Q_1)_{j_1+1}\infty_0 Q_2)\cdots)_{j_1+\cdots+j_{k-1}+1}\infty_0 Q_k \quad (28)$$

when the conditions for every sewing procedure in the right-hand side of (28) are satisfied. The identity element I is the element of $K_2 = K(1)$ which is the equivalence class of the standard sphere $\mathbb{C} \cup \{\infty\}$ with ∞ the negatively oriented puncture, 0 the only positively oriented puncture, and with standard local coordinates vanishing at ∞ and 0. For $\sigma \in S_j$ and $Q \in K(j)$, $\sigma(Q)$ is defined to be the conformal equivalence class of spheres with $j+1$ tubes obtained from members of the class Q by permuting the orderings of their positively oriented punctures using σ. Thus S_j acts on $K(j)$.

THEOREM 29. *The moduli space $K = \{K(j) \mid j \in \mathbb{N}\}$ has the structure of an associative analytic $\mathbb{C}^\times$-rescalable partial operad.*

We give a brief description of the structures on K. From the definition it is easy to see that K is a partial operad. Let $\mathbf{0}$ be the element of H with all components equal to 0. It can be verified that the subset $\{(\mathbf{0}, a, \mathbf{0}) \in H \times (\mathbb{C}^\times \times H) = K(1) \mid a \in \mathbb{C}^\times\}$ of $K(1)$ is a rescaling group for K, isomorphic to $\mathbb{C}^\times$. In fact, with this structure, K is a $\mathbb{C}^\times$-rescalable partial operad. It is also easy to see that K is associative, with associative element the equivalence class of the sphere $\mathbb{C} \cup \{\infty\}$ with ∞ the negatively oriented puncture, 1 and 0 the first and second positively oriented punctures, respectively, and with the standard local coordinates vanishing at these punctures. There are also natural topologies and infinite-dimensional complex manifold structures on $K(j)$, $j \in \mathbb{N}$. In the proof of the analyticity of the partial operad K, we have to prove that the sewing operations are analytic as maps between infinite-dimensional complex manifolds. In [**H1**] the Fischer-Grauert theorem [**FG**] in the deformation theory of complex manifolds is used to prove the analyticity of certain functions and then the convergence of certain series. The analyticity of the sewing operation is proved using this method.

The solution described in [**H2**] of the central charge problem in the geometric interpretation of vertex operator algebras can be reformulated in terms of the following notion of "$\mathbb{C}$-extension" of the operad K:

DEFINITION 30. An *extension by* $\mathbb{C}$ (or $\mathbb{C}$-*extension*) of the operad K is an analytic $\mathbb{C}^\times$-rescalable partial operad $\tilde{K}$ together with a morphism $\pi : \tilde{K} \longrightarrow K$ of $\mathbb{C}^\times$-rescalable partial operads satisfying the following axioms:

(i) For any $j \in \mathbb{N}$ the triple $(\tilde{K}(j), K(j), \pi)$ is a holomorphic vector bundle with fiber isomorphic to $\mathbb{C}$;

(ii) For any $\tilde{Q} \in \tilde{K}(k)$, $\tilde{Q}_1 \in \tilde{K}(j_1), \ldots, \tilde{Q}_k \in \tilde{K}(j_k)$, $k, j_1, \ldots, j_k \in \mathbb{N}$, the substitution $\gamma_{\tilde{K}}(\tilde{Q}; \tilde{Q}_1, \ldots, \tilde{Q}_k)$ in $\tilde{K}$ exists if (and only if)

$$\gamma_K(\pi(\tilde{Q}); \pi(\tilde{Q}_1), \ldots, \pi(\tilde{Q}_k))$$

exists;

(iii) Let $Q \in K(k)$, $Q_1 \in K(j_1), \dots, Q_k \in K(j_k)$, $k, j_1, \dots, j_k \in \mathbb{N}$, such that $\gamma(Q; Q_1, \dots, Q_k)$ exists. The map from the Cartesian product of the fibers over Q, $Q_1, \dots, Q_k$ to the fiber over $\gamma_K(Q; Q_1, \dots, Q_k)$ induced from the substitution map of $\tilde{K}$ is multilinear and gives an isomorphism from the tensor product of the fibers over Q, $Q_1, \dots, Q_k$ to the fiber over $\gamma_K(Q; Q_1, \dots, Q_k)$.

A $\mathbb{C}$-extension of K is also clearly associative; any element of the fiber over the associative element of K is an associative element for the $\mathbb{C}$-extension of K.

EXAMPLE 31. The determinant line bundle over the moduli space of spheres with boundaries induces a line bundle over K. We still call it the determinant line bundle and denote it $\tilde{K}^1$. For any complex number c, the line bundle $\tilde{K}^1$ raised to the complex power c is a well-defined line bundle over K which we denote $\tilde{K}^c$. For any $c \in \mathbb{C}$, $\tilde{K}^c$ is a $\mathbb{C}$-extension of K.

Following [**H2**], we can use the methods of Segal's and Mumford's work on one-dimensional modular functors [**Se**] to prove:

THEOREM 32. *Any $\mathbb{C}$-extension of K is a complex power of the determinant line bundle over K.*

6. Vertex operator algebras as meromorphic associative (pseudo-)algebras associated to $\mathbb{C}$-extensions of the operad K

For every $c \in \mathbb{C}$ we have an associative analytic $\mathbb{C}^\times$-rescalable partial operad $\tilde{K}^c$. It is natural to consider $\tilde{K}^c$-associative algebras (in the sense of Section 4). But since it is not easy to verify that a $\tilde{K}^c$-associative pseudo-algebra is a $\tilde{K}^c$-associative algebra, it is natural to ask what kinds of conditions will guarantee that a $\tilde{K}^c$-associative pseudo-algebra is a $\tilde{K}^c$-associative algebra. It turns out that the positive energy axiom and the analyticity axiom in [**H1**] and [**H2**] are such conditions.

We consider the operad $\tilde{K}^c$ for a given $c \in \mathbb{C}$. From the properties of the determinant line bundle, we know that there is a natural connection on $\tilde{K}^c$. Moreover, this connection is flat over $\overline{K}(j) = M^{j-1} \times \{\mathbf{0}\} \times (\{1\} \times \{\mathbf{0}\})^j \subset K(j)$, $j \geq 1$, where $1 \in \mathbb{C}$ and as above, $\mathbf{0} \in H$ is the sequence with all entries 0. Note that $\overline{K}(j)$ can be identified with M^{j-1}.

Since an equivalence class of irreducible modules for $\mathbb{C}^\times$ is determined by an integer n such that $a \in \mathbb{C}^\times$ acts on modules in this class as scalar multiplication by a^{-n}, any completely reducible module for $\mathbb{C}^\times$ is of the form $V = \coprod_{n \in \mathbb{Z}} V_{(n)}$ where $V_{(n)}$ is the sum of the $\mathbb{C}^\times$-submodules in the class corresponding to the integer n. In particular, the vector space of a $\tilde{K}^c$-associative pseudo-algebra is of this form. Note that for a $\tilde{K}^c$-associative pseudo-algebra we have $\dim V_{(n)} < \infty$ by definition.

DEFINITION 33. A $\tilde{K}^c$-associative pseudo-algebra (V, W, ν) is *meromorphic* if the following axioms are satisfied:

(i) $V_{(n)} = 0$ for n sufficiently small.

(ii) For any $v' \in V'$, $v_1, \dots, v_j \in V$, $\langle v', \nu(\cdot)(v_1, \dots, v_j) \rangle$ is analytic as a function on $\tilde{K}^c(j)$.

(iii) Given any $v_1, \dots, v_j \in V$ and $v' \in V'$ and any flat section ϕ of the restriction of the line bundle $\tilde{K}^c(j)$ to $\overline{K}(j)$, $\langle v', \nu(\phi(\cdot))(v_1, \dots, v_j)\rangle$ is a meromorphic function on $\overline{K}(j) = M^{j-1}$ with $z_i = 0$ and $z_i = z_k$, $i < k$, as the only possible poles, and for fixed $v_i, v_k \in V$ there is an upper bound for the orders of the pole $z_i = z_k$ of the functions

$$\langle v', \nu(\phi(\cdot))(v_1, \dots, v_{i-1}, v_i, v_{i+1}, \dots, v_{k-1}, v_k, v_{k+1}, \dots, v_j)\rangle$$

for all $v_1, \dots, v_{i-1},\ v_{i+1}, \dots, v_{k-1},\ v_{k+1}, \dots, v_j \in V,\ v' \in V'$.

PROPOSITION 34. *Any meromorphic $\tilde{K}^c$-associative pseudo-algebra is a $\tilde{K}^c$-associative algebra.*

The proof of this proposition uses the convergence and divergence properties of the meromorphic functions of the form in the definition above in certain regions and the $\mathbb{C}^\times$-rescalability of $\tilde{K}^c$.

We call a meromorphic $\tilde{K}^c$-associative (pseudo-)algebra a *vertex associative algebra with central charge* or *rank* c. *Morphisms* (respectively, *isomorphisms*) of vertex associative algebras are morphisms (respectively, isomorphisms) of the underlying $\tilde{K}^c$-associative algebras. The main theorem in [**H1**] and [**H2**] can now be reformulated using the language we have developed in this paper as follows:

THEOREM 35. *The category of vertex operator algebras with central charge (or rank) c is isomorphic to the category of vertex associative algebras with central charge (or rank) c.*

Here we give a brief description of the functor from the category of vertex operator algebras with central charge c to the category of vertex associative algebras with central charge c. Let $(V, Y, \mathbf{1}, \omega)$ be a vertex operator algebra with central charge c (see [**FLM**] or [**FHL**] for the definition). The $\mathbb{Z}$-graded vector space V is naturally a completely reducible $\mathbb{C}^\times$-module. The module W for the Virasoro algebra generated by $\mathbf{1}$ is a $\mathbb{Z}$-graded subspace of V and therefore is a $\mathbb{C}^\times$-submodule of V. In [**H2**] and [**H4**], a certain section of the line bundle $\tilde{K}^c$ over K is chosen. Here we denote this section ϕ. For an element $Q = (z, A^{(0)}, a_0^{(1)}, A^{(1)}, a_0^{(2)}, A^{(2)}) \in M^1 \times H \times (\mathbb{C}^\times \times H) \times (\mathbb{C}^\times \times H) = K(2)$, any element of the fiber over Q is of the form $\lambda\phi(Q)$ where $\lambda \in \mathbb{C}$. We define $\nu(\lambda\phi(Q))$ by

$$\begin{aligned}\nu(\lambda\phi(Q))(v_1, v_2) &= \lambda e^{-\sum_{i>0} A_i^{(0)} L(-i)} Y(e^{-\sum_{i>0} A_i^{(1)} L(i)} (a_0^{(1)})^{-L(0)} v_1, x) \cdot \\ &\quad \cdot e^{-\sum_{i>0} A_i^{(2)} L(i)} (a_0^{(2)})^{-L(0)} v_2 \big|_{x=z} \qquad (29)\end{aligned}$$

for any $v_1, v_2 \in V$, where $L(i)$, $i \in \mathbb{Z}$, are the coefficients of the vertex operator associated to ω, i.e., $Y(\omega, x) = \sum_{i\in\mathbb{Z}} L(i) x^{-i-2}$. For $\tilde{Q} \in \tilde{K}^c(j)$, $j \neq 2$, we can define $\nu(\tilde{Q})$ in a similar way. The triple (V, W, ν) is the vertex associative algebra corresponding to $(V, Y, \mathbf{1}, \omega)$.

References

[BPZ] A. A. Belavin, A. M. Polyakov and A. B. Zamolodchikov, Infinite conformal symmetries in two-dimensional quantum field theory, *Nucl. Phys.* **B241** (1984), 333–380.

[B] R. E. Borcherds, Vertex algebras, Kac-Moody algebras, and the Monster, *Proc. Natl. Acad. Sci. USA* **83** (1986), 3068–3071.

[FG] W. Fischer and H. Grauert, Lokal-triviale Familien kompakter komplexer Mannigfaltigkeiten, *Nachr. Akad. Wiss. Göttingen Math.-Phys.Kl.* 1965, 89–94.

[FHL] I. B. Frenkel, Y.-Z. Huang and J. Lepowsky, On axiomatic approaches to vertex operator algebras and modules, preprint, 1989; *Memoirs Amer. Math. Soc.* **104**, No. 494, 1993.

[FLM] I. B. Frenkel, J. Lepowsky and A. Meurman, *Vertex operator algebras and the Monster*, Pure and Appl. Math., Vol. 134, Academic Press, Boston, 1988.

[FS] D. Friedan and S. Shenker, The analytic geometry of two-dimensional conformal field theory, *Nucl. Phys.* **B281** (1987), 509–545.

[H1] Y.-Z. Huang, *On the geometric interpretation of vertex operator algebras*, Ph.D. thesis, Rutgers University, 1990; Operads and the geometric interpretation of vertex operator algebras, I, to appear.

[H2] Y.-Z. Huang, Geometric interpretation of vertex operator algebras, *Proc. Natl. Acad. Sci. USA* **88** (1991), 9964–9968.

[H3] Y.-Z. Huang, Applications of the geometric interpretation of vertex operator algebras, *Proc. 20th International Conference on Differential Geometric Methods in Theoretical Physics, New York, 1991*, ed. S. Catto and A. Rocha, World Scientific, Singapore, 1992, Vol. 1, 333–343.

[H4] Y.-Z. Huang, Operads and the geometric interpretation of vertex operator algebras, II, in preparation.

[HL1] Y.-Z. Huang and J. Lepowsky, Toward a theory of tensor products for representations of a vertex operator algebra, *Proc. 20th International Conference on Differential Geometric Methods in Theoretical Physics, New York, 1991*, ed. S. Catto and A. Rocha, World Scientific, Singapore, 1992, Vol. 1, 344–354.

[HL2] Y.-Z. Huang and J. Lepowsky, Vertex operator algebras and operads, *The Gelfand Mathematical Seminars, 1990–1992*, ed. L. Corwin, I. Gelfand and J. Lepowsky, Birkhäuser, Boston, 1993, 145-161.

[HL3] Y.-Z. Huang and J. Lepowsky, Tensor products of modules for a vertex operator algebra and vertex tensor categories, *Lie Theory and Geometry, In Honor of Bertram Kostant*, ed. R. Brylinski, J.-L. Brylinski, V. Guillemin and V. Kac, Birkhäuser, Boston, 1994, to appear.

[ML] S. Mac Lane, Natural associativity and commutativity, *Rice Univ. Studies* **49** (1963), 28–46.

[M1] J. P. May, *The geometry of iterated loop spaces*, Lecture Notes in Mathematics, No. 271, Springer-Verlag, 1972.

[M2] J. P. May, *E_∞ ring spaces and E_∞ ring spectra*, Lecture Notes in Mathematics, No. 577, Springer-Verlag, 1977.

[Se] G. Segal, The definition of conformal field theory, preprint, 1988.

[St1] J. D. Stasheff, Homotopy associativity of H-spaces, I, *Trans. Amer. Math. Soc.* **108** (1963), 275–292.

[St2] J. D. Stasheff, Homotopy associativity of H-spaces, II, *Trans. Amer. Math. Soc.* **108** (1963), 293–312.

[Ste] R. Steiner, A canonical operad pair, *Math. Proc. Camb. Phil. Soc.* **86** (1979), 443-449.

[V] C. Vafa, Conformal theories and punctured surfaces, *Phys. Lett.* **B199** (1987), 195–202.

DEPARTMENT OF MATHEMATICS, UNIVERSITY OF PENNSYLVANIA, PHILADELPHIA, PA 19104
Current address: Department of Mathematics, Rutgers University, New Brunswick, NJ 08903
E-mail address: yzhuang@math.rutgers.edu

SCHOOL OF MATHEMATICS, INSTITUTE FOR ADVANCED STUDY, PRINCETON, NJ 08540
Current address: Department of Mathematics, Rutgers University, New Brunswick, NJ 08903
E-mail address: lepowsky@math.rutgers.edu

Contemporary Mathematics
Volume **175**, 1994

Torus Actions, Moment Maps, and the Symplectic Geometry of the Moduli Space of Flat Connections on a Two-Manifold*

Lisa C. Jeffrey†
Downing College
Cambridge CB2 1DQ, UK
and
Jonathan Weitsman‡§
Isaac Newton Institute for Mathematical Sciences
20 Clarkson Road
Cambridge CB2 0EH, UK

Abstract

We summarize recent work ([W],[JW91a], [JW92]) on the symplectic geometry of the moduli space of flat SU(2) connections on a two-manifold. In this work we exploit the existence in these moduli spaces of Hamiltonian torus actions. Using these torus actions and the images of the corresponding moment maps we find a simple description of the moduli space; this description can be used to compute symplectic volumes and other quantities arising in the geometry and topology of the moduli space.

*Lecture given by J.W. at the AMS Summer Conference on Conformal Field Theory, Topological Field theory, and Quantum Groups, Mt. Holyoke College, June, 1992, and at the NATO Advanced Research Workshop on Low Dimensional Topology and Quantum Field Theory, Cambridge, UK, September, 1992

†Address after September 1, 1993: Department of Mathematics, Princeton University, Princeton, NJ 08544

‡Address after January 1, 1993: Department of Mathematics, Columbia University, New York, NY 10027

§Supported in part by NSF Mathematical Sciences Postdoctoral Research Fellowship 88-07291

1 Introduction

In this talk we will review some recent work on the structure of some moduli spaces associated to two-manifolds. We will be particularly interested in the structure of these spaces as symplectic manifolds. The moduli spaces we consider can also be viewed as Kähler varieties once a Riemann surface structure is put on the underlying two-manifold, and the study of the Kähler geometry of the resulting projective varieties has been a topic of much interest since the 1960's (see for example [AB]). In our work we show that the study of these spaces as *symplectic* varieties reveals a good deal of structure not obviously present in the Kähler setting.

Let us describe more explicitly the spaces in question. We consider a compact, connected, oriented two-manifold Σ^g of genus g, and associate to it the space of conjugacy classes of representations $\bar{\mathcal{S}}_g = Hom(\pi_1(\Sigma^g), G)/G$, where G is a compact simple Lie group which in this talk will be $G = SU(2)$. The space $\bar{\mathcal{S}}_g$ contains an open dense set $\mathcal{S}_g$ which is a symplectic manifold; we denote the symplectic form on $\mathcal{S}_g$ by ω.

The space $\bar{\mathcal{S}}_g$ makes its appearance in various areas of mathematics. In *gauge theory*, $\bar{\mathcal{S}}_g$ appears as the moduli space of flat connections on the trivial principal G-bundle on the two-manifold Σ^g. In *topology* $\bar{\mathcal{S}}_g$ appears in relation with the Casson invariant of homology three-spheres; this invariant is given roughly by the intersection number of two Lagrangian subvarieties in $\bar{\mathcal{S}}_g$. In *algebraic geometry* $\bar{\mathcal{S}}_g$ appears once a conformal structure is chosen on the underlying two-manifold Σ^g; the space $\bar{\mathcal{S}}_g$ is then the moduli space of semistable holomorphic vector bundles on the corresponding Riemann surface, with rank 2, degree zero, and fixed determinant. The final role of $\bar{\mathcal{S}}_g$ and the one most closely related to the topic of this conference is its appearance in connection with *topological field theory.* For $\bar{\mathcal{S}}_g$ is the classical phase space of Chern-Simons gauge theory, and as such is conjectured to be related to the Witten-Reshetikhin-Turaev (WRT) invariants of three manifolds. To relate our work to that of the other speakers, we will phrase our results about the moduli space in connection with this topic.

Recall that a $2+1$ dimensional topological field theory would assign to every two-manifold Σ^g a Hilbert space $\mathcal{H}(\Sigma^g)$, and to every three-manifold M bounding Σ^g an element $v(M) \in \mathcal{H}(\Sigma^g)$. The WRT topological invariants arise from a family of topological field theories, one for each positive integer level k. In these theories the Hilbert space $\mathcal{H}(\Sigma^g)$ is naturally isomorphic to a vector space with a basis naturally identified with a family of marked trivalent graphs; these graphs are obtained from a *pants decomposition* Γ of Σ^g as follows. We decompose Σ^g into $(2g-2)$ pairs of pants P_γ, $\gamma = 1, \ldots, 2g-2$; this gives $(3g-3)$ boundary circles C_i, $i = 1, \ldots, 3g-3$. We can associate to this decomposition a trivalent graph given by assigning a vertex to each pair of pants and an edge connecting two vertices to every boundary circle shared by the two corresponding pairs of pants. A *marked trivalent graph* corresponding to the pants decomposition Γ is a labeling of each edge of the graph by a real number;

we denote the real number assigned to the edge associated to the circle C_i by x_i. A marked trivalent graph will be called *admissible* if for every $i = 1, \dots, 3g-3$ we have

$$0 \leq x_i \leq 1 \tag{A1}$$

and if in addition, for each pair of pants P_γ with boundary $C_{i_1(\gamma)} \cup C_{i_2(\gamma)} \cup C_{i_3(\gamma)}$ we have

$$0 \leq x_{i_1(\gamma)} + x_{i_2(\gamma)} + x_{i_3(\gamma)} \leq 2 \tag{A2}$$

$$|x_{i_1(\gamma)} - x_{i_2(\gamma)}| \leq x_{i_3(\gamma)} \leq x_{i_1(\gamma)} + x_{i_2(\gamma)}. \tag{A3}$$

A marked trivalent graph will be called *integral* if for every $i = 1, \dots, 3g-3$ and every $\gamma = 1, \dots, 2g-2$ we have

$$x_i \in \frac{1}{k}\mathbb{Z} \tag{I1}$$

$$x_{i_1(\gamma)} + x_{i_2(\gamma)} + x_{i_3(\gamma)} \in \frac{1}{k}2\mathbb{Z}. \tag{I2}$$

Then the Hilbert space assigned by the level k WRT topological field theory to a surface Σ^g of genus g is naturally isomorphic to the vector space with basis given by generators corresponding to the set $\mathcal{D}(g, k, \Gamma)$ of integral admissible markings of the trivalent graph associated to *any* pants decomposition Γ of the surface Σ^g. In particular the dimension of the Hilbert space is equal to the number $D(g, k)$ of such markings (which is in fact independent of the pants decomposition; see [MS]).

The moduli space $\mathcal{S}_g$ enters this picture in the following way. On $\bar{\mathcal{S}}_g$ there exists a line bundle $\mathcal{L}$ with connection of curvature ω (see *e.g.* [RSW]). If we choose a Riemann surface structure on Σ^g and work with the corresponding Kähler structure on $\bar{\mathcal{S}}_g$, this line bundle acquires the structure of a hermitian holomorphic line bundle. We may then form a vector space by *geometric quantization* from the space $\bar{\mathcal{S}}_g$ and this Kähler polarization; this quantization is none other than the space $H^0(\bar{\mathcal{S}}_g, \mathcal{L}^k)$ of holomorphic sections of the line bundle $\mathcal{L}^k$ for any $k \in \mathbb{Z}$. Then the following theorem holds:

Theorem (Verlinde Dimension Formula): ([V];[Don],[K],[NR],[Sz], [BSz],[Th]) $dim H^0(\bar{\mathcal{S}}_g, \mathcal{L}^k) = D(g, k)$.

The purpose of this lecture is to show how the geometry of $\bar{\mathcal{S}}_g$ is mirrored in this remarkable combinatorial formula. In brief, we will show that the symplectic variety $\bar{\mathcal{S}}_g$ is "well approximated" by the toric variety constructed from the convex polyhedron described by the inequalities $(A1) - (A3)$. Morally, then, we would expect the holomorphic sections of a line bundle on this toric variety to be associated to characters of the torus, and hence to the integral points of the polyhedron—that is, to points satisfying $(I1) - (I2)$. The rest of this talk will be devoted to explaining the terms used above, to constructing the torus actions which give the toric variety structure, and to a precise delineation of the words "well approximated." We will then prove the following result:

Theorem 1.1: ([W],[JW91a],[JW92]) Fix a pants decomposition Γ of Σ^g. There exists a dense open set $U \subset \bar{\mathcal{S}}_g$ which is symplectically diffeomorphic to a noncompact toric variety. The convex body associated to this toric variety (via the image of the moment map of the torus action) has as its closure the convex polyhedron described by the inequalities $(A1) - (A3)$.

It is important to note what this theorem does *not* say. We do not claim that the natural complex structure on the toric variety has anything to do with any Kähler structure coming from a Riemann surface; in other words we do not show that the torus actions preserve the complex structures on $U \subset \bar{\mathcal{S}}_g$ coming from the conformal structure on Σ^g. Thus we cannot directly compute the dimension of the space of holomorphic sections by counting integral points. Instead we must content ourselves with computing quantities such as the symplectic volume of the space $\bar{\mathcal{S}}_g$ which can be deduced from the toric structure on U. In one of our papers [JW92], however, we explain how knowledge of such volumes can be combined with techniques in the literature to compute the dimension of this space of holomorphic sections.

A second remark is directed at physicists, who are accustomed to different terminology for the toric structure we describe. In physics language our result is that the symplectic manifold U is a classical integrable system. Now a classical integrable system may be quantized using the Bohr-Sommerfeld rules; the allowable quantized momenta (expected for a compact phase space) are just given by the integral points of the polyhedron $(A1) - (A3)$, and are therefore in correspondence with the integral admissible markings of the trivalent graph corresponding to the pants decomposition Γ. In [JW91a] we show that this Bohr-Sommerfeld procedure can indeed be justified despite the fact that the integrable system structure does not extend from U to all of $\bar{\mathcal{S}}_g$; and our result shows that the quantization of this space in our real polarization is isomorphic to the quantization in a Kähler polarization, which is the space of holomorphic sections of $\mathcal{L}^k$.

The basic plan of the talk will be as follows. In the next section we will explain in a brief way the basic ideas of the geometry of Hamiltonian torus actions on symplectic manifolds. We do so with some trepidation given the many excellent texts on this subject (see for example [GS84]). Our description is tailored to explain the meaning of the words used in Theorem 1.1 above, and the reader should refer to those texts for an overview of this beautiful theory. We then proceed in section 3 to show how these ideas can be applied to the moduli space $\bar{\mathcal{S}}_g$, and to construct the S^1 actions on the toric variety of Theorem 1.1. We follow this in section 4 with a description of the polyhedron, which completes the proof of Theorem 1.1. In section 5 we show how our result can be applied to calculate the symplectic volume of $\mathcal{S}_g$. An application of our methods to the construction of three-manifold invariants is described in [JW91b].

2 Symplectic manifolds, torus actions, and toric varieties

Let (M^{2m}, ω) be a $2m$–dimensional, compact, connected symplectic manifold. Suppose we are given a smooth action of the circle group S^1 on M. This action is called *Hamiltonian* if the vector field X which generates this action is symplectically dual to an exact one-form; that is, if there exists a function $\mu : M \to \mathbb{R}$ such that

$$d\mu|_p(v) = \omega|_p(v, X|_p) \tag{2.1}$$

for any tangent vector v to M at any point $p \in M$.

The function $\mu : M \to \mathbb{R}$ associated to a Hamiltonian circle action in this way is called the *moment map* of this circle action. More generally, we may consider several (say j) commuting circle actions on M. Then the corresponding moment maps may be combined to form a moment map $\mu : M \to \mathbb{R}^j$. The following result characterizes the image of the moment map associated to a Hamiltonian torus action on a compact symplectic manifold.

Convexity Theorem:([A], [GS]) The image $\mu(M)$ of the moment map is a convex polyhedron in $\mathbb{R}^j$.

A particularly interesting case of this situation is where the torus acts effectively and has half the dimension of the manifold; that is where $j = m$. In this case the fibres of the moment map of the torus action on M are the orbits of this action. In other words every point in M is uniquely specified by giving the values of the moment map along with the values of a collection of parameters describing the location of the point along the torus orbit. In classical mechanics the components of the moment map are known as *action variables*, while the parameters describing the orbits can be chosen to be *angle variables.* Since generically the orbits will be diffeomorphic to $(S^1)^m$, these angle variables will be given on an open dense set $V \subset M$ by functions $\phi_i : V \to S^1, i = 1, \dots, m$. In terms of these functions the symplectic form ω may be written

$$\omega|_V = \sum_i d\mu_i \wedge d\phi_i. \tag{2.2}$$

Thus a symplectic manifold of dimension $2m$ endowed with a Hamiltonian $(S^1)^m$ action is completely determined by the image of the moment map μ; the manifold is given by the fibering above $\mu(M)$, while the symplectic form is determined by the "globalized" Darboux formula (2.2). A manifold of this type is symplectomorphic to a *toric variety*; that is, to the closure of the orbit of a linear $(\mathbb{C}^*)^m$ action in some complex projective space $\mathbb{C}P^N$, equipped with the symplectic form inherited from $\mathbb{C}P^N$ (see for example [Del]).

We pause to give some simple examples of this situation. The first is just the two sphere S^2, endowed with the usual coordinates θ, ϕ and the symplectic form $\omega = d\cos\theta \wedge d\phi$. The circle action given by rotation about the $z-$axis has as its generating vector field $\frac{\partial}{\partial\phi}$. Hence the moment map $\mu : S^2 \to \mathbb{R}$ must satisfy

$$d\mu(v) = \omega(v, \frac{\partial}{\partial\phi}) = d\cos\theta(v);$$

thus $\mu(\theta, \phi) = \cos\theta$ and the moment map is just projection onto the axis of rotation. The image of S^2 under this map is the interval, a convex polyhedron!

Exercise: Show that the convex polyhedron given by the image of the moment map of the linear $(S^1)^m$ action on $\mathbb{C}P^m$ is the $m-$simplex.

In light of the results summarized in this section, we can rephrase our task in proving the main theorem of this lecture. It is to construct on an open dense set in $\bar{\mathcal{S}}_g$ the action of a $3g-3$ dimensional torus, and to compute the image of the corresponding moment map. As we shall see, this moment map will extend naturally to a continuous (but not differentiable!) function defined on all of $\bar{\mathcal{S}}_g$, whose image will be the convex polyhedron described by $(A1)-(A3)$.

3 Torus Actions on the Moduli Space

We now come to the main focus of this work, which is the construction of the torus action on $\bar{\mathcal{S}}_g$. We wish to construct $3g-3$ commuting Hamiltonian circle actions on this space. Equivalently, we may give the moment maps of those circle actions; these will be functions $\mu_i : \bar{\mathcal{S}}_g \to \mathbb{R}, i = 1, \ldots, 3g-3$. The functions μ_i are defined as follows.

Recall that a *pair of pants* is a copy of the two holed disc, or alternatively, the three holed sphere. Let us fix a decomposition Γ of the surface Σ^g into pants $P_\gamma, \gamma = 1, \ldots, 2g-2$. Each of the boundary circles $C_i, i = 1, \ldots, 3g-3$ of the pairs of pants will then give rise to a function $\mu_i : \bar{\mathcal{S}}_g \to \mathbb{R}^{3g-3}$. To give a formula for μ_i, we recall that $\bar{\mathcal{S}}_g$ is the quotient of $Hom(\pi_1(\Sigma^g), G)$ by the conjugation action of G. So we may define a function on $\bar{\mathcal{S}}_g$ by giving a conjugation invariant function $\tilde{\mu}_i : Hom(\pi_1(\Sigma^g), G) \to \mathbb{R}$.

To define such a function we note that given a choice of basepoints, arcs, and orientations, each circle C_i gives a homotopy class $[C_i] \in \pi_1(\Sigma^g)$. Given an element $\rho \in Hom(\pi_1(\Sigma^g), G)$, that is, a representation of $\pi_1(\Sigma^g)$ into G, we may evaluate ρ on $[C_i]$, to give an element of $G = SU(2)$. The trace of $\rho([C_i])$ is conjugation invariant; and being the trace of an $SU(2)$ matrix, it is equal to $2\cos\theta_i$ for some angle $0 \leq \theta_i \leq \pi$. Thus we may define

$$\tilde{\mu}_i : Hom(\pi_1(\Sigma^g), G) \to \mathbb{R}$$

by

$$\tilde{\mu}_i(\rho) = \frac{1}{\pi} \cos^{-1} \frac{1}{2} tr\ \rho([C_i]). \tag{3.1}$$

We have the following lemma.

Lemma 3.1: The function $\tilde{\mu}_i$ is independent of the choices of basepoints, arcs, and orientations, and descends to a function $\mu_i : \bar{\mathcal{S}}_g \to \mathbb{R}$.

We claim that the function μ_i is the moment map for a densely defined circle action on $\bar{\mathcal{S}}_g$. The formula (3.1) for $\tilde{\mu}_i$ shows why this cannot hold everywhere in $\bar{\mathcal{S}}_g$; for where $\rho([C_i]) = \pm 1$, the function $\tilde{\mu}_i = \frac{1}{\pi} \cos^{-1} \frac{1}{2} tr\ \rho([C_i])$ will have infinite derivative, so that the vector field X generating the putative S^1 action (according to formula (2.1)) would be ill-defined.

Away from these bad points, however, the following theorem of W. Goldman applies:

Theorem 3.2:([G]) Let $U_i = \mu_i^{-1}((0,1))$. The function $\mu_i|_{U_i}$ is the moment map for a Hamiltonian circle action of period 1 on U_i.

We thus obtain a circle action in $\bar{\mathcal{S}}_g$ for each boundary circle in the pants decomposition Γ. In order to obtain a torus action in $\bar{\mathcal{S}}_g$, we must show that these circle actions commute. This is in fact true, and is essentially due to the fact that the corresponding boundary circles are *disjoint*. We summarize this as follows.

Theorem 3.3:([G]) Let $U = \cap_{i=1}^{3g-3} U_i \subset \bar{\mathcal{S}}_g$. The functions $\mu_i|_U$ combine to give a function $\mu : U \to \mathbb{R}^{3g-3}$ which is the moment map for an $(S^1)^{(3g-3)}$ action on U.

Note that Theorem 3.3 proves the existence of the toric variety structure on the dense open set $U \subset \bar{\mathcal{S}}_g$. This is the main part of the proof of Theorem 1.1. The image of the moment map will be computed in the next section.

Note that circle actions will commute if the corresponding generating vector fields commute. In view of the formula (2.1), this amounts to saying that the functions μ_i form a Poisson commuting family of functions on U. Geometrically, the orbit of any point $x \in U$ under the torus action spans a torus T_x with $\omega|_{T_x} = 0$; this torus is of dimension *dim* $T_x = 3g - 3$. Hence these tori foliate $U \subset \bar{\mathcal{S}}_g$ by *Lagrangian* submanifolds. Such a foliation of a symplectic manifold is called a *real polarization.*

4 The Image of the Moment Map

We have seen that the function $\mu : U \subset \bar{\mathcal{S}}_g \to \mathbb{R}$ gives a moment map for a $(S^1)^{(3g-3)}$ action on U. Since dim $U = 6g - 6$, this shows that U is fibred by tori over the image $\mu(U)$; in order to understand the structure of U as a symplectic manifold, we have to compute the image $\mu(U) \subset \mathbb{R}^{(3g-3)}$, or, equivalently, its closure $\mu(\bar{S}_g)$. This is the purpose of this section.

To do this we must find which $(3g-3)$–tuples of real numbers $(x_1, \ldots, x_{(3g-3)})$ occur as

$$x_i = \frac{1}{\pi} \cos^{-1} \frac{1}{2} tr\ \rho([C_i]) \tag{4.1}$$

where C_i is a curve in the fixed pants decomposition Γ of Σ^g, and ρ is a representation of $\pi_1(\Sigma^g)$ in $G = SU(2)$. In order to answer this question, it is useful to observe that it is enough to check whether the real numbers $x_{i_1(\gamma)}, x_{i_2(\gamma)}, x_{i_3(\gamma)}$ associated to each pair of pants come via equation (4.1) from a representation of the fundamental group $\pi_1(P_\gamma)$ of the pair of pants—and to do so for each pair of pants. The easiest way to see this is via gauge theory, and we refer the reader to [JW91a] for details. We state the result as a lemma.

Lemma 4.1: Let there be given a decomposition of the surface Σ^g into pairs of pants $P_\gamma, \gamma = 1, \ldots, 2g-2$. Label the boundary circles of the pants decomposition by $C_i, i = 1, \ldots, 3g-3$. Let $(x_1, \ldots, x_{(3g-3)}) \in \mathbb{R}^{(3g-3)}$. Suppose that each pair of pants P_γ has as its boundary $C_{i_1(\gamma)} \cup C_{i_2(\gamma)} \cup C_{i_3(\gamma)}$.

Suppose that for each $\gamma = 1, \ldots, 2g-2$, there exist $g_1(\gamma), g_2(\gamma), g_3(\gamma) \in G$ with

$$x_{i_j(\gamma)} = \frac{1}{\pi} \cos^{-1} \frac{1}{2} tr\ g_i(\gamma), i = 1, 2, 3$$

and with

$$g_1(\gamma) g_2(\gamma) g_3(\gamma) = 1.[1]$$

Then there exists a representation $\rho \in Hom(\pi_1(\Sigma^g), G)$ such that (4.1) holds.

Remark: This representation is *not* unique, and not even unique up to conjugation; in fact, if it corresponds to a point in U, the circle actions will give a $(3g-3)$ dimensional torus' worth of such representations.

The lemma reduces the task of finding the image of μ to a computation in $SU(2)$; we must find which real numbers x_1, x_2, x_3 occur via equation (4.1) as traces of matrices g_1, g_2, g_3 which multiply to give the identity. The answer to this is the following lemma.

Lemma 4.2: Let $g_1, g_2, g_3 \in SU(2)$ and suppose that $g_1 g_2 g_3 = 1$. Let $x_i = \frac{1}{\pi} \cos^{-1} \frac{1}{2} tr\ g_i, i = 1, 2, 3$.

Then

$$0 \leq x_i \leq 1 \tag{4.2a}$$

$$|x_1 - x_2| \leq x_3 \leq x_1 + x_2 \tag{4.2b}$$

$$x_1 + x_2 + x_3 \leq 2. \tag{4.2c}$$

[1]This is of course just the condition that $g_1(\gamma), g_2(\gamma), g_3(\gamma)$ give a representation of $\pi_1(P_\gamma)$ in G.

Conversely, every triple of real numbers satisfying (4.2a-c) will occur in this way.

Note that the conditions (4.2a-c) require that x_1, x_2, x_3 form a spherical triangle.

We can now read off the image of the moment map $\mu(\bar{\mathcal{S}}_g)$. Combining Lemmas 4.1 and 4.2 we see that

Theorem 4.3: The image $\mu(\bar{\mathcal{S}}_g)$ of the map μ is the polyhedron given by the inequalities (A1),(A2),(A3) of Section 1.

Combining Theorems 3.3 and 4.3 we see that we have proved Theorem 1.1 of the Introduction.

5 Application: The Symplectic Volume

We now show how Theorem 1.1 can be used to calculate symplectic invariants of the moduli space $\bar{\mathcal{S}}_g$. As we noted in the introduction, the torus action on U does not extend to all of $\bar{\mathcal{S}}_g$; furthermore, this action does not preserve the complex structure coming from a Riemann surface. Thus we cannot use the standard theorems on holomorphic sections of line bundles on toric varieties (see [Oda]) to prove the Verlinde formula. Instead we content ourselves with the computation of the volume of the moduli space.

Let us return for a moment to the general case of a symplectic manifold (M^{2m}, ω). Its symplectic volume is the integral

$$vol\ (M) = \frac{1}{m!} \int_M \omega^m. \tag{5.1}$$

Suppose now that M is symplectomorphic to a toric variety with an effective torus action corresponding to a moment map $\mu : M \to \mathbb{R}^m$. Then by equation (2.2), there exists a dense open set $V \subset M$ with

$$\omega|_V = \sum_i d\mu_i \wedge d\phi_i$$

for some functions $\phi_i : V \to (S^1)^m$, which parametrize the orbits of the $(S^1)^m$ action. Thus (5.1) becomes

$$vol\ (M) = \int_V \bigwedge_{i=1}^m d\mu_i \wedge \bigwedge_{i=1}^m d\phi_i.$$

We may however perform the integrals over the orbit; we have normalized the volume of the orbits to 1, so that

$$vol\ (M) = \int_{\mu(V)} \bigwedge_{i=1}^m d\mu_i;$$

in other words,

$$vol\ (M) = vol\ (\mu(M)). \tag{5.2}$$

Equation (5.2) may be looked upon as a simple case of the Duistermaat-Heckman theorem [DH].

Now $\mu(M)$ is a convex polyhedron in Euclidean space, and we may compute its volume by counting integral points. More precisely suppose that we are given some lattice $\Lambda \subset \mathbb{R}^m$. This lattice will be called *volume-approximating* if a fundamental domain for the lattice action on $\mathbb{R}^m$ has volume 1. Then the volume of any polyhedron P is given by

$$vol\ (P) = lim_{k\to\infty}\frac{1}{k^m}\#(\Lambda \cap kP) \tag{5.3}$$

where the symbol # is used to denote the number of points in a finite set, and where kP is the polyhedron given by expanding the polyhedron $P \in \mathbb{R}^m$ by a factor of k (that is, the image of P under the self map of $\mathbb{R}^m$ given by componentwise multiplication by k.)

We now combine equations (5.2) and (5.3); if we are given a volume-approximating lattice $\Lambda \subset \mathbb{R}^m$, we have

$$vol\ (M) = lim_{k\to\infty}\frac{1}{k^m}\#(\Lambda \cap k\mu(M)). \tag{5.4}$$

Let us apply this relation to the toric variety $U \subset \bar{\mathcal{S}}_g$. For our integral lattice, we choose the lattice Λ given by conditions $(I1), (I2)$ of the introduction.[2]

$$vol\ (\mathcal{S}_g) = lim_{k\to\infty}\frac{1}{k^m}\#(\Lambda \cap k\mu(U)) = lim_{k\to\infty}\frac{1}{k^m}D(g,k). \tag{5.5}$$

Equation (5.5) appears in the literature in various guises; see for example [Wit]. In our approach the appearance of the volume as the leading term in the large-k asymptotics of the Verlinde dimension $D(g,k)$ is a simple consequence of Theorem 1.1, which is an expression of the symplectic geometry behind this formula.

6 References

[A] M. F. Atiyah. Convexity and Commuting Hamiltonians. *Bull. Lond. Math. Soc.* **14**, 1 (1981)

[AB] M.F. Atiyah, R. Bott. The Yang Mills Equations over Riemann Surfaces. *Phil. Trans. Roy. Soc. London* **A 308** 523 (1982).

[2]There is a slight difference between our setting and that of equation (5.4) in that the torus action corresponding to the moment map μ on U is not effective. There does exist, however, an effective Hamiltonian action on U of a quotient of this torus by a finite group. The same methods then apply, except that the lattice Λ must be substituted for the volume-approximating lattice given by the sole condition $(I1)$; see [JW91a].

[BSz] A. Bertram, A. Szenes. Hilbert Polynomials of Moduli Spaces of Rank 2 Vector Bundles II. Harvard preprint (1991).

[Del] T. Delzant. Hamiltoniens Periodiques et Images Convexes de l'Application Moment. *Bull. Soc. Math. France* **116**, 315 (1988)

[Don]S.K. Donaldson. Gluing Techniques in the Cohomology of Moduli Spaces. Oxford preprint (1992).

[DH] J.J. Duistermaat, G. Heckman. On the Variation in the Cohomology of the Symplectic Form of the Reduced Phase-Space. *Inv. Math.* **69**, 259 (1982)

[G] W. Goldman. Invariant Functions on Lie Groups and Hamiltonian Flows of Surface Group Representations. *Inv. Math.* **85**, 263 (1986).

[GS] V. Guillemin, S. Sternberg. Convexity Properties of the Moment Mapping. *Inv. Math.* **67**, 491 (1982)

[GS84] V. Guillemin, S. Sternberg. Symplectic Techniques in Physics. Cambridge University Press, 1984.

[JW91a] L.C. Jeffrey, J. Weitsman. Bohr-Sommerfeld Orbits in the Moduli Space of Flat Connections and the Verlinde Dimension Formula. IAS preprint IASSNS-HEP-91/82; *Commun. Math Phys.,* (1992)

[JW91b] L.C. Jeffrey, J. Weitsman. Half Density Quantization of the Moduli Space of Flat Connections and Witten's Semiclassical Manifold Invariants. IAS preprint IASSNS-HEP-91/94; *Topology,* to appear.

[JW92] L. C. Jeffrey, J. Weitsman. Toric Structures on the Moduli Space of Flat Connections on a Riemann Surface: Volumes and the Moment Map. Institute for Advanced Study Preprint IASSNS-HEP-92/25; *Adv. Math.,* to appear.

[K] F. Kirwan. The Cohomology Rings of Moduli Spaces of Bundles over Riemann Surfaces. *J. Amer. Math. Soc.,* to appear.

[MS] G. Moore, N. Seiberg. Classical and Quantum Conformal Field Theory. *Commun. Math. Phys.* **123**, 77 (1989).

[NR] M. S. Narasimhan, T. R. Ramadas. Factorization of Generalized Theta Functions. Tata Institute Preprint, 1991.

[Oda] T. Oda.Convex Bodies and Algebraic Geometry. New York: Springer Verlag, 1988.

[RSW] T. R. Ramadas, I. M. Singer, J. Weitsman. Some Comments on Chern-Simons Gauge Theory. *Commun. Math. Phys.* **126**, 409 (1989)

[Sz] A. Szenes. Hilbert Polynomials of Moduli Spaces of Rank 2 Vector Bundles I. Harvard preprint (1991).

[Th] M. Thaddeus. Conformal Field Theory and the Cohomology of the Moduli Space of Stable Bundles. *J. Diff. Geom.*, to appear (1991).

[V] E. Verlinde. Fusion Rules and Modular Transformations in 2d Conformal Field Theory. *Nucl. Phys.* **B300**, 351 (1988).

[W] J. Weitsman. Real Polarization of the Moduli Space of Flat Connections on a Riemann Surface. *Commun. Math. Phys.* **145**, 425 (1992).

[Wit] E. Witten. On Quantum Gauge Theories in Two Dimensions, *Commun. Math. Phys.* **140**, 153 (1991).

Contemporary Mathematics
Volume **175**, 1994

Vertex Operator Superalgebras and Their Representations*†

Victor Kac and Weiqiang Wang

0 Introduction

Vertex operator algebras (VOA) were introduced in physics by Belavin, Polyakov and Zamolodchikov [BPZ] and in mathematics by Borcherds [B]. For a detailed exposition of the theory of VOAs see [FLM] and [FHL]. In a remarkable development of the theory, Zhu [Z] constructed an associative algebra $A(V)$ corresponding to a VOA V and established a 1-1 correspondence between the irreducible representations of V and those of $A(V)$. Furthermore, Frenkel and Zhu [FZ] defined an $A(V)$-module $A(M)$ for any V-module M and then described the fusion rules in terms of the modules $A(M)$. An important feature of these constructions is that $A(V)$ and $A(M)$ can usually be computed explicitly. For example, they enabled Frenkel and Zhu to prove the rationality and compute the fusion rules of VOAs associated to the representations of affine Kac-Moody algebras with a positive integral level. They also allowed one of the authors [W] to prove the rationality and compute the fusion rules of VOAs associated to the minimal series representations of the Virasoro algebra. (Independently, Dong, Mason and Zhu [DMZ] proved the rationality for the *unitary* minimal series of the Virasoro algebra and calculated the fusion rules in the case of central charge $c = \frac{1}{2}$).

In this paper we generalize Frenkel-Zhu's construction to vertex operator superalgebras (SVOA) and then discuss in detail several interesting classes of SVOAs. We present explicit formulas for the "top" singular vectors and defining relations for the integrable representations of the affine Kac-Moody superalgebras. These formulas are not only crucial for the theory of the associated SVOAs and their modules, but also of independent interest.

We organize this paper in the following way. In Subsec.1.1 we present definitions of vertex operator superalgebras and their modules, emphasizing the existence of the Neveu-Schwarz element in the so-called $N = 1$ (NS-type) SVOAs. We define in Subsec.1.2 an associative algebra $A(V)$ corresponding to a SVOA V and establish a bijective correspondence between the irreducible representations of V and the irreducible representations of $A(V)$. In Subsec.1.3 we define an $A(V)$-module $A(M)$ for every V-module M and then describe the fusion rules in terms of modules $A(M)$. Needless to say that, if we view a VOA as a SVOA with zero odd part, then our construction reduces to Frenkel-Zhu's original one.

In Subsec.2.1 we construct $N = 1$ SVOAs $M_{k,0}$ and $L_{k,0}$ corresponding to the representations of an affine Kac-Moody superalgebra $\hat{\hat{\mathfrak{g}}}$. In [KT], the minimal representation $L(h^\vee \Lambda_0)$ of $\hat{\hat{\mathfrak{g}}}$ was realized in a Fock space F of a certain infinite-

*1991 Mathematics Subject Classification. Primary 17b65; Secondary 17A70, 17B67.
†This paper is in the final form, and no version of it will be submitted for publication elsewhere.

dimensional Clifford algebra contained in $\hat{\hat{\mathfrak{g}}}$. Kac and Todorov [KT] proved that any unitary highest weight representation of $\hat{\hat{\mathfrak{g}}}$ is of the form $L(\Lambda + h^\vee \Lambda_0) = F \otimes \bar{L}(\Lambda)$, where $\bar{L}(\Lambda)$ is the irreducible unitary highest weight representation of the affine Kac-Moody algebra $\hat{\mathfrak{g}}$. Explicit formulas for the "top" singular vectors of the Verma module $M(\Lambda + h^\vee \Lambda_0)$ of $\hat{\hat{\mathfrak{g}}}$ and the defining relations of $L(\Lambda + h^\vee \Lambda_0)$ are presented in detail in the Appendix. With the help of the theory developed in Sec.1, we prove in Subsec.2.2 that the SVOA $L_{k,0}$ is rational for positive integral k and that the representations and fusion rules for the SVOA $L_{k,0}$ are in 1-1 correpondence with those for the VOA $\bar{L}_{k,0}$.

In Subsection 3.1 we construct $N = 1$ SVOAs M_c and V_c corresponding to the representations of the Neveu-Schwarz algebra. We then discuss the rationality and the fusion rules of V_c.

In Sec. 4 we construct the SVOAs generated by charged and neutral free fermionic fields. We prove that such an SVOA is rational and has a unique irreducible representation, namely itself.

Acknowledgement. We thank Shun-Jen Cheng for useful discussions.

1 General constructions and theorems

1.1 Definitions

For a rational function $f(z, w)$, with possible poles only at $z = w, z = 0$ and $w = 0$, we denote by $\iota_{z,w} f(z, w)$ the power series expansion of $f(z, w)$ in the domain $| z |>| w |$. Set $\mathbb{Z}_+ = \{0, 1, 2, \ldots\}$, $\mathbb{N} = \{1, 2, 3, \ldots\}$.

A superalgebra is an algebra V with a $\mathbb{Z}_2$-gradation $V = V_{\bar{0}} \oplus V_{\bar{1}}$. Elements in $V_{\bar{0}}$ (*resp.*$V_{\bar{1}}$) are called even (*resp.* odd). Let $\tilde{a}$ be 0 if $a \in V_{\bar{0}}$, and 1 if $a \in V_{\bar{1}}$. The general principle to extend identities in VOAs to SVOAs is the usual one: if in certain formulas of VOAs there are some monomials of vertex operators with interchanged terms, then in the corresponding formulas in SVOAs every interchange of neighboring terms, say a and b, is accompanied by multiplication of the monomial by the factor $(-1)^{\tilde{a}\tilde{b}}$.

Definition 1.1 *A* vertex operator superalgebra *is a $\frac{1}{2}\mathbb{Z}_+$-graded vector space $V = \bigoplus_{n \in \frac{1}{2}\mathbb{Z}_+} V_n$ with a sequence of linear operators $\{a(n) \mid n \in \mathbb{Z}\} \subset End\ V$ associated to every $a \in V$, whose generating series $Y(a, z) = \sum_{n\in\mathbb{Z}} a(n) z^{-n-1} \in (End\ V)\left[\left[z, z^{-1}\right]\right]$, called the* vertex operators *associated to a, satisfy the following axioms:*

Axiom A1 $Y(a, z) = 0$ *iff* $a = 0$.

Axiom A2 *There is a* vacuum *vector, which we denote by 1, such that*

$$Y(1, z) = I_V \ (I_V \text{ is the identity of } End V).$$

Axiom A3 *There is a special element $\omega \in V$ (called the* Virasoro element*), whose vertex operator we write in the form*

$$Y(\omega, z) = \sum_{n\in\mathbb{Z}} \omega(n) z^{-n-1} = \sum_{n\in\mathbb{Z}} L_n z^{-n-2},$$

such that

$$L_0 \mid_{V_n} = nI \mid_{V_n},$$

$$Y(L_{-1}a, z) = \frac{d}{dz}Y(a, z) \text{ for every } a \in V, \tag{1.1}$$

$$[L_m, L_n] = (m-n)L_{m+n} + \delta_{m+n,0}\frac{m^3-m}{12}c, \tag{1.2}$$

where c is some constant in $\mathbb{C}$, which is called the rank *of V.*

Axiom A4 *The* Jacobi identity *holds, i.e.*

$$\begin{aligned} &Res_{z-w}\big(Y(Y(a, z-w)b, w)\iota_{w,z-w}((z-w)^m z^n)\big) \\ &= Res_z\left(Y(a,z)Y(b,w)\iota_{z,w}(z-w)^m z^n\right) \\ &\quad - (-1)^{\tilde{a}\tilde{b}} Res_z\left(Y(b,w)Y(a,z)\iota_{w,z}(z-w)^m z^n\right) \end{aligned}$$

for any $m, n \in \mathbb{Z}$.

An element $a \in V$ is called *homogeneous* of degree n if a is in V_n. In this case we write $\deg a = n$.

Define a natural $\mathbb{Z}_2$-gradation of V by letting

$$V_{\bar{0}} = \bigoplus_{n \in \mathbb{Z}_+} V_n, \quad V_{\bar{1}} = \bigoplus_{n \in \frac{1}{2}+\mathbb{Z}_+} V_n.$$

$V = V_{\bar{0}} + V_{\bar{1}}$. $V_{\bar{0}}$, (*resp.* $V_{\bar{1}}$) is called the even (*resp.*, odd) part of V. Elements in $V_{\bar{0}}$ (*resp.* $V_{\bar{1}}$) are called even (*resp.* odd).

We now introduce the notion of an $N = 1$ SVOA.

Definition 1.2 *V is called an $N = 1$* (NS-type) SVOA *if axiom (A3) is replaced by the following stronger axiom:*

Axiom A3′ *There is a special element $\tau \in V$ (called the* Neveu-Schwarz element*), whose corresponding vertex operator we write in the form*

$$Y(\tau, z) = \sum_{n \in \mathbb{Z}} \tau(n) z^{-n-1} = \sum_{n \in \mathbb{Z}} G_{n+\frac{1}{2}} z^{-n-2},$$

such that the element $\omega := \frac{1}{2}G_{-\frac{1}{2}}\tau$ satisfies (A3), and the commutation relations

$$\begin{aligned} \left[G_{m+\frac{1}{2}}, L_n\right] &= \left(m + \frac{1}{2} - \frac{n}{2}\right) G_{m+n+\frac{1}{2}}, \\ \left[G_{m+\frac{1}{2}}, G_{n-\frac{1}{2}}\right]_+ &= 2L_{m+n} + \frac{1}{3}m(m+1)\delta_{m+n,0}c, \quad m, n \in \mathbb{Z} \end{aligned}$$

also hold.

We list some properties of *SVOAs* which are anologous to those in the VOA case. For more detail see [FLM].

$$[a(n), Y(b,z)]_{\mp} = \sum_{i \geq 0} \binom{n}{i} z^{n-i} Y(a(i)b, z), \tag{1.3}$$

$$[L_0, Y(a,z)] = \left(z\frac{d}{dz} + \deg a\right) Y(a,z),$$

$$[L_{-1}, Y(a,z)] = \frac{d}{dz}Y(a,z), \tag{1.4}$$

$$a(n)V_m \subset V_{m+\deg a-n-1}, \tag{1.5}$$

$$\begin{aligned} Y(a,z)1 &= e^{zL_{-1}}a, \\ Y(a,z)b &= (-1)^{\tilde{a}\tilde{b}}e^{zL_{-1}}Y(b,-z)a, \\ a(n)1 &= 0, \quad \text{for } n \geq 0, \\ a(-n-1)1 &= \frac{1}{n!}L_{-1}^n a \quad \text{for } n \geq 0. \end{aligned}$$

Moreover, $N = 1$ SVOAs have the extra property that:

$$\left[G_{-\frac{1}{2}}, Y(a,z)\right]_{\mp} = Y(G_{-\frac{1}{2}}a, z).$$

Definition 1.3 *Given an SVOA* V, a representation of V (or V-module) *is a* $\frac{1}{2}\mathbb{Z}_+$*-graded vector space* $M = \bigoplus_{n\in\frac{1}{2}\mathbb{Z}_+} M_n$ *and a linear map*

$$\begin{aligned} V &\longrightarrow (End\ M)\left[\left[z, z^{-1}\right]\right], \\ a &\longmapsto Y_M(a,z) = \textstyle\sum_{n\in\mathbb{Z}} a(n)z^{-n-1}, \end{aligned}$$

satisfying

Axiom R1 $a(n)M_m \subset M_{m+\deg a-n-1}$ *for every homogeneous element* a.

Axiom R2 $Y_M(1,z) = I_M$, *and setting* $Y_M(\omega,z) = \sum_{n\in\mathbb{Z}} L_n z^{-n-2}$, *we have*

$$\begin{aligned} [L_m, L_n] &= (m-n)L_{m+n} + \delta_{m+n,0}\frac{m^3-m}{12}c, \\ Y_M(L_{-1}a, z) &= \frac{d}{dz}Y_M(a,z) \ \textit{for every } a \in V. \end{aligned}$$

Axiom R3 *The* Jacobi identity *holds, i.e.*

$$\begin{aligned} &Res_{z-w}\left(Y_M\left(Y(a,z-w)b,w\right)\iota_{w,z-w}\left((z-w)^m z^n\right)\right) \\ &\quad = Res_z\left(Y_M(a,z)Y_M(b,w)\iota_{z,w}(z-w)^m z^n\right) \\ &\qquad - (-1)^{\tilde{a}\tilde{b}}Res_z\left(Y_M(b,w)Y_M(a,z)\iota_{w,z}(z-w)^m z^n\right) \end{aligned}$$

for any $m, n \in \mathbb{Z}$.

Definition 1.4 *Given an* $N = 1$ *SVOA* V, M *is called* a representation of V *if axiom (R2) is replaced by the following stronger axiom:*

Axiom R2′ *Set* $Y_M(\tau,z) = \sum_{n\in\mathbb{Z}} G_{n+\frac{1}{2}}z^{-n-2}$ *and* $\omega := \frac{1}{2}G_{-\frac{1}{2}}\tau$. *Then* ω *satisfies (R2), and the commutation relations*

$$\begin{aligned} \left[G_{m+\frac{1}{2}}, L_n\right] &= \left(m+\frac{1}{2}-\frac{n}{2}\right)G_{m+n+\frac{1}{2}}, \\ \left[G_{m+\frac{1}{2}}, G_{n-\frac{1}{2}}\right]_+ &= 2L_{m+n} + \frac{1}{3}m(m+1)\delta_{m+n,0}c, \quad m,n \in \mathbb{Z} \end{aligned}$$

also hold.

The notions of submodules, quotient modules, submodules generated by a subset, direct sums, irreducible modules, completely reducible modules, etc., can be introduced in the usual way. As a module over itself, V is called the *adjoint module*. A submodule of the adjoint module is called an *ideal* of V. Given an ideal I in V such that $1 \notin I$, $\omega \notin I$, the quotient V/I admits a natural SVOA structure.

Definition 1.5 *A SVOA is called* rational *if it has finitely many irreducible modules and every module is a direct sum of irreducibles.*

We will now extend the definition of intertwining operators and fusion rules of representations of VOAs ([FHL]) to SVOAs.

For simplicity, we will only define an intertwining operator for V-modules $M^i = \oplus_{n\in\frac{1}{2}\mathbb{Z}_+} M^i(n)$, $i = 1, 2, 3$, satisfying $L_0 \mid_{M^i(n)} = (h_i + n) I \mid_{M^i(n)}$, for some complex numbers h_1, h_2, h_3. We define a $\mathbb{Z}_2$-gradation of M^i by letting $\tilde{v} = 0$ if $v \in M^i(n)$, $n \in \mathbb{Z}$; $\tilde{v} = 1$ if $v \in M^i(n)$, $n \in \frac{1}{2} + \mathbb{Z}$.

Definition 1.6 *Under the above assumptions, an intertwining operator of type* $\binom{M^3}{M^1\ M^2}$ *is a linear map*

$$I(\cdot, z) : v \mapsto \sum_{k\in I} v(n) z^{-n-1+(h_3-h_1-h_2)}, \quad v \in M^1, \quad v(n) \in Hom_{\mathbb{C}}\left(M^2, M^3\right)$$

satisfying

Axiom I1 *For homogeneous* $v \in M^1$,

$$I(L_{-1}v, z) = \frac{d}{dz} I(v, z) \ \ \textit{for every } v \in M^1,$$

Axiom I2 *For any* $a \in V, v \in M^1$, *and* $m, n \in \mathbb{Z}$,

$$\begin{aligned} &Res_{z-w}\left(I(Y(a, z-w)v, w)\iota_{w,z-w}\left((z-w)^m z^n\right)\right) \\ = \ &Res_z\left(Y(a, z)I(v, w)\iota_{z,w}(z-w)^m z^n\right) \\ &-(-1)^{\tilde{a}\tilde{v}} Res_z\left(I(v, w)Y(a, z)\iota_{w,z}(z-w)^m z^n\right). \end{aligned}$$

We denote by $I\binom{M^3}{M^1\ M^2}$ the vector space of intertwining operators of type $\binom{M^3}{M^1\ M^2}$.

An immediate consequence of this definition is that for homogeneous $v \in M^1$,

$$v(n) M^2_m \subset M^3_{m+\deg v-n-1},$$

where $\deg v = k$ means that $v \in M^1_k$.

We now assume that V is a rational SVOA and $\{M^i, i \in J\}$ is the complete set of the irreducible modules of V. Denote by N^{ij}_k the dimension of the vector space $I\binom{M^k}{M^i\ M^j}$. We define the fusion rules as the formal product rules

$$M^i \times M^j = \sum_{k\in J} N^{ij}_k M^k.$$

1.2 The associative algebra A(V) and related theorems

Definition 1.7 *We define bilinear maps $*: V\times V \to V$, $\circ : V \times V \to V$ as follows. For homogeneous a, b, let*

$$a * b = \begin{cases} Res_z\left(Y(a,z)\frac{(z+1)^{\deg a}}{z}b\right), & \text{if } a,b \in V_{\bar{0}},\\ 0, & \text{if } a \text{ or } b \in V_{\bar{1}}.\end{cases}$$

$$a \circ b = \begin{cases} Res_z\left(Y(a,z)\frac{(z+1)^{\deg a}}{z^2}b\right), & \text{for } a \in V_{\bar{0}}\\ Res_z\left(Y(a,z)\frac{(z+1)^{\deg a-\frac{1}{2}}}{z}b\right), & \text{for } a \in V_{\bar{1}}.\end{cases}$$

Extend to $V \times V$ bilinearity, denote by $O(V) \subset V$ the linear span of elements of the form $a \circ b$, and by $A(V)$ the quotient space $V/O(V)$.

Remark 1.1 *1) $O(V)$ is a $\mathbb{Z}_2$-graded subspace of V.*

2) If $a \in V_{\bar{1}}$, then

$$a \circ 1 = Res_z\left(Y(a,z)\frac{(z+1)^{\deg a-\frac{1}{2}}}{z}1\right) = a.$$

Hence $O(V) = O_{\bar{0}}(V) + V_{\bar{1}}$, where $O_{\bar{0}}(V) = O(V) \cap V_{\bar{0}}$. Thus $A(V) = V_{\bar{0}}/O_{\bar{0}}(V)$. Denote by $O_e(V)$ (resp. $O_d(V)$) the linear span of the elements $a \circ b$ for $a, b \in V_{\bar{0}}$ (resp. $V_{\bar{1}}$). The intersection $O_e(V) \cap O_d(V)$ need not be empty.

It is convenient to introduce an equivalence relation $\sim$ as follows. For $a, b \in V$, $a \sim b$ means $a - b \equiv 0 \bmod O(V)$. For $f, g \in End\ V$, $f \sim g$ means $f \cdot c \sim g \cdot c$ for any $c \in V$. Let $[a]$ to denote the image of a in V under the projection of V onto $A(V)$.

Lemma 1.1 *1) $L_{-1}a + L_0 a \sim 0$ if $a \in V_{\bar{0}}$.*

2) For every homogeneous element $a \in V$, and $m \geq n \geq 0$, one has

$$Res_z\left((Y(a,z)\frac{(z+1)^{\deg a+n}}{z^{2+m}}\right) \sim 0, \quad \text{if } a \in V_{\bar{0}}.$$

$$Res_z\left(Y(a,z)\frac{(z+1)^{\deg a+n-\frac{1}{2}}}{z^{1+m}}\right) \sim 0, \quad \text{if } a \in V_{\bar{1}}.$$

3) For any homogeneous element $a, b \in V_{\bar{0}}$, one has

$$a * b \sim Res_z\left(Y(b,z)\frac{(z+1)^{\deg b-1}}{z}a\right).$$

Proof. Noting that $V_{\bar{0}}$ is a vertex operator algebra, we see that 1), the first part of 2) and 3) are the same as Lemma 2.1.1, 2.1.2 and 2.1.3 in [Z]. The proof of the second part of 2) is similar to that of the first part. □

The following theorem is an analog of Theorem 2.1.1 in [Z].

Theorem 1.1 *1) $O(V)$ is a two-sided ideal of V under the multiplication $*$. Moreover, the quotient algebra $(A(V), *)$ is associative.*

2) $[\mathbf{1}]$ is the unit element of the algebra $A(V)$.

3) $[\omega]$ is in the center of $A(V)$.

4) $A(V)$ has a filtration $A_0(V) \subset A_1(V) \subset \cdots$, where $A_n(V)$ is the image of $\oplus_{i\in\frac{1}{2}\mathbb{Z}_+, i\le n} V_i$.

Sketch of a proof. To prove 1), it is enough to prove the following relations:

$$\begin{aligned} O_{\bar{0}}(V) * V &\subset O(V),\\ V_{\bar{0}} * O_{\bar{0}}(V) &\subset O(V),\\ (a*b)*c - a*(b*c) &\in O(V). \end{aligned}$$

By the definition of the operation $*$ and Remark 1.1, it suffices to prove that for homogeneous a, b, c one has

$$(a\circ b)*c \in O(V) \text{ for } a,b,c\in V_{\bar{0}}, \tag{1.6}$$

$$a*(b\circ c) \in O(V) \text{ for } a,b,c\in V_{\bar{0}}, \tag{1.7}$$

$$(a\circ b)*c \in O(V) \text{ for } a,b\in V_{\bar{1}}, \tag{1.8}$$

$$a*(b\circ c) \in O(V) \text{ for } a\in V_{\bar{0}}, b,c\in V_{\bar{1}}, \tag{1.9}$$

$$(a*b)*c - a*(b*c) \in O(V) \text{ for } a,b,c\in V_{\bar{0}}. \tag{1.10}$$

The proofs of (1.6), (1.7) and (1.10) are the same as in the VOA cases (see the proof of Theorem 2.1.1 in [Z]).

To prove (1.8), for $a, b \in V_{\bar{1}}$, $c \in V$ homogeneous, we have

$$\begin{aligned}
&(a\circ b)*c\\
&= Res_z\left(Y(a,z)\frac{(z+1)^{\deg a-\frac{1}{2}}}{z}b\right)*c\\
&= \sum_{i=0}^{\deg a-\frac{1}{2}}\binom{\deg a-\frac{1}{2}}{i}(a(i-1)b)*c\\
&= \sum_{i=0}^{\deg a-\frac{1}{2}}\binom{\deg a-\frac{1}{2}}{i}Res_w\left(Y(a(i-1)b,w)\frac{(w+1)^{\deg a+\deg b-i}}{w}c\right)\\
&= \sum_{i=0}^{\deg a-\frac{1}{2}}\binom{\deg a-\frac{1}{2}}{i}Res_w Res_{z-w}\\
&\quad\times\left(Y(Y(a,z-w)b,w)(z-w)^{i-1}\frac{(w+1)^{\deg a+\deg b-i}}{w}c\right)\\
&= Res_w Res_{z-w}\left(Y(Y(a,z-w)b,w)\frac{(z+1)^{\deg a-\frac{1}{2}}(w+1)^{\deg b+\frac{1}{2}}}{w(z-w)}c\right)\\
&= Res_z Res_w\left(Y(a,z)Y(b,w)\frac{(z+1)^{\deg a-\frac{1}{2}}(w+1)^{\deg b+\frac{1}{2}}}{w(z-w)}c\right)
\end{aligned}$$

$$
\begin{aligned}
&+ Res_w Res_z \left(Y(b,w)Y(a,z) \frac{(z+1)^{\deg a - \frac{1}{2}}(w+1)^{\deg b + \frac{1}{2}}}{w(z-w)} c \right) \\
= & \sum_{i \in \mathbb{Z}_+} Res_z Res_w \left(Y(a,z)Y(b,w) z^{-1-i} w^i \frac{(z+1)^{\deg a - \frac{1}{2}}(w+1)^{\deg b + \frac{1}{2}}}{w} c \right) \\
&- \sum_{i \in \mathbb{Z}_+} Res_w Res_z \left(Y(b,w)Y(a,z) w^{-1-i} z^i \frac{(z+1)^{\deg a - \frac{1}{2}}(w+1)^{\deg b + \frac{1}{2}}}{w} c \right).
\end{aligned}
$$

By Lemma 1.1 the right hand side of the last identity is in $O(V)$.

To prove (1.9), for $a \in V_{\bar{0}}, b, c \in V_{\bar{1}}$ homogeneous, we have

$$
\begin{aligned}
&a * (b \circ c) - b \circ (a * c) \\
= & Res_z \left(Y(a,z) \frac{(z+1)^{\deg a}}{z} \right) Res_w \left(Y(b,w) \frac{(w+1)^{\deg b - \frac{1}{2}}}{w} c \right) \\
&- Res_w \left(Y(b,w) \frac{(w+1)^{\deg b - \frac{1}{2}}}{w} \right) Res_z \left(Y(a,z) \frac{(z+1)^{\deg a}}{z} c \right) \\
= & Res_w Res_{z-w} \left(Y(Y(a, z-w)b, w) \frac{(z+1)^{\deg a}}{z} \frac{(w+1)^{\deg b - \frac{1}{2}}}{w} c \right) \\
= & \sum_{i=0}^{\deg a} \sum_{j \in \mathbb{Z}_+} \binom{\deg a}{i} Res_w \left(Y(a(i+j)b, w)(-1)^j \frac{(w+1)^{\deg a + \deg b - i - \frac{1}{2}}}{w^{j+2}} c \right).
\end{aligned}
$$

Since $\deg(a(i+j)b) = \deg a + \deg b - i - j - 1$, and $a(i+j)b \in V_{\bar{1}}$, by Lemma 1.1, the right-hand side of the last identity is in $O(V)$. The second term of the left-hand side is also in $O(V)$ by definition. Then so is the first term.

The proof of statements 2), 3) and 4) is the same as in the VOA case. (For details see the proof of Theorem 2.1.1 in [Z]). □

The following proposition follows from the definition of $A(V)$.

Proposition 1.1 *Let I be an ideal of V with the $\mathbb{Z}_2$-gradation $I_{\bar{0}} \oplus I_{\bar{1}}$ consistent with that of V. Assume $1 \notin I, \omega \notin I$. Then the associative algebra $A(V/I)$ is isomorphic to $A(V)/[I]$, where $[I]$ is the image of I in $A(V)$.*

For any homogeneous $a \in V_{\bar{0}}$ we define $o(a) = a(\deg a - 1)$ and extend this map linearly to $V_{\bar{0}}$. It follows from (1.5) that $o(a)M_n \subset M_n$. In particular, $o(a)$ maps M_0 into itself. We may assume that $M_0 \neq 0$ without loss of generality.

Theorem 1.2 *Let $M = \bigoplus_{n \in \frac{1}{2}\mathbb{Z}_+} M_n$ be a V-module. Then M_0 is an $A(V)$-module defined as follows: for $[a] \in A(V)$, let $a \in V_{\bar{0}}$ be a preimage of $[a]$. Then $[a]$ acts on M_0 as $o(a)$.*

Proof. An equivalent way to state this theorem is that for $a, b \in V_{\bar{0}}$, $o(a)o(b) \mid_{M_0} = o(a * b) \mid_{M_0}$, and for $c \in O(V) = O_d(V) + O_e(V)$, $o(c) \mid_{M_0} = 0$. We only need to prove that $o(c) \mid_{M_0} = 0$ for $c \in O_d(V)$ since $V_{\bar{0}}$ is a vertex operator algebra and so the rest of the statements above holds (For details see Theorem 2.1.2 and its proof in [Z]).

Given $a, b \in V_{\bar{1}}$ homogeneous, we have

$$\begin{aligned}
& o(a \circ b) \\
= \;& o\left(Res_z\left(Y(a,z)\frac{(z+1)^{\deg a-\frac{1}{2}}}{z}b\right)\right) \\
= \;& \sum_{i=0}^{\deg a-\frac{1}{2}} \binom{\deg a-\frac{1}{2}}{i} o\left(a(i-1)b\right) \\
= \;& \sum_{i=0}^{\deg a-\frac{1}{2}} \binom{\deg a-\frac{1}{2}}{i} \left(a(i-1)b\right)\left(\deg a+\deg b-i-1\right) \\
= \;& Res_w Res_{z-w} \sum_{i=0}^{\deg a-\frac{1}{2}} \binom{\deg a-\frac{1}{2}}{i} \times \\
& \times \left(Y((a,z-w)b,w)(z-w)^{i-1}w^{\deg a+\deg b-i-1}\right) \\
= \;& Res_w Res_{z-w}\left(Y((a,z-w)b,w)\frac{z^{\deg a-\frac{1}{2}}w^{\deg b-\frac{1}{2}}}{z-w}\right) \\
= \;& Res_z Res_w\left(Y(a,z)Y(b,w)\frac{z^{\deg a-\frac{1}{2}}w^{\deg b-\frac{1}{2}}}{z-w}\right) \\
& + Res_w Res_z\left(Y(b,w)Y(a,z)\frac{z^{\deg a-\frac{1}{2}}w^{\deg b-\frac{1}{2}}}{z-w}\right) \\
= \;& \sum_{i\in\mathbb{Z}_+} Res_z Res_w\left(Y(a,z)Y(b,w)z^{\deg a-i-\frac{3}{2}}w^{\deg b-\frac{1}{2}+i}\right) \\
& - \sum_{i\in\mathbb{Z}_+} Res_w Res_z\left(Y(b,w)Y(a,z)z^{\deg a+i-\frac{1}{2}}w^{\deg b-i-\frac{3}{2}}\right) \\
= \;& \sum_{i\in\mathbb{Z}_+} a(\deg a-i-\frac{3}{2})b(\deg b+i-\frac{1}{2}) \\
& - \sum_{i\in\mathbb{Z}_+} b(\deg b-i-\frac{3}{2})a(\deg a+i-\frac{1}{2}).
\end{aligned}$$

The right-hand side of the above identities acting on M_0 is 0 since

$$a(\deg a+i-\frac{1}{2})\mid_{M_0} = b(\deg b+i-\frac{1}{2})\mid_{M_0} = 0.$$

□

Theorem 1.3 *Given an $A(V)$-module (W,π), there exists a V-module $M = \bigoplus_{n\in\frac{1}{2}\mathbb{Z}_+} M_n$ such that the $A(V)$-modules M_0 and W are isomorphic. Moreover, this gives a bijective correspondence between the set of irreducible $A(V)$-modules and the set of irreducible V-modules.*

Sketch of a proof. First we have the following recurrent formula for n-correlation functions on $\langle M_0, (M_0)^*\rangle$ for a given V-module $M = \oplus_{i\in\frac{1}{2}\mathbb{Z}_+} M_i$, where M_0^* is the dual space of M_0. (The proof is similar to that of Lemma 2.2.1 in [Z].) Given $v \in M_0$, $v' \in M_0^*$, and homogeneous $a_1 \in V$, we have

$$\langle v',\, Y(a_1,z_1)Y(a_2,z_2)\cdots Y(a_m,z_m)v\rangle$$
$$= \begin{cases} \displaystyle\sum_{k=2}^{m}\sum_{i\in\mathbb{Z}_+}(-1)^{(\tilde{a}_2+\cdots+\tilde{a}_{k-1})}F_{\deg a_1-\frac{1}{2},i}(z_1,z_k) \\ \qquad \times \langle v',Y(a_2,z_2)\cdots Y(a_1(i)a_k,z_k)\cdots Y(a_m,z_m)v\rangle \text{ if } a_1\in V_{\bar{1}}, \\ \displaystyle\sum_{k=2}^{m}\sum_{i\in\mathbb{Z}_+}F_{\deg a_1,i}(z_1,z_k)\times \\ \qquad \times \langle v',Y(a_2,z_2)\cdots Y(a_1(i)a_k,z_k)\cdots Y(a_m,z_m)v\rangle \\ \qquad + z_1^{-\deg a_1}\langle a_1(\deg a_1-1)^*v',Y(a_2,z_2)\cdots Y(a_m,z_m)v\rangle \text{ if } a_1\in V_{\bar{0}}, \end{cases}$$

where $F_{\deg a,i}$ is defined by

$$\begin{aligned} F_{n,i}(z,w) &= \sum_{j\in\mathbb{Z}_+}\binom{n+j}{i}z^{-n-j}w^{n+j-i} \\ &= \iota_{z,w}\left(z^{-n}\frac{1}{i!}\left(\frac{d^i}{dw^i}\right)\frac{w^n}{z-w}\right). \end{aligned}$$

This recurrent formula means that the n-correlation functions on $\langle M_0,(M_0)^*\rangle$ are determined by the $A(V)$-module structure on M_0. The completion of the proof of this theorem is similar to that in Theorem 2.2.1 in [Z]. □

Remark 1.2 *Thus we have a functor from the category of V-modules to the category of $A(V)$-modules which is bijective on the sets of irreducibles.*

1.3 Fusion rules

In this subsection, to generalize the construction of [FZ], we define a bimodule $A(M)$ of $A(V)$ for every V-module M. We then give a description of the fusion rules in terms of $A(M)$. The proofs are only sketched.

Definition 1.8 *For a V-module M, we define bilinear operations $a * v$ and $v * a$, for $a \in V$ homogeneous and $v \in M$, as follows*

$$a * v = Res_z\left(Y(a,z)\frac{(z+1)^{\deg a}}{z}v\right),\ \textit{for } a\in V_{\bar{0}}, \tag{1.11}$$

$$v * a = Res_z\left(Y(a,z)\frac{(z+1)^{\deg a-1}}{z}v\right),\ \textit{for } a\in V_{\bar{0}}, \tag{1.12}$$

$$a * v = 0,\quad v * a = 0,\ \textit{for } a\in V_{\bar{1}}$$

and extend linearly to V. We also define $O(M) \subset M$ to be the linear span of elements of the forms

$$Res_z\left(Y(a,z)\frac{(z+1)^{\deg a}}{z^2}v\right),\ \textit{for } a\in V_{\bar{0}} \textit{ and}$$

$$Res_z\left(Y(a,z)\frac{(z+1)^{\deg a-\frac{1}{2}}}{z}v\right),\ \textit{for } a\in V_{\bar{1}}.$$

Let $A(M)$ be the quotient space $M/O(M)$.

We have the following theorem which is an analogue of Theorem 1.5.1 in [Z].

Theorem 1.4 *$A(M)$ is an $A(V)$-bimodule with the left action of $A(V)$ defined by (1.11) and the right action by (1.12). Moreover the left and right action of $A(V)$ commute with each other.*

Sketch of a proof. By a similar argument to Lemma 1.1, we see that

$$Res_z\left(Y(a,z)\frac{(z+1)^{\deg a+n}}{z^{2+m}}v\right) \in O(M), \text{ for } a\in V_{\bar{0}},$$

$$Res_z\left(Y(a,z)\frac{(z+1)^{\deg a+n-\frac{1}{2}}}{z^{1+m}}v\right) \in O(M), \text{ for } a\in V_{\bar{1}},$$

for $m\geq n\geq 0$, $v\in M$.

Recall that $O(V)=O_d(V)+O_e(V)$. To prove the theorem, we need to check that

$$O_d(V)*v\subset O(M),\qquad v*O_d(V)\subset O(M), \tag{1.13}$$

$$O_e(V)*v\subset O(M),\qquad v*O_e(V)\subset O(M), \tag{1.14}$$

$$a*O(M)\subset O(M),\qquad O(M)*a\subset O(M), \tag{1.15}$$

$$(a*b)*v-a*(b*v) \in O(M), \tag{1.16}$$

$$(v*a)*b-v*(a*b) \in O(M), \tag{1.17}$$

$$(a*v)*b-a*(v*b) \in O(M). \tag{1.18}$$

The proof of (1.13) is similar to that of Theorem 1.1. The proofs of (1.14), (1.15), (1.16), (1.17), and (1.18) are similar to those in [Z]. □

Consider left V-modules $M^i=\oplus_{n\in\frac{1}{2}\mathbb{Z}_+}M^i(n)$, $i=1,2,3$. Note that $M^2(0)$ is a left module over $A(V)$, $\left(M^3(0)\right)^*$ is a right module over $A(V)$, and $A(M^1)$ is a bimodule over $A(V)$. Hence we can consider the tensor product $M^3(0)^*\otimes_{A(V)} A(M^1)\otimes_{A(V)} M^2(0)$ of $A(V)$-modules.

The following theorem is an analogue of Theorems 1.5.2 and 1.5.3 in [FZ].

Theorem 1.5 *Let V be a rational VOA and $M^i=\sum_{n\in\frac{1}{2}\mathbb{Z}_+}M^i(n)(i=1,2,3)$ be V-modules, satisfying $L_0\mid_{M^i(n)}=(h_i+n)I\mid_{M^i(n)}$, for some complex numbers h_1,h_2,h_3.*

1) *Let $I(\cdot,z)$ be an intertwining operator of type $\binom{M^3}{M^1\ M^2}$. Then $\langle v_3',o(v_1)v_2\rangle$ defines a linear functional f_I on $M^3(0)^*\otimes_{A(V)} A(M^1)\otimes_{A(V)} M^2(0)$, where $v_3'\in M^3(0)^*$, $v_1\in M^1$, $v_2\in M^2$,*

2) *The map $I\mapsto f_I$ given in 1) defines an isomorphism of vector spaces $I\binom{M^3}{M^1\ M^2}$ and $\left(M^3(0)^*\otimes_{A(V)} A(M^1)\otimes_{A(V)} M^2(0)\right)^*$ if M^i $(i=1,2,3)$ are irreducible.*

Proof. The argument is similar to that in Theorem 1.3. □

As a consequence, we obtain the following proposition, which is an anologue of Proposition 1.5.4 in [FZ].

Proposition 1.2 *1) Given a V-module M and a submodule M^1 of M, then the image $A(M^1)$ of M^1 in $A(M)$, is a submodule of $A(V)$-bimodule $A(M)$, and the quotient $A(M)/A(M^1)$ is isomorphic to the bimodule $A(M/M^1)$ corresponding to the quotient V-module M/M^1.*

2) If I is an ideal of V, $1 \notin I$, $\omega \notin I$, and $I \cdot M \subset M^1$, then $A(V/I)$-bimodule $A(M)/A(M^1)$ is isomorophic to the $A(M/M^1)$.

Remark 1.3 *One can also consider the pre-SVOA (i.e., the SVOA which may not admit a Virasoro element). Similarly to the VOA case, one can still define the associative algebra $A(V)$ and the $A(V)$-module $A(M)$ for any V-module M [L]. Theorems 1.3 and 1.5 are valid for the pre-SVOAs.*

2 SVOA associated to representations of affine Kac-Moody superalgebras

2.1 SVOA structures on $M_{k,0}$ and $L_{k,0}$

In this subsection, we construct the SVOAs associated to representations of affine Kac-Moody superalgebras which are analogous to the construction of VOAs associated to representations of affine algebras [FZ]. First let us recall some basic notions of affine Kac-Moody (super)algebras. Given a simple finite-dimensional Lie algebra $\mathfrak{g}$ of rank l over $\mathbb{C}$, we fix a Cartan subalgebra $\mathfrak{h}$, a root system $\Delta \subset \mathfrak{h}^*$ and a set of positive roots $\Delta_+ \subset \Delta$. Let $\mathfrak{g} = \mathfrak{h} \bigoplus \left(\bigoplus_{\alpha\in\Delta} \mathfrak{g}_\alpha\right)$ be the root space decomposition of $\mathfrak{g}$. Let e_i, f_i, h_i $(i = 1, \ldots, l)$ be the corresponding Chevalley generators. Denote by θ the highest root and normalize the Killing form

$$(\, , \,) : \mathfrak{g} \times \mathfrak{g} \rightarrow \mathbb{C}$$

by the condition $(\theta, \theta) = 2$. Let σ be the antilinear anti-involution of $\mathfrak{g}$. We choose $f_\theta \in \mathfrak{g}_{-\theta}$ so that $(f_\theta, \sigma(f_\theta)) = 1$, and set $e_\theta = \sigma(f_\theta)$. We denote by r_α the reflection with respect to $\alpha \in \Delta$ in the Weyl group $W \in GL(\mathfrak{h})$ of $\mathfrak{g}$.

The affine Kac-Moody superalgebra (of NS type) is then defined by

$$\hat{\hat{\mathfrak{g}}} = \mathfrak{g} \bigotimes \mathbb{C}\left[t, t^{-1}, \xi\right] \bigoplus \mathbb{C}\mathbf{k} \bigoplus \mathbb{C}d$$

with the following commutation relations

$$
\begin{aligned}
&(2.1) & [a(m), b(n)] &= [a,b](m+n) + m\delta_{m+n,0}(a,b)\mathbf{k}, \\
&(2.2) & \left[\bar{a}(m), \bar{b}(n)\right]_+ &= \delta_{m+n+1,0}(a,b)\mathbf{k}, \\
&(2.3) & \left[a(m), \bar{b}(n)\right] &= \overline{[a,b]}(m+n), \\
&(2.4) & [\mathbf{k}, a(m)] &= 0, \\
&(2.5) & [d, a(m)] &= ma(m), \\
&(2.6) & [d, \bar{a}(m)] &= (m + \frac{1}{2})\bar{a}(m),
\end{aligned}
$$

where $a, b \in \mathfrak{g}$, $m, n \in \mathbb{Z}$, $a(m) := a \otimes t^m$, $\bar{a}(m) := a \otimes \xi t^m$.

Let

$$\begin{aligned}
\bar{\mathfrak{g}} &= \mathfrak{g}\bigotimes\xi, \\
\hat{\hat{\mathfrak{g}}}_+ &= \mathfrak{g}\bigotimes t\mathbb{C}[t]\bigoplus\bar{\mathfrak{g}}\bigotimes\mathbb{C}[t], \\
\hat{\hat{\mathfrak{g}}}_- &= \mathfrak{g}\bigotimes t^{-1}\mathbb{C}\left[t^{-1}\right]\bigoplus\bar{\mathfrak{g}}\bigotimes t^{-1}\mathbb{C}\left[t^{-1}\right].
\end{aligned}$$

Then $\hat{\hat{\mathfrak{g}}}_+$ and $\hat{\hat{\mathfrak{g}}}_-$ are subalgebras of $\hat{\hat{\mathfrak{g}}}$ and

$$\hat{\hat{\mathfrak{g}}} = \hat{\hat{\mathfrak{g}}}_+\bigoplus\hat{\hat{\mathfrak{g}}}_-\bigoplus\mathfrak{g}\bigoplus\mathbb{C}\mathbf{k}\bigoplus\mathbb{C}d.$$

Here we identify $\mathfrak{g}\otimes 1$ with $\mathfrak{g}$. We let

$$\hat{\mathfrak{h}} = \mathfrak{h}\bigoplus\mathbb{C}\mathbf{k}\bigoplus\mathbb{C}d,$$

and extend the Killing form on $\mathfrak{h}$ to $\hat{\mathfrak{h}}$ by letting $(\mathbf{k}, d) = 1$, $(\mathbf{k}, \mathbf{k}) = 0$, $(d, d) = 0$, $(\mathbb{C}\mathbf{k} + \mathbb{C}d, \mathfrak{h}) = 0$. We identify $\hat{\mathfrak{h}}^*$ with $\hat{\mathfrak{h}}$ using this bilinear form on $\hat{\mathfrak{h}}$.

Given a $\mathfrak{g}$-module V and a complex number k, we can define the induced module $\tilde{V}_k$ over $\hat{\hat{\mathfrak{g}}}$ as follows: V can be viewed as a module over $\hat{\hat{\mathfrak{g}}}_+ + \mathfrak{g} + \mathbb{C}\mathbf{k} + \mathbb{C}d$ by letting $(\hat{\hat{\mathfrak{g}}}_+ \bigoplus \mathbb{C}d)V = 0$ and $\mathbf{k} = (k + h^\vee)\, I\,|_V$. Then we let

$$\tilde{V}_k = \mathfrak{U}(\hat{\hat{\mathfrak{g}}}) \bigotimes_{\mathfrak{U}\left(\hat{\hat{\mathfrak{g}}}_+ + \mathfrak{g} + \mathbb{C}\mathbf{k} + \mathbb{C}d\right)} V.$$

Here and further $\mathfrak{U}(\mathfrak{A})$ denotes the universal enveloping algebra of a Lie (super)algebra $\mathfrak{A}$. In particular for any $\lambda \in \mathfrak{h}^*$, we let $L(\lambda)$ be the irreducible highest weight $\mathfrak{g}$-module with highest weight λ, and denote the $\hat{\hat{\mathfrak{g}}}$-module $\tilde{L}(\lambda)_k$ by $M_{k,\lambda}$. Let $J_{k,\lambda}$ be the maximal proper submodule of the $\hat{\hat{\mathfrak{g}}}$-module $M_{k,\lambda}$. Denote $M_{k,\lambda}/J_{k,\lambda}$ by $L_{k,\lambda}$. Note that if $\lambda = 0$, $L(0)$ is the trivial $\mathfrak{g}$-module $\mathbb{C}$ and $M_{k,0} \cong \mathfrak{U}\left(\hat{\hat{\mathfrak{g}}}_-\right)$ as $\hat{\hat{\mathfrak{g}}}_-$-modules.

Define a $\frac{1}{2}\mathbb{Z}$-gradation of $\hat{\hat{\mathfrak{g}}}$ by the eigenvalues of $-d$:

$$\deg \mathbf{k} = 0, \quad \deg a(n) = -n, \quad \deg \bar{a}(n) = -n - \frac{1}{2}, \quad a \in \mathfrak{g}.$$

This induces $\frac{1}{2}\mathbb{Z}$-gradations of $\mathfrak{U}(\hat{\hat{\mathfrak{g}}}), \mathfrak{U}(\hat{\hat{\mathfrak{g}}}_-)$ and $\frac{1}{2}\mathbb{Z}_+$-gradations of $M_{k,\lambda}$ if we let the degree of the highest weight of $L(\lambda)$ to be zero. We denote the gradation decompositions by

$$\begin{aligned}
\mathfrak{U}(\hat{\hat{\mathfrak{g}}}) &= \bigoplus_{n\in\frac{1}{2}\mathbb{Z}} \mathfrak{U}(\hat{\hat{\mathfrak{g}}})(n), \\
\mathfrak{U}(\hat{\hat{\mathfrak{g}}}_-) &= \bigoplus_{n\in\frac{1}{2}\mathbb{Z}} \mathfrak{U}(\hat{\hat{\mathfrak{g}}}_-)(n), \\
M_{k,\lambda} &= \bigoplus_{n\in\frac{1}{2}\mathbb{Z}_+} M_{k,\lambda}(n),
\end{aligned}$$

where $M_{k,\lambda}(n) = \mathfrak{U}\left(\hat{\hat{\mathfrak{g}}}_-\right)(n)L(\lambda)$.

Define a topological completion $\hat{\mathfrak{U}}\left(\hat{\hat{\mathfrak{g}}}\right)$ of $\mathfrak{U}\left(\hat{\hat{\mathfrak{g}}}\right)$ as follows. Let

$$\mathfrak{U}\left(\hat{\hat{\mathfrak{g}}}\right)_n^m = \sum_{i\leq m, i\in\frac{1}{2}\mathbb{Z}} \mathfrak{U}\left(\hat{\hat{\mathfrak{g}}}\right)_{n-i} \mathfrak{U}\left(\hat{\hat{\mathfrak{g}}}\right)_i, \quad \text{for } m \in \frac{1}{2}\mathbb{Z}.$$

It is easy to see that

$$\mathfrak{U}\left(\hat{\hat{\mathfrak{g}}}\right)_n^{m+\frac{1}{2}} \subset \mathfrak{U}\left(\hat{\hat{\mathfrak{g}}}\right)_n^m, \quad \bigcap_{m\in\frac{1}{2}\mathbb{Z}} \mathfrak{U}\left(\hat{\hat{\mathfrak{g}}}\right)_n^m = 0, \quad \bigcup_{m\in\frac{1}{2}\mathbb{Z}} \mathfrak{U}\left(\hat{\hat{\mathfrak{g}}}\right)_n^m = \mathfrak{U}\left(\hat{\hat{\mathfrak{g}}}\right)_n.$$

We take $\{\mathfrak{U}\left(\hat{\hat{\mathfrak{g}}}\right)_n^m, m \in \frac{1}{2}\mathbb{Z}\}$ for a fundamental neighborhood system of $\mathfrak{U}\left(\hat{\hat{\mathfrak{g}}}\right)_n$, and denote the corresponding completion by $\tilde{\mathfrak{U}}\left(\hat{\hat{\mathfrak{g}}}\right)_n$. We let

$$\tilde{\mathfrak{U}}\left(\hat{\hat{\mathfrak{g}}}\right) = \bigoplus_{n\in\frac{1}{2}\mathbb{Z}} \tilde{\mathfrak{U}}\left(\hat{\hat{\mathfrak{g}}}\right)_n.$$

Let $\langle \mathbf{k} - (k + h^\vee)\rangle$ be the two-sided ideal of the associative superalgebra $\tilde{\mathfrak{U}}\left(\hat{\hat{\mathfrak{g}}}\right)$ generated by the element $\mathbf{k} - (k + h^\vee)$. Denote $\tilde{\mathfrak{U}}\left(\hat{\hat{\mathfrak{g}}}\right)/\langle \mathbf{k} - (k + h^\vee)\rangle$ by $\tilde{\mathfrak{U}}\left(\hat{\hat{\mathfrak{g}}}, k\right)$.

A $\hat{\hat{\mathfrak{g}}}$-module M is called *restricted* if for any fixed $v \in M$, $x(n)v = 0$ for $n \gg 0$. For example, $\hat{V}_k$ is a restricted module. The action of $\hat{\hat{\mathfrak{g}}}$ on any restricted module can be extended to $\tilde{\mathfrak{U}}\left(\hat{\hat{\mathfrak{g}}}\right)$ naturally.

Let $\tilde{\mathfrak{U}}\left(\hat{\hat{\mathfrak{g}}}, k\right)[[z, z^{-1}]]$ be the space of power series of z, z^{-1} with coefficients in $\tilde{\mathfrak{U}}\left(\hat{\hat{\mathfrak{g}}}, k\right)$. An element $b(z) = \sum_{n\in\mathbb{Z}} b(n)z^{-n-1}$ in $\tilde{\mathfrak{U}}\left(\hat{\hat{\mathfrak{g}}}, k\right)[[z, z^{-1}]]$ is called *regular* if every $b(n)$ is homogeneous in $\tilde{\mathfrak{U}}\left(\hat{\hat{\mathfrak{g}}}, k\right)$ and $\deg(b(n)) = -n + N_b$, where $N_b \in \frac{1}{2}\mathbb{Z}$ is a constant independent of n. We say that the regular element is odd if $N_b \in \frac{1}{2} + \mathbb{Z}$, even if $N_b \in \mathbb{Z}$. We define $\tilde{b}$ to be 1, if $b(z)$ is odd and 0 if $b(z)$ even. We denote by $\tilde{\mathfrak{U}}\left(\hat{\hat{\mathfrak{g}}}, k\right)\langle z\rangle$ the subspace linearly spanned by the regular elements in $\tilde{\mathfrak{U}}(\hat{\hat{\mathfrak{g}}}, k)[[z, z^{-1}]]$.

Recall the $\frac{1}{2}\mathbb{Z}_+$-gradation $M_{k,0} = \oplus_{n\in\frac{1}{2}\mathbb{Z}_+} M_{k,0}(n)$. Let $1 \in M_{k,0}(0)$ be the vacuum element of $M_{k,0}$. Define $Y(1, z) = I\,|_{M_{k,0}}$. We have

$$M_{k,0}(0) = \mathbb{C}\cdot 1, \quad M_{k,0}(\frac{1}{2}) = \bar{\mathfrak{g}}(-1)\cdot 1 \cong \bar{\mathfrak{g}}, \quad M_{k,0}(1) = \mathfrak{g}(-1)\cdot 1 \cong \mathfrak{g}.$$

For $a \in \mathfrak{g} \subset M_{k,0}$, $\bar{a} \in \bar{\mathfrak{g}} \subset M_{k,0}$, we define

$$a(z) = \sum_{n\in\mathbb{Z}} a(n)z^{-n-1}, \bar{a}(z) = \sum_{n\in\mathbb{Z}} \bar{a}(n)z^{-n-1}.$$

It is clear that $a(z), \bar{a}(z) \in \tilde{\mathfrak{U}}\left(\hat{\hat{\mathfrak{g}}}, k\right)\langle z\rangle$. $a(z)$ is even while $\bar{a}(z)$ is odd.

Definition 2.1 *For $b(z) = \sum_{n\in\mathbb{Z}} b(n)z^{-n-1}$, and $a, \bar{a} \in M_{k,0}$, we define*

$$\begin{aligned} a(n)\bullet b(z) &= Res_w\left(a(w)b(z)\iota_{w,z}(w-z)^n - b(z)a(w)\iota_{z,w}(w-z)^n\right), \\ \bar{a}(n)\bullet b(z) &= Res_w\left(\bar{a}(w)b(z)\iota_{w,z}(w-z)^n - (-1)^{\tilde{b}}b(z)\bar{a}(w)\iota_{z,w}(w-z)^n\right). \end{aligned}$$

By direct calculation, the following proposition which is an anologue of Proposition 2.2.1 in [FZ] follows from Definition 2.1.

Proposition 2.1 *The definition above gives* $\tilde{\mathfrak{U}}\left(\hat{\bar{\mathfrak{g}}}, k\right)\langle z\rangle$ *the structure of a* $\hat{\bar{\mathfrak{g}}}$-*module, where* $\mathbf{k} \in \hat{\bar{\mathfrak{g}}}$ *acts as* $(k+h^{\vee}) I$.

As a consequence of Definition 2.1, we have the following.

Corollary 2.1 *For* $a \in \mathfrak{g}, \bar{a} \in \bar{\mathfrak{g}}$, *we have*

$$\begin{aligned} a(n) \bullet 1 &= \begin{cases} 0, & n \geq 0 \\ \dfrac{1}{(-n-1)!}\left(\dfrac{d}{dz}\right)^{-n-1} a(z), & n < 0, \end{cases} \\ \bar{a}(n) \bullet 1 &= \begin{cases} 0, & n \geq 0 \\ \frac{1}{(-n-1)!}\left(\frac{d}{dz}\right)^{-n-1}\bar{a}(z), & n < 0. \end{cases} \end{aligned}$$

By Propositions 2.1 and Corollary 2.1, we have a well-defined homomorphism of $\hat{\bar{\mathfrak{g}}}$-modules from $M_{k,0}$ to $\tilde{\mathfrak{U}}\left(\hat{\bar{\mathfrak{g}}}, k\right)\langle z\rangle$:

$$\begin{aligned} Y(\ ,z) &: a_1(-i_1)\cdots a_n(-i_n)\bar{b}_1(-j_1)\cdots\bar{b}_m(-j_m)1 \\ &\longmapsto a_1(-i_1)\bullet\cdots\bullet a_n(-i_n)\bullet\bar{b}_1(-j_1)\bullet\cdots\bullet\bar{b}_m(-j_m)\bullet 1. \end{aligned}$$

Moreover, $Y(a(-1)1,z) = a(z)$, $Y(\bar{a}(-1)1,z) = \bar{a}(z)$ for $a \in \mathfrak{g}$, $\bar{a} \in \bar{\mathfrak{g}}$.

Since $M_{k,0}$ is a $\tilde{\mathfrak{U}}\left(\hat{\bar{\mathfrak{g}}}, k\right)$-module, we have a map from $\tilde{\mathfrak{U}}\left(\hat{\bar{\mathfrak{g}}}, k\right)$ to $End(M_{k,0})$. Now we have a series of maps

$$M_{k,0} \longrightarrow \tilde{\mathfrak{U}}\left(\hat{\bar{\mathfrak{g}}}, k\right)\langle z\rangle \subset \tilde{\mathfrak{U}}\left(\hat{\bar{\mathfrak{g}}}, k\right)\left[\left[z, z^{-1}\right]\right] \longrightarrow End\ (M_{k,0})\left[\left[z, z^{-1}\right]\right].$$

We still denote the composition of these maps by

$$Y(\ ,z) : M_{k,0} \to End\ (M_{k,0})\left[\left[z, z^{-1}\right]\right].$$

For $b \in M_{k,0}$, we call $Y(b,z)$ the vertex operator of b.

We use small letters $a, b, c, \ldots$ to denote the index among $1, 2, \ldots, \dim\mathfrak{g}$. Choose a basis $\{u_a\}$ of $\mathfrak{g}$ satisfying $(u_a, u_b) = \frac{1}{2}\delta_{ab}$, $[u_a, u_b] = if_{abc}u_c$, where f_{abc} is anti-symmetric in a, b, c and real valued (these notations agree with those in [KS]). Here and below we assume, as usual, summation over repeated indices.

Theorem 2.1 $(M_{k,0}, 1, \omega, \tau, Y(\ ,z))$ *is an* $N=1$ *SVOA of rank* c_k *provided that* $k \neq -h^{\vee}$, *where*

$$\begin{aligned} c_k &= \frac{\dim\mathfrak{g}}{2} + \frac{k\dim\mathfrak{g}}{k+h^{\vee}}, \\ \tau &= \frac{2}{k+h^{\vee}}u_a(-1)\bar{u}_a(-1)1 + \frac{4i}{3(k+h^{\vee})^2}f_{abc}\bar{u}_a(-1)\bar{u}_b(-1)\bar{u}_c(-1)1, \\ \omega &= \frac{1}{k+h^{\vee}}\left\{u_a(-1)u_a(-1)1 + \bar{u}_a(-2)\bar{u}_a(-1)1\right\} \\ &\quad + \frac{2i}{3(k+h^{\vee})^2}f_{abc}\bar{u}_a(-1)\bar{u}_b(-1)u_c(-1)1. \end{aligned}$$

The fact that the components of the fields

$$Y(\tau, z) = \sum_{n\in\mathbb{Z}} G_{n+\frac{1}{2}} z^{-n-2}, \quad Y(\omega, z) = \sum_{n\in\mathbb{Z}} L_n z^{-n-2},$$

satisfy the commutation relations of the Neveu-Schwarz algebra with the central charge c_k is ensured by Theorem 4 in [KT]. The rest of the proof of the above theorem is similar to Theorem 2.4.1 in [FZ].

It follows from [KT] that $[L_0, a(m)] = -ma(m)$, $[L_0, \bar{a}(m)] = -(m+\frac{1}{2})\bar{a}(m)$. Thus $L_0 + d$, which commutes with all $a(m), \bar{a}(m) \in \hat{\hat{\mathfrak{g}}}$. We call the generalized Casimir operator the element $\Omega = 2\,(k+h^\vee)\,(L_0+d)$. It follows from [KT] that

$$\Omega(v) = (\lambda + 2\rho, \lambda)v \tag{2.7}$$

if v is a singular vector of weight λ.

Let $J_{k,0}$ be the maximal proper submodule of $M_{k,0}$. It is easy to see that if $k \neq -h^\vee$ then $1 \notin J_{k,0}$, $\tau \notin J_{k,0}$ and hence the quotient $L_{k,0} = M_{k,0}/J_{k,0}$ is also a *SVOA*. To understand what $J_{k,0}$ is, we need to find the formulas for singular vectors of $M_{k,0}$. This is done in the Appendix (Sec.5).

Remark 2.1 *One may construct the $N = 2$ SVOA (i.e., the SVOA which admits vertex operators whose Fourier components satisfy the $N = 2$ superconformal algebra) from the ($N = 1$) affine Kac-Moody superalgebra [KS].*

2.2 Rationality and fusion rules of the SVOA $L_{k,0}$

Lemma 2.1 *The associative algebra $A(M_{k,0})$ is canonically isomorphic to $\mathfrak{U}(\mathfrak{g})$.*

Proof. By the definition of $A(M_{k,0})$ and Lemma 1.1 we have

$$[c] * [a(-1)1] = [a(-1)c],$$

where $a \in \mathfrak{g}$, $c \in M_{k,0}$. Hence

$$[a_m(-1)1] * \cdots * [a_1(-1)1] = [a_1(-1)\cdots a_n(-1)1].$$

Therefore we have a homomorphism of associative algebras

$$F : \mathfrak{U}(\mathfrak{g}) \longrightarrow A(M_{k,0}) \tag{2.8}$$

given by

$$a_m \cdots a_1 \mapsto [a_1(-1)\cdots a_n(-1)1].$$

It is clear that

$$\begin{aligned}(a(-n-2) + a(-n-1))c &= Res_z\left(Y(a(-1)1, z)\frac{z+1}{z^{n+2}}c\right),\\ \bar{b}(-n-1))c &= Res_z\left(Y(\bar{b}(-1)1, z)\frac{1}{z^{n+1}}c\right).\end{aligned}$$

By Lemma 1.1, we have

$$O'(M_{k,0}) \subset O(M_{k,0}),$$

where

$$O'(M_{k,0}) = \{(a(-n-2)+a(-n-1))c, \quad \bar{b}(-n-1)c \text{ for } n \geq 0\}.$$

Then it follows that

$$[a_1(-i_1-1)\cdots a_m(-i_m-1)] = (-1)^{i_1+\cdots+i_m}[a_1(-1)\cdots a_n(-1)1]$$

for $i_1, \ldots i_m \geq 0$. So F is an epimorphism. To show that F is indeed an isomorphism, we still need to show that

$$O'(M_{k,0}) = O(M_{k,0}). \tag{2.9}$$

However this is standard (see the proof of a similar fact in Appendix of [W]). □

Lemma 2.2 *If k is a positive integer, then the map (2.8) induces an isomorphism from $\mathfrak{U}(\mathfrak{g})/\langle e_\theta^{k+1}\rangle$ onto $A(L_{k,0})$, where $\langle e_\theta^{k+1}\rangle$ is the two-sided ideal of $\mathfrak{U}(\mathfrak{g})$ generated by e_θ^{k+1}.*

Proof. It follows from Theorem 5.3 in the Appendix that the SVOA $M_{k,0}$ is isomorphic to

$$M(\Lambda + h^\vee \Lambda_0)/\langle f_i 1, i = 1, \ldots, l\rangle,$$

with Λ given by $\lambda_i = \Lambda(h_i) = 0$, $i = 1, \ldots, l$, $\lambda_0 = \Lambda(\mathbf{k}) = k$. Then the SVOA $L_{k,0}$ is isomorphic to $M_{k,0}/\langle v_k\rangle$, where v_k is defined by Theorem 5.2. By Remark 5.3 and the identity (2.9), we see that under the isomorphism (2.8), v_{λ_0} corresponds to $e_\theta^{k+1} \in \mathfrak{U}(\mathfrak{g})$. Hence the lemma follows from Proposition 1.1. □

Lemma 2.3 *If $x \in \mathfrak{g}$, and $N \in \mathbb{N}$, then the algebra $\mathfrak{U}(\mathfrak{g})/\langle x^N\rangle$ is finite dimensional and semisimple.*

Proof. Let G be the adjoint group of $\mathfrak{g}$. Since G is generated by $exp(ad\, y)$, $y \in \mathfrak{g}$, the ideal $\langle x^N\rangle$ is G-invariant, hence it contains all elements $g(x)^N, g \in G$. Since $\mathfrak{g}$ is simple, it coincides with the linear span of the orbit $G(x)$, hence $u_i^N \in \langle x^N\rangle$ for some basis $\{u_i\}$ of $\mathfrak{g}$. It follows that $\dim \mathfrak{U}(\mathfrak{g})/\langle x^N\rangle \leq N^{\dim \mathfrak{g}}$.

Since any finite-dimensional representation of $\mathfrak{U}(\mathfrak{g})$ is semisimple, it follows that any representation of $\mathfrak{U}(\mathfrak{g})/\langle x^N\rangle$ is semisimple. Hence the latter algebra is semisimple. □

Theorem 2.2 *For any positive integral k, the SVOA $L_{k,0}$ is rational. Moreover, $L_{k,\lambda}$, for $\lambda \in \mathfrak{h}^*$ dominant integrable with $\langle\lambda, \theta\rangle \leq k$, are precisely all the irreducible $L_{k,0}$-modules.*

Proof. The second part of this theorem follows from Theorem 1.3 and Lemma 2.2 because by Lemma 2.3, $L_{k,\lambda}$, for $\lambda \in \mathfrak{h}^*$ dominant integrable with $\langle\lambda, \theta\rangle \leq k$, are all the irreducible modules of $\mathfrak{U}(\mathfrak{g})/\langle e_\theta^{k+1}\rangle$. Any $L_{k,0}$-module M is a restricted module over $\hat{\hat{\mathfrak{g}}}$. Hence any $\hat{\hat{\mathfrak{g}}}$-submodule of M is also an $L_{k,0}$-submodule of M. To prove the complete reducibility of any $L_{k,0}$-module, we only need to prove the following.

Lemma 2.4 *Given λ, $\mu \in P_+$ such that $\langle\lambda, \theta\rangle \leq k, \langle\mu, \theta\rangle \leq k$, any short exact sequence of $\hat{\hat{\mathfrak{g}}}$-modules*

$$0 \longrightarrow L_{k,\lambda} \stackrel{\iota}{\longrightarrow} M \stackrel{\pi}{\longrightarrow} L_{k,\mu} \longrightarrow 0$$

splits.

Proof. Let $Q_+ = \Sigma_i \mathbb{Z}_+ \alpha_i$. First let us define a partial order in P_+ as follows: $\lambda > \mu$ iff $\lambda - \mu \in Q_+$ and $\lambda \neq \mu$. Without loss of generality, we may assume that $\lambda \not> \mu$. Otherwise we can apply the contragredient functor to the short exact sequence to reverse it. Let v_μ be the vacuum vector of $L_{k,\mu}$. Pick a vector $v'_{\mu'} \in M$ of weight μ such that $\pi\left(v'_\mu\right) = v_\mu$. We claim that v'_μ is a singular vector of M, i.e. $e_i v'_\mu = 0$ for any i. Indeed, if $e_i v'_\mu \neq 0$ for some i, then

$$\pi\left(e_i v'_\mu\right) = e_i \pi\left(v'_\mu\right) = e_i v_\mu = 0.$$

So

$$e_i v'_\mu = \iota(u) \tag{2.10}$$

for some nonzero $u \in L_{k,\lambda}$, since the short sequence is exact. Comparing the weights of both sides of equation (2.10), we have $\lambda - \beta = \mu + \alpha_i$ for nonzero $\alpha_i, \beta \in Q_+$. It follows that $\lambda = \mu + \alpha_i + \beta > \mu$, which is a contradiction.

Denote by M' the submodule of M generated by the singular vector v'_μ. It suffices to show that the module M' is irreducible. But this follows in the same way as in Chapter 11 of [K] by making use of formula (2.7). □

Of course the Lie subalgebra

$$\hat{\mathfrak{g}} = \mathfrak{g} \bigotimes \mathbb{C}\left[t, t^{-1}\right] \bigoplus \mathbb{C}\mathbf{k} \bigoplus \mathbb{C}d$$

of $\hat{\bar{\mathfrak{g}}}$ is the usual affine Kac-Moody algebra. As usual, we let

$$\begin{aligned} \hat{\mathfrak{g}}_+ &= \mathfrak{g} \bigotimes t\mathbb{C}[t], \\ \hat{\mathfrak{g}}_- &= \mathfrak{g} \bigotimes t^{-1}\mathbb{C}\left[t^{-1}\right], \\ \hat{\mathfrak{n}}_\pm &= \hat{\mathfrak{g}}_\pm \bigoplus \bigoplus_{\alpha \in \Delta_+} \mathfrak{g}_{\pm\alpha}. \end{aligned}$$

Given $\lambda \in P_+$ and $k \in \mathbb{Z}_+$, consider the irreducible $\mathfrak{g}$-module $\bar{L}(\lambda)$ as a module over $\hat{\mathfrak{g}}_+ \bigoplus \mathfrak{g} \bigoplus \mathbb{C}\mathbf{k} \bigoplus \mathbb{C}d$ by letting $(\hat{\mathfrak{g}}_+ \bigoplus \mathbb{C}d)\bar{L}(\lambda) = 0$ and $\mathbf{k} = kI\,|_V$. Let

$$\bar{M}_{k,\lambda} = \mathfrak{U}(\hat{\mathfrak{g}}) \bigotimes_{\mathfrak{U}(\hat{\mathfrak{g}}_+ \bigoplus \mathfrak{g} \bigoplus \mathbb{C}\mathbf{k} \bigoplus \mathbb{C}d)} \bar{L}(\lambda),$$

and $\bar{L}_{k,\lambda} = \bar{M}_{k,\lambda}/\bar{J}_{k,\lambda}$, where $\bar{J}_{k,\lambda}$ is the unique maximal $\hat{\mathfrak{g}}$-submodule of $\bar{M}_{k,\lambda}$.

The associative algebra of the VOA $\bar{L}_{k,0}$ was computed in [FZ]. Comparing with our results, we see that the associative algebras $A\left(L_{k,0}\right)$ and $A\left(\bar{L}_{k,0}\right)$ are the same. And so the irreducible modules of the SVOA $L_{k,0}$ are canonically in 1–1 correspondence with those of the VOA $\bar{L}_{k,0}$. One can calculate the fusion rules using the $A(L_{k,0})$-modules similarly to Section 3.2 in [FZ] and find that the fusion rules for the modules of the SVOA $L_{k,0}$ are canonically in 1–1 correspondence with those for the VOA $\bar{L}_{k,0}$ (see the statements in Theorem 3.2.3 and Corollary 3.2.1 in [FZ]).

3 Neveu-Schwarz SVOAs

3.1 SVOA structure on $M_{c,0}$ and $L_{c,0}$

Let us recall first that the Neveu-Schwarz algebra is the Lie superalgebra

$$\mathfrak{NS} = \bigoplus_{n\in\mathbb{Z}} \mathbb{C}L_n \bigoplus \bigoplus_{m\in\frac{1}{2}+\mathbb{Z}} \mathbb{C}G_m \bigoplus \mathbb{C}C$$

with commutation relations $(m, n \in \mathbb{Z})$:

$$\begin{aligned} [L_m, L_n] &= (m-n)L_{m+n} + \delta_{m+n,0}\frac{m^3-m}{12}C, \\ \left[G_{m+\frac{1}{2}}, L_n\right] &= \left(m+\frac{1}{2}-\frac{n}{2}\right)G_{m+n+\frac{1}{2}}, \\ \left[G_{m+\frac{1}{2}}, G_{n-\frac{1}{2}}\right]_+ &= 2L_{m+n} + \frac{1}{3}m(m+1)\delta_{m+n,0}C, \\ [L_m, C] = 0, & \qquad \left[G_{m+\frac{1}{2}}, C\right] = 0. \end{aligned}$$

The $\mathbb{Z}_2$-gradation is given by $\tilde{L_n} = \tilde{C} = \bar{0}$, $\tilde{G_n} = \bar{1}$ (so that the even part is the Virasoro algebra). Set

$$\mathfrak{NS}_\pm = \bigoplus_{n\in\mathbb{N}} \mathbb{C}L_{\pm n} \bigoplus \bigoplus_{m\in\frac{1}{2}+\mathbb{Z}_+} \mathbb{C}G_{\pm m}.$$

Given complex numbers c and h, the Verma module $M_{c,h}$ over $\mathfrak{NS}$ is the free $\mathfrak{U}(\mathfrak{NS}_-)$-module generated by 1, such that $\mathfrak{NS}_+1 = 0$, $L_0 1 = h\cdot 1$ and $C\cdot 1 = c\cdot 1$. There exists a unique maximal proper submodule $J_{c,h}$ of $M_{c,h}$. Denote the quotient $M_{c,h}/J_{c,h}$ by $L_{c,h}$. Recall that $v \in M_{c,h}$ is called a singular vector if $\mathfrak{NS}_+1 = 0$ and v is an eigenvector of L_0. For example, $G_{-\frac{1}{2}}1$ is a singular vector of $M_{c,0}$ for any c. Denote $M_{c,0}/\langle G_{-\frac{1}{2}}1\rangle$ by M_c, where $\langle G_{-\frac{1}{2}}1\rangle$ is the submodule of $M_{c,0}$ generated by the singular vector $G_{-\frac{1}{2}}1$. For simplicity we denote $L_{c,0}$ by V_c.

It is well known that

$$L_{-i_1}L_{-i_2}\cdots L_{-i_m}G_{-j_1}G_{-j_2}\cdots G_{-j_n},$$

for $i_1 \geq \cdots \geq i_m \geq 1$, $j_1 > \cdots > j_n \geq \frac{1}{2}$, $i_1\cdots i_m \in \mathbb{N}$, and $j_1\cdots j_n \in \frac{1}{2}+\mathbb{Z}_+$ is a basis of $\mathfrak{U}(\mathfrak{NS}_-)$. There is a natural gradation on $M_{c,0}$, M_c and V_c given by the eigenspace decomposition of L_0:

$$\deg L_{-i_1}L_{-i_2}\cdots L_{-i_m}G_{-j_1}G_{-j_2}\cdots G_{-j_n}1 = i_1+i_2+\cdots+i_m+j_1+\cdots+j_n.$$

We can define $\tilde{\mathfrak{U}}(\mathfrak{NS}, c)$, the completion of $\mathfrak{U}(\mathfrak{NS})$, as in Section 2. The action of $\mathfrak{NS}$ on any restricted module of $\mathfrak{NS}$ can be extended to $\tilde{\mathfrak{U}}(\mathfrak{NS})$ naturally. In particular, $\tilde{\mathfrak{U}}(\mathfrak{NS}, c)$ acts on $M_{c,h}$ and M_c. We can also define the notions of even and odd regular elements in $\tilde{\mathfrak{U}}(\mathfrak{NS}, c)\left[\left[z, z^{-1}\right]\right]$. Denote by $\tilde{\mathfrak{U}}(\mathfrak{NS}, c)\langle z\rangle$ the linear span of regular elements in $\tilde{\mathfrak{U}}(\mathfrak{NS}, c)\left[\left[z, z^{-1}\right]\right]$. Set

$$\begin{aligned} L(z) &= \sum_{n\in\mathbb{Z}} L_n z^{-n-2}, \\ G(z) &= \sum_{n\in\mathbb{Z}} G_{n+\frac{1}{2}} z^{-n-2}. \end{aligned}$$

Clearly $L(z)$ is even while $G(z)$ is odd. For a regular element $b(z) \in \tilde{\mathfrak{U}}(\mathfrak{NS}, c)$ $[[z, z^{-1}]]$ we define

$$\begin{aligned}(3.1)\qquad L_n \bullet b(z) &= Res_w \left(L(w)b(z)\iota_{w,z}(w-z)^{n+1}\right.\\ &\qquad \left. - b(z)L(w)\iota_{z,w}(w-z)^{n+1}\right),\end{aligned}$$

$$\begin{aligned}(3.2)\qquad G_{n+\frac{1}{2}} \bullet b(z) &= Res_w \Big(G(w)b(z)\iota_{w,z}(w-z)^{n+1}\\ &\qquad - (-1)^{\tilde{b}} b(z)G(w)\iota_{z,w}(w-z)^{n+1}\Big),\end{aligned}$$

where $\tilde{b} = 0$ if $b(z)$ is even and $\tilde{b} = 1$ if $b(z)$ odd.

Claim 3.1. (3.1) and (3.2) define a $\mathfrak{NS}$-module structure on $\tilde{\mathfrak{U}}(\mathfrak{NS}, c)\langle z\rangle$ with central charge c.

Claim 3.2.

$$L_n \bullet 1 = \begin{cases} 0, & n \geq -1 \\ \dfrac{1}{(-n-2)!}\left(\dfrac{d}{dz}\right)^{-n-2} L(z), & n < -1, \end{cases}$$

$$G_{n+\frac{1}{2}} \bullet 1 = \begin{cases} 0, & n \geq -1 \\ \dfrac{1}{(-n-2)!}\left(\dfrac{d}{dz}\right)^{-n-2} G(z), & n < -1, \end{cases}$$

From Claim 3.2 we have a well-defined homomorphism of $\mathfrak{NS}$-modules

$$(3.3)\qquad Y(\ ,z): M_{k,0} \rightarrow \tilde{\mathfrak{U}}(\mathfrak{NS}, c)\langle z\rangle$$

$$\begin{aligned}&L_{-i_1}L_{-i_2}\cdots L_{-i_m}G_{-j_1}G_{-j_2}\cdots G_{-j_n}1\\ \mapsto\ &L_{-i_1}\bullet L_{-i_2}\bullet\cdots\bullet L_{-i_m}\bullet G_{-j_1}\bullet G_{-j_2}\bullet\cdots\bullet G_{-j_n}\bullet 1.\end{aligned}$$

In particular, if we set $\tau = G_{-3/2}1$, and $\omega = L_{-2}1$, we have $Y(\tau, z) = G(z)$, and $Y(\omega, z) = L(z)$.

Since $\tilde{\mathfrak{U}}(\mathfrak{NS}, c)$ acts on $M_{c,h}$, we have a map from $\tilde{\mathfrak{U}}(\mathfrak{NS}, c)$ to $End\ (M_{k,0})$. Then (3.3) induces a linear map

$$Y(\ ,z): M_{k,0} \longrightarrow End\ (M_{k,0})\left[\left[z, z^{-1}\right]\right].$$

Thus we have

Theorem 3.1 *$(M_c, f1, \tau, Y(\cdot, z))$ is an $N = 1$ SVOA.*

3.2 Rationality and fusion rules of $V_{c_{p,q}}$

Lemma 3.1 *There exists an isomorphism of associative algebras, $F : A(M_c) \cong \mathbb{C}[x]$, given by $[\omega]^n \mapsto x^n$, where $\mathbb{C}[x]$ is the polynomial algebra on one generator x.*

Proof. Set

$$M_c = M_c^0 + M_c^1,$$

where M_c^0 (*resp.* M_c^1) is the even (*resp.* odd) part of M_c. By Lemma 1.1 we have

$$(3.4)((L_{-m-3} + 2L_{-m-2} + L_{-m-1})\,b) = \mathrm{Res}_z \left(Y(\omega, z)\frac{(z+1)^2}{z^{2+n}} b\right) \in O(M_c),$$

for every $m \geq 0, b \in M_c^0$.

$$(3.5) \qquad ((G_{-n-1} + G_{-n})\, b) \;=\; \mathrm{Res}_z \left(Y(\tau, z) \frac{(z+1)}{z^{1+n-\frac{1}{2}}} b \right) \in O(M_c),$$

for every $n \in \frac{1}{2} + \mathbb{Z}_+, b \in M_c^1$. It follows by induction that

$$(3.6) \qquad L_{-m} \;\sim\; (-1)^m \left((m-1)\left(L_{-2} + L_{-1}\right) + L_0 \right),$$

for every $m \geq 1$.

$$(3.7) \qquad G_{-n} \;\sim\; (-1)^{n-\frac{1}{2}} G_{-\frac{1}{2}}, \text{ for every } n \in \frac{1}{2} + \mathbb{Z}_+.$$

By Lemma 1.1, we have

$$(3.8) \qquad [b] * [\omega] \;=\; [(L_{-2} + L_{-1})b], b \in M_c^0$$

Using (3.4) and (3.5), it is easy to show by induction on $m + n$ that

$$[L_{-i_1} L_{-i_2} \cdots L_{-i_m} G_{-j_1} G_{-j_2} \cdots G_{-j_{2n}} 1] \;=\; P([\omega])$$

for some $P(x) \in \mathbb{C}[x]$. Since the elements

$$L_{-i_1} L_{-i_2} \cdots L_{-i_m} G_{-j_1} G_{-j_2} \cdots G_{-j_{2n}} 1$$

for $i_1 \geq \cdots \geq i_m \geq 1$, $j_1 > \cdots > j_{2n} \geq \frac{1}{2}$, $i_1 \cdots i_m \in \mathbb{N}$, $j_1 \cdots j_{2n} \in \frac{1}{2} + \mathbb{Z}_+$, span M_c, the homomorphism of associative algebras

$$F : \mathbb{C}[x] \to A(M_c)$$

given by $x^n \mapsto [\omega]^n$ is surjective. (This homomorphism is well defined since $[\omega]$ is in the center of $A(M_c)$.)

To prove that F is also injective, it suffices to show that $O(M_c)$ is the linear span of the elements of the form (3.4) and (3.5), i.e.

$$O(M_c) = \left\{ \left(L_{-n-3} + 2L_{-n-2} + L_{-n-1}\right) b, \left(G_{-n-3/2} + G_{-n-3/2}\right) bn \geq 0 b \in M_c \right\}.$$

This can be proved in a standard way (for a proof of a similar fact see Appendix of [W]). □

Set

$$c_{p,q} \;=\; \frac{3}{2}\left(1 - \frac{2(p-q)^2}{pq}\right),$$

$$h_{p,q}^{r,s} \;=\; \frac{(sp - rq)^2 - (p-q)^2}{8pq}.$$

Whenever we mention $c_{p,q}$ again, we always assume that p, $q \in \{2, 3, 4, \ldots\}$, $p - q \in 2\mathbb{Z}$, and that $(p-q)/2$ and q are relatively prime to each other. The submodule structure of a Verma module over the Neveu-Schwarz algebra [A] is very similar to that for the Virasoro algebras [FF]. From the results of [A], we have the following lemma which is an analogue of the results in [FF] (also see Lemma 4.2 of [W]).

Lemma 3.2 *1) $J_{c,0}$ is generated by the singular vector $G_{-\frac{1}{2}}1$ if $c \neq c_{p,q}$.*

2) $J_{c,0}$ is generated by two singular vectors if $c = c_{p,q}$. One of them is $G_{-\frac{1}{2}}1$. The other, denoted by $v_{p,q}$ has degree $\frac{1}{2}(p-1)(q-1)$.

From this lemma we immediately derive an analogue of Corollary 4.1 in [W].

Corollary 3.1 *If $c \neq c_{p,q}$, then V_c is not rational.*

Proof. See proof of Corollary 4.1 in [W]. □

From now on, we always assume that $c = c_{p,q}$ and that $h^{r,s} = h^{r,s}_{p,q}$. It follows from Lemma 3.2 that $V_c = M_c/\langle v_{p,q}\rangle$, where $\langle v_{p,q}\rangle$ denotes the submodule of M_c generated by $v_{p,q}$. Then we have

Proposition 3.1 *One has:*

$$A(V_c) \cong \mathbb{C}[x]/\langle F_{p,q}(x)\rangle,$$

where $\deg F_{p,q} = \frac{1}{4}(p-1)(q-1)$ *if p, q are odd;* $\deg F = \frac{1}{4}(p-1)(q-1) + \frac{1}{4}$ *if p, q are even.*

Proof. If p, q are odd, $v_{p,q}$ is an even element of degree $\frac{1}{2}(p-1)(q-1)$ which corresponds to a polynomial $F_{p,q}$ of degree $\frac{1}{4}(p-1)(q-1)$; if p, q are even, $v_{p,q}$ is an odd element of degree $\frac{1}{2}(p-1)(q-1)$. From the definition of the associative algebra $A(V_c)$, it is $G_{-\frac{1}{2}}v_{p,q}$ which corresponds to $F_{p,q}$ of degree $\frac{1}{4}(p-1)(q-1)+\frac{1}{4}$. □

We expect the following conjecture, which is an analogue of Theorem 4.2 in [W], to be true.

Conjecture 3.1 *The vertex operator superalgebra $V_{c_{p,q}}$ is rational. Moreover, the minimal series modules $L_{c,h_{r,s}}, 0 < r < p, 0 < s < q, r-s \in 2\mathbb{Z}$ are all the irreducible representations of V_c.*

Remark 3.1 *If $p-q = 2$, $V_{c,h^{r,s}}$ is unitary. In this case we can prove Theorem 3.1 by using the well-known GKO construction [KW] and Theorem 2.2 (see [DMZ] for a similar proof in the Virasoro algebra case). It follows that Conjecture 3.1 holds at least in the cases $p-q=2$.*

From the argument of Lemma 3.1, we see that

$$A(V_c) = H_0(\mathfrak{S}, V_c),$$

where $\mathfrak{S} = \{L_{-n-2} + 2L_{-n-1} + L_{-n}, G_{-n-1} + G_{-n}, n > 0\}$ is a locally nilpotent subalgebra of $\mathfrak{N}\mathfrak{S}$. This conjecture can probably be proved as in [W], by calculating the coinvariants $H_0(\mathfrak{S}, V_c)$. It is easy to see by Lemma 3.1 that

$$A(L_{c,h^{r,s}}) = H_0(\mathfrak{S}, L_{c,h^{r,s}}).$$

Then by applying Theorem 1.5, and Proposition 1.2, we can obtain the fusion rules for the $V_{c_{p,q}}$-modules $L_{c,h_{r,s}}$, $0 < r < p$, $0 < s < q$, $r-s \in 2\mathbb{Z}$ if the coinvariants $H_0(\mathfrak{S}, L_{c,h^{r,s}})$ are calculated.

To support our conjecture, we present some examples.

Example 1. Consider the case $(p,q) = (5,3)$, $c_{5,3} = 7/10$, $h^{1,1} = 0$, $h^{2,2} = 1/10$. It is easy to check that the singular vector $v_{5,3}$ is given by

$$v_{5,3} = 3L_{-4}1 + 10L_{-2}^2 1 - 15G_{-5/2}G_{-3/2}1.$$

Using (3.6), (3.7) and (3.8), we have

$$F_{5,3}(x) = 10\left(x^2 - \frac{1}{10}x\right)$$

which gives the values of $h^{1,1}$ and $h^{2,2}$.

Example 2. Let $(p,q) = (8,2)$, $c_{8,2} = -\frac{21}{4}$, $h^{1,1} = 0, h^{3,1} = -\frac{1}{4}$. This is a non-unitary case. The singular vector $v_{8,2}$ is given by

$$v_{8,2} = 3G_{-7/2}1 - 4L_{-2}G_{-3/2}1.$$

Since $v_{8,2}$ is an odd element, we consider $G_{-\frac{1}{2}}v_{8,2}$ in order to get the polynomial $F_{8,2}(x)$. Using (3.6), (3.7) and (3.8), we get

$$G_{-\frac{1}{2}}v_{8,2} \sim -8x\left(x + \frac{1}{4}\right)$$

which gives the values of $h^{1,1}$ and $h^{3,1}$.

4 SVOAs generated by free fermionic fields

The free fermionic fields are

$$\Phi^a(z) = \sum_{i\in\frac{1}{2}+\mathbb{Z}} \phi_i^a z^{-i-\frac{1}{2}}, \quad \text{(neutral)}$$

$$\Psi^{a,\pm}(z) = \sum_{i\in\frac{1}{2}+\mathbb{Z}} \psi_i^{a,\pm} z^{-i-\frac{1}{2}}, \quad \text{(charged)}$$

with the following nontrivial commutation relations

$$\left[\phi_i^a, \phi_j^b\right]_+ = \delta_{a,b}\delta_{i,j}$$

$$\left[\psi_i^{a,+}, \psi_j^{b,-}\right]_+ = \delta_{a,b}\delta_{i,j},$$

where $a, b = 1, \ldots, l$.

It is easy to see that from a pair of charged free fermionic fields $\Psi^\pm(z)$ one can construct two neutral free fermionic fields $\Phi(z)$ by letting

$$\Phi^1(z) = \frac{1}{\sqrt{2}}\left(\Psi^+(z) + \Psi^-(z)\right),$$

$$\Phi^2(z) = \frac{i}{\sqrt{2}}\left(\Psi^+(z) - \Psi^-(z)\right),$$

and vise versa. Hence we only need to consider the SVOAs generated by neutral free fermionic fields. Let F be the Fock space defined by $\phi_{i>0}^a|0\rangle = 0$, $a = 1, 2, \ldots, l$. F is a SVOA with the Virasoro element $\omega = \frac{1}{2}\sum_{a=1}^l \phi_{-3/2}^a\phi_{-1/2}^a 1$, and central charge $c = \frac{l}{2}$. Denote by $\mathfrak{a}$ the Lie algebra linearly spanned by $\left\{\phi_{n+\frac{1}{2}}^a,\ a = 1, \ldots, l,\ n \in \mathbb{Z}\right\}$. $\mathfrak{U}(\mathfrak{a})$ admits a natural gradation by letting $\deg \phi_{n+\frac{1}{2}}^a = -n - \frac{1}{2}$. Then, as in Subsection 2.1, we can define the completion $\tilde{\mathfrak{U}}(\mathfrak{a})$ of $\mathfrak{U}(\mathfrak{a})$. Similarly, we can define the notion of an (even or odd) *regular* vector in $\tilde{\mathfrak{U}}(\mathfrak{a})\left[\left[z, z^{-1}\right]\right]$. Let $\tilde{\mathfrak{U}}(\mathfrak{a})\langle z\rangle$ be the linear span of all regular vectors. Define $Y\left(\phi^a_{-1/2}1, z\right) = \Phi^a(z)$. For any regular element $b(z) \in \tilde{\mathfrak{U}}(\mathfrak{a})\langle z\rangle$ we define an action

$$\phi^a_{n+\frac{1}{2}} \bullet b(z) \quad = \quad Res_w\left(\Phi^a(w)b(z)\iota_{w,z}(w-z)^n - (-1)^{\tilde{b}}b(z)\Phi^a(w)\iota_{z,w}(w-z)^n\right).$$

It is easy to see that $\phi^a_{n+\frac{1}{2}} \bullet b(z)$ is also a regular vector. Then we define the vertex operator associated to any $v = \phi^{a_1}_{n_1+\frac{1}{2}} \cdots \phi^{a_s}_{n_s+\frac{1}{2}} 1 \in F$ by

$$Y(v,z) \quad = \quad \phi^{a_1}_{n_1+\frac{1}{2}} \bullet \cdots \phi^{a_s}_{n_s+\frac{1}{2}} \bullet 1.$$

This definition turns out to be the same as that defined by the normal ordering product [T].

Theorem 4.1 *F is a rational SVOA. Moreover, F has a unique irreducible representation, namely F itself.*

Proof. First we calculate the associative algebra $A(F)$. By Lemma 1.1, we have

$$\phi^a_{-\frac{1}{2}-n}v \quad = \quad Res_z\left(Y(\phi^a, z)\frac{1}{z^{1+n}}\right) \in O(F), \quad n \geq 0, \quad v \in F.$$

Since

$$\left\{\phi^a_{-\frac{1}{2}-n}v,\ 1 \leq a \leq ln \geq 0,\ v \in F\right\} \quad = \quad \bigoplus_{n \in \frac{1}{2}\mathbb{N}} F_n,$$

we have $\bigoplus_{n\in\frac{1}{2}\mathbb{N}} F_n \subset O(F)$.

On the other hand, it is easy to check by definition of $O(F)$ that $O(F) \subset \bigoplus_{n\in\frac{1}{2}\mathbb{N}} F_n$. Thus $O(F) = \bigoplus_{n\in\frac{1}{2}\mathbb{N}} F_n$, and so $A(F) = F/O(F) \cong \mathbb{C}$. Hence there exists a unique representation of the associative algebra $A(F) \cong \mathbb{C}$, i.e., $\mathbb{C}$ itself. By Theorem 1.3, there exists a unique representation of F, i.e., F itself.

The complete reducibility of modules of F follows from a similar argument to the proof of Lemma 2.4. So F is rational. □

Remark 4.1 *The above SVOA F is not an $N = 1$ SVOA. To construct the $N = 1$ SVOAs one needs to add some bosonic fields. For example, one can see that the Fock space of one free bosonic field and one free neutral fermionic field is an $N = 1$ SVOA of rank $\frac{3}{2}$. This is just the special case of the SVOA associated to the affine Kac-Moody superalgebra corresponding to the 1-dimensional Lie algebra $\mathfrak{g}$.*

5 Appendix: Singular vectors and defining relations for the integrable representations of affine Kac-Moody superalgebras

We continue using the notation on affine (super)algebras introduced in Subsection 2.1.

Recall the triangular decomposition $\hat{\mathfrak{g}} = \hat{\mathfrak{n}_+} \bigoplus \hat{\mathfrak{h}} \bigoplus \hat{\mathfrak{n}}_-$, where

$$\hat{\mathfrak{n}}_\pm = \hat{\mathfrak{g}}_\pm \bigoplus \left(\bigoplus_{\alpha \in \Delta_+} \mathfrak{g}_{\pm\alpha} \right).$$

Recall that for $\Lambda \in \hat{\mathfrak{h}}^*$ we have the Verma module $\bar{M}(\Lambda) = \mathfrak{U}(\hat{\mathfrak{g}}) \bigotimes_{\mathfrak{U}(\hat{\mathfrak{h}}+\hat{\mathfrak{n}}_+)} \mathbb{C}_\Lambda$ over $\hat{\mathfrak{g}}$, where $\mathbb{C}_\Lambda$ is the 1-dimensional $\mathfrak{U}\left(\hat{\mathfrak{h}} + \hat{\mathfrak{n}}_+\right)$-module defined by $h \longmapsto \Lambda(h)$, $\hat{\mathfrak{n}}_+ \longmapsto 0$. Note that $\bar{M}_{k,\lambda}$, where $k = \Lambda(\mathbf{k})$ and $\lambda = \Lambda\mid_{\mathfrak{h}}$, is a quotient module of $\bar{M}(\Lambda)$ and that $\bar{L}(\lambda)$ is the quotient of $\bar{M}(\Lambda)$ by the maximal submodule.

Similarly we have the triangular decomposition $\hat{\hat{\mathfrak{g}}} = \hat{\hat{\mathfrak{n}}}_- \bigoplus \mathfrak{h} \bigoplus \hat{\hat{\mathfrak{n}}}_+$, where $\hat{\hat{\mathfrak{n}}}_\pm = \hat{\hat{\mathfrak{g}}}_\pm \bigoplus \bigoplus_{\alpha\in\Delta_+} \mathfrak{g}_{\pm\alpha}$, and for a given $\Lambda \in \hat{\mathfrak{h}}^*$, we define the Verma module $M(\Lambda)$ over $\hat{\hat{\mathfrak{g}}}$, so that $M_{k,\lambda}$ is a quotient of $M(\Lambda)$ and $L(\Lambda)$ is the irreducible quotient of $M(\Lambda)$.

Set $e_0 = e_{-\theta}(1)$, $f_0 = e_\theta(-1)$, and $h_0 = \alpha_0 = \mathbf{k} - \theta$. Given $\Lambda \in \hat{\mathfrak{h}}^*$, let $\lambda_i = \Lambda(h_i)$. Let $P_+ = \left\{\Lambda \in \hat{\mathfrak{h}}^* \mid \Lambda(h_i) \in \mathbb{Z}_+\right\}$.

It is well known [K] that if $\Lambda \in P_+$, then $\{f_i^{\lambda_i+1}1, i = 0, 1, \ldots, l\}$ are the singular vectors of $\bar{M}(\Lambda)$ which generate the maximal proper submodule of $\bar{M}(\Lambda)$, denoted by $\bar{\langle} f_i^{\lambda_i+1}1, i = 0, 1, \ldots, l \bar{\rangle}$, i.e.,

$$\bar{L}(\Lambda) \cong \bar{M}(\Lambda)/\bar{\langle} f_i^{\lambda_i+1}1, i = 0, 1, \ldots, l \bar{\rangle}$$

and that the $\hat{\mathfrak{g}}$-modules $\bar{L}(\Lambda)$, $\Lambda \in P_+$, are all the unitary highest weight modules. For $\hat{\hat{\mathfrak{g}}}$ the situation is similar. In more detail, there exists a unique hermitian form $H(\cdot,\cdot)$ on the Verma module $M(\Lambda)$ satisfying

$$H(1,1) = 1,$$

$$a(n)^* = \sigma(a)(-n), \tag{5.1}$$

$$\bar{a}(n)^* = \overline{\sigma(a)}(-n-1), \tag{5.2}$$

where $*$ denotes the adjoint operator with respect to the hermitian form $H(\cdot,\cdot)$. Then $L(\Lambda) = M(\Lambda)/Ker\ H$. The $\hat{\hat{\mathfrak{g}}}$-module is called unitary if the form H on $L(\Lambda)$ is positive definite. It is known that there exists a unique unitary highest weight $\hat{\hat{\mathfrak{g}}}$-module of level $h^\vee$, called the minimal representation F which is given by the Fock space realization of the infinite dimensional Clifford algebra (2.2) and as a $\hat{\mathfrak{g}}$-module is isomorphic to $L(h^\vee d)$ [KT]. Furthermore [KT], any unitary highest weight representation of $\hat{\hat{\mathfrak{g}}}$ is of the form $L(\Lambda + h^\vee d)$, where $\Lambda \in P_+$, and that one has an isomorphism as $\hat{\mathfrak{g}}$-modules:

$$(\Lambda + h^\vee d) \cong F \bigotimes \bar{L}(\Lambda).$$

From the construction of the minimal representation, we also see that as $\hat{\mathfrak{g}}$-modules

$$M\left(\Lambda + h^\vee d\right) \cong F \bigotimes \bar{M}(\Lambda). \tag{5.3}$$

It is clear that $\left\{f_i^{\lambda_i+1}1, i = 1, \ldots, l\right\}$ are the singular vectors of $M\left(\Lambda + h^\vee d\right)$. By comparing the character formulas of both sides of (5.3), we see that there also exists a unique singular vector of weight $\Lambda + h^\vee d - (\lambda_0 + 1)\,\alpha_0$ in $M\left(\Lambda + h^\vee d\right)$. To get an explicit formula for this singular vector, we need to introduce the following notion of special roots.

Definition 5.1 *A root* α *in* Δ_+ *is called* special *if* $\theta - \alpha$ *is also a root.*

Denote by $\mathbb{S}$ the set of all special roots. The following is an equivalent way to define the set $\mathbb{S}$:

Remark 5.1 *The set* $\mathbb{S}$ *is also characterized by the property:* $r_\theta(\alpha) = \alpha - \theta$, *if* $\alpha \in \mathbb{S}$; $r_\theta(\alpha) = \alpha$, *if* $\alpha \in \Delta - (\mathbb{S} \cup \{\theta\})$. *Also we have:*

$$\mathbb{S} \cup \{\theta\} = \{\alpha \in \Delta_+ | r_\theta(\alpha) \in -\Delta_+\}. \tag{5.4}$$

Lemma 5.1 *The number of special roots is* $2(h^\vee - 2)$.

Proof. Choose the shortest $w \in W$ such that $w(\alpha_i) = \theta$ for some simple root α_i of $\mathfrak{g}$. It is not difficult to see that $l(w) = h^\vee - 2$. Pick a reduced expression $w = r_{i_1} \cdots r_{i_{h^\vee-2}}$. We claim that the expression

$$r_\theta = r_{i_1} \cdots r_{i_{h^\vee-2}} r_i r_{i_{h^\vee-2}} \cdots r_{i_1}, \tag{5.5}$$

is reduced, i.e., $l(r_\theta) = 2h^\vee - 3$. Indeed, first from the expression (5.5) of r_θ we see that $l(r_\theta) \le 2h^\vee - 3$. Let

$$\beta_1 = \alpha_{i_1}, \beta_2 = r_{i_1}(\alpha_{i_2}), \ldots, \beta_{h^\vee-2} = r_{i_1} \cdots r_{i_{h^\vee-3}} \left(\alpha_{i_{h^\vee-2}}\right)$$

and let

$$\gamma_s := \theta - \beta_s = w r_i r_{i_{h^\vee-2}} \cdots r_{i_{s+1}}(\alpha_{i_s}).$$

It is easy to see (using Remark 5.1) that $\{\beta_1, \ldots, \beta_{h^\vee-2}, \gamma_1, \ldots \gamma_{h^\vee-2}\} \subset \mathbb{S}$. Hence $l(r_\theta) \ge 2h^\vee - 3$. Then the lemma follows since $l(w) = \#\left\{\alpha \in \Delta_+ | w^{-1}(\alpha) \in -\Delta_+\right\}$ for any $w \in W$. □

Remark 5.2 *It follows from the above proof that*

$$\mathbb{S} = \{\beta_1, \ldots, \beta_{h^\vee-2}, \gamma_1, \ldots \gamma_{h^\vee-2}\}. \tag{5.6}$$

The sum of two elements from $\mathbb{S} \cup \{\theta\}$ *is a root if and only if one of them is* β_i *and the other is* γ_i.

Lemma 5.2 *Assume that* $\delta_i \in \Delta_+ - \{\theta\}$, $i = 1, 2, \ldots, p$ *satisfy*

$$\sum_{i=1}^{I} \delta_i = p\theta + \sum_k \eta_k \text{ for some } p \in \mathbb{N} \text{ and } \eta_k \in \Delta_+. \tag{5.7}$$

Then $I \ge 2p$, *and at least* $2p\delta_i$*'s are contained in* $\mathbb{S}$.

Proof. Assume that there are $q\delta_i$'s which are contained in $\mathbb{S}$. By applying r_θ to both sides of the equation (5.7), it follows from Remark 5.1 that

$$\sum_{i=1}^{I} \delta_i - q\theta = -p\theta + \sum_k r_\theta(\eta_k). \tag{5.8}$$

By subtracting (5.8) from (5.7), we have

$$(q - 2p)\theta = \sum_k (\eta_k - r_\theta(\eta_k)). \tag{5.9}$$

By Remark (5.1), $\eta_k - r_\theta(\eta_k)$ is in $Q_+ = \sum_{\alpha \in \Delta_+} \mathbb{Z}_+ \alpha$. It follows that $q \ge 2p$. □

Theorem 5.1 *The element* $v_{\lambda_0} = \bar{e}_{-\theta}(1)\bar{e}_{-\theta}\Pi_{\alpha\in S}\bar{e}_{-\alpha}\cdot e_\theta(-1)^{\lambda_0+h^\vee+1}1$ *is a singular vector in* $M(\Lambda + h^\vee d)$ *of weight* $\Lambda + h^\vee d - (\lambda_0+1)\alpha_0$. *(Here and further,* $\bar{e}$ *stands for* $\bar{e}(0)$*, where* $e \in \mathfrak{g}$.*)*

A different arrangement of order in $\Pi_{\alpha\in S}\bar{e}_{-\alpha}$ only makes a difference in the sign of v_{λ_0}. In the following proof, for convenience we use $x\left(m+\frac{1}{2}\right)$ to denote $\bar{x}(m)$ in $\hat{\hat{\mathfrak{g}}}$, $m \in \mathbb{Z}$.

Proof. It follows from Lemma 5.1 that the weight of v_{λ_0} is $\mu = \Lambda + h^\vee d - (\lambda_0+1)\alpha_0$. To prove that v_{λ_0} is a singular vector, it suffices to prove that for every homogeneous element $w \in M^s(\Lambda + h^\vee d)_\mu$, we have $H(v_{\lambda_0}, w) = 0$ and that $v_{\lambda_0} \neq 0$. The latter statement will follow from another formula for v_{λ_0} given by Theorem 5.2. Here we prove the former one.

By (5.2), it is enough to show that

$$H\left(e_\theta(-1)^{\lambda_0+h^\vee+1}1, \tilde{w}\right) = 0, \tag{5.10}$$

where

$$\tilde{w} = \Pi_{\alpha\in S}e_\alpha\left(-\frac{1}{2}\right)e_\theta\left(-\frac{3}{2}\right)e_\theta\left(-\frac{1}{2}\right)\cdot w. \tag{5.11}$$

Any homogeneous element w in $M(\Lambda + h^\vee d)_\mu$ is of the form

$$\Pi_{i=1}^{I}e_{\gamma_i}(-m_i)\Pi_{j=1}^{J}e_\theta(-n_j)\Pi_{k=1}^{K}e_{-\eta_k}(-l_k)\cdot 1,$$

where $\eta_k \in \Delta_+$, $\gamma_i \in \Delta_+ - \{\theta\}$, $K, I, J \in \mathbb{Z}_+, l_k \in \frac{1}{2}\mathbb{Z}_+$, m_i, $n_j \in \frac{1}{2}\mathbb{N}$, and

$$-\sum_k \eta_k + \sum_i \gamma_i + J\theta = (\lambda_0+1)\theta, \tag{5.12}$$

$$\sum_k l_k + \sum_i m_i + \sum_j n_j = \lambda_0 + 1. \tag{5.13}$$

Case 1) $n_j = \frac{1}{2}$ for some j.

Note that $e_\theta\left(-\frac{1}{2}\right)$ commutes up to a sign with elements of the form $e_\theta(-n_j)$ or $e_{\gamma_i}(-m_i)$. Since $e_\theta^2\left(-\frac{1}{2}\right) = 0$, we have $\tilde{w} = 0$ by (5.11). (5.10) is satisfied automatically.

Case 2) All $n_j \geq 1$.

On one hand, since all $m_i \geq \frac{1}{2}$, we have the inequality

$$\frac{I}{2} \leq \lambda_0 + 1 - J \tag{5.14}$$

since by (5.13) we have

$$\frac{I}{2} = \sum_i \frac{1}{2} \leq \sum_i m_i = (\lambda_0+1) - \sum_j n_j - \sum_k l_k \leq d + 1 - \sum_j 1 = d + 1 - J.$$

And the equality in (5.14) holds iff

$$m_i = \frac{1}{2}, \quad n_j = 1, \quad l_k = 0. \tag{5.15}$$

for all i, j, k.

On the other hand, we rewrite (5.12) as

$$\sum_{i=1}^{I} \gamma_i = (\lambda_0 + 1 - J)\,\theta + \sum_k \eta_k.$$

By Lemma 5.2,

$$I \geq 2(\lambda_0 + 1 - J). \tag{5.16}$$

Comparing (5.14) and (5.16), we obtain that $I = 2\,(\lambda_0 + 1 - J)$ and then (5.15) holds. Furthermore at least $2\,(\lambda_0 + 1 - J)$ γ_i's are in $\mathbb{S}$. Now we divide case 2) into two subcases:

Subcase 2.1) $J < \lambda_0 + 1$.

Then at least one γ_{i_0} is in $\mathbb{S}$, i.e. w can be expressed in the form $e_{\gamma_{i_0}}\left(-\frac{1}{2}\right) w'$. Then $\tilde{w} = 0$ since $e^2_{\gamma_{i_0}}\left(-\frac{1}{2}\right) = 0$.

Subcase 2.2) $J = \lambda_0 + 1$.

Under this assumption we have $I = K = 0$. $w = e_\theta(-1)^{\lambda_0+1}1$. Then w' can be expressed in the form $e_\theta(-1)w''$ since $e_\theta(-1)$ commutes with $\Pi_{\alpha\in\mathbb{S}} e_\alpha\left(-\frac{1}{2}\right)$ $e_\theta\left(-\frac{3}{2}\right) e_\theta\left(-\frac{1}{2}\right)$. By (5.2), we see that (5.10) is equivalent to

$$H\left(e_{-\theta}(1)e_\theta(-1)^{\lambda_0+h^\vee+1}1, w''\right) = 0.$$

However it is well known that $e_{-\theta}(1)e_\theta(-1)^{\lambda_0+h^\vee+1}1 = 0$. □

We can choose $e_{\pm\alpha} \in \mathfrak{g}_{\pm\alpha}, \alpha \in \mathbb{S} \cup \{\theta\}$, in such a way that

$$[e_\alpha, e_{-\alpha}] = -\alpha \tag{5.17}$$

$$[e_{-\gamma_i}, e_\theta] = e_{\beta_i} \tag{5.18}$$

$$[e_{-\beta_i}, e_\theta] = -e_{\gamma_i} \tag{5.19}$$

Indeed, we pick $e_{-\gamma_i}$ and e_θ arbitrarily, and define e_{β_i} by the formula (5.18). Then (5.17) fixes e_{γ_i} and $e_{-\beta_i}$. The formula (5.19) holds automatically since

$$([e_{-\beta_i}, e_\theta], e_{-\gamma_i}) = (e_\theta, [e_{-\gamma_i}, e_{-\beta_i}]) = -(e_{\beta_i}, e_{-\beta_i}) = 1.$$

In the above notations, the singular vector v_{λ_0} can be written as (cf. (5.6))

$$v_{\lambda_0} = \bar{e}_{-\theta}(1)\bar{e}_{-\theta} \prod_{i=1}^{h^\vee-2} (\bar{e}_{-\beta_i}\bar{e}_{-\gamma_i}) \cdot e_\theta(-1)^{\lambda_0+h^\vee+1}1.$$

Note that v_{λ_0} is independent of the order of $\prod_{i=1}^{h^\vee-2} (\bar{e}_{-\beta_i}\bar{e}_{-\gamma_i})$. Now we rewrite this formula of singular vectors in terms of a PBW basis. We introduce a combinatorial symbol $[m]_n = m(m-1)\cdots(m-n+1)$.

Theorem 5.2 *One has:*

$$\begin{aligned} v_{\lambda_0} = \sum_{s=0}^{h^\vee-2} \sum_{(i_1,\ldots,i_s)} &(k+h^\vee)^{h^\vee-s-2} [\lambda_0 + h^\vee + 1]_{2(h^\vee-2)-s} \\ &\times \big((k+h^\vee)\,[\lambda_0 - s + 3]_2\, e_\theta(-1)^{\lambda_0-s+1} \\ &+ [\lambda_0 - s + 3]_3\, e_\theta(-1)^{\lambda_0-s}\bar{e}_\theta(-1)\bar{h}_\theta(-1) \\ &+ [\lambda_0 - s + 3]_4\, e_\theta(-1)^{\lambda_0-s-1}\bar{e}_\theta(-1)\bar{e}_\theta(-2)\big)\, Q_{i_1}\cdots Q_{i_s}\cdot 1, \end{aligned}$$

where $Q_i = \bar{e}_{\beta_i}(-1)\bar{e}_{\gamma_i}(-1)$, *and the sum* $\sum_{(i_1,\ldots,i_s)}$ *is taken over all subsets of the set* $\{1,\ldots,h^\vee-2\}$.

Proof. We assume that whenever some negative power of $e_\theta(-1)$ appears in the following, the corresponding monomial term is zero.

It is not hard to prove by induction that

$$\begin{aligned} &\bar{e}_{-\beta_i}\bar{e}_{-\gamma_i}\cdot e_\theta(-1)^n \\ &= n\,(k+h^\vee)\,e_\theta(-1)^{n-1} + n(n-1)e_\theta(-1)^{n-2}\bar{e}_{\beta_i}(-1)\bar{e}_{\gamma_i}(-1) \\ &\quad + e_\theta(-1)^n\bar{e}_{-\beta_i}\bar{e}_{-\gamma_i} - ne_\theta(-1)^{n-1}\bar{e}_{\beta_i}(-1)\bar{e}_{-\beta_i} \\ &\quad - ne_\theta(-1)^{n-1}\bar{e}_{\alpha_i}(-1)\bar{e}_{-\alpha_i}. \end{aligned} \tag{5.20}$$

It follows that

$$\begin{aligned} &\bar{e}_{-\beta_i}\bar{e}_{-\gamma_i}\cdot e_\theta(-1)^n\cdot 1 \\ &= n\,(k+h^\vee)\,e_\theta(-1)^{n-1}\cdot 1 + n(n-1)e_\theta(-1)^{n-2}\bar{e}_{\beta_i}(-1)\bar{e}_{\gamma_i}(-1)\cdot 1. \end{aligned} \tag{5.21}$$

Using (5.20) and (5.21), we get by induction that

$$\begin{aligned} v' &:= \prod_{i=1}^{h^\vee-2}\bar{e}_{-\beta_i}\bar{e}_{-\gamma_i}\cdot e_\theta(-1)^{\lambda_0+h^\vee+1}1 \\ &= \sum_{s=0}^{h^\vee-2}\sum_{(i_1,\ldots,i_s)}(k+h^\vee)^{h^\vee-s-2}[\lambda_0+h^\vee+1]_{(2(h^\vee-2)-s)} \\ &\qquad \times e_\theta(-1)^{\lambda_0-s+3}Q_{i_1}\cdots Q_{i_s}\cdot 1. \end{aligned}$$

Since

$$[\bar{e}_{-\theta}, Q_i] = 0$$

and

$$[\bar{e}_{-\theta}, e_\theta(-1)^n] = n(n-1)e_\theta(-1)^{n-2}\bar{e}_{-\theta}(-2) + ne_\theta(-1)^{n-1}\bar{h}_\theta(-1),$$

we have

$$\begin{aligned} \bar{e}_{-\theta}v' &= \sum_{s=0}^{h^\vee-2}\sum_{(i_1,\ldots,i_s)}(k+h^\vee)^{h^\vee-s-2}[\lambda_0+h^\vee+1]_{(2(h^\vee-2)-s)}\times \\ &\qquad \times\left([\lambda_0-s+3]_2\,e_\theta(-1)^{\lambda_0-s+1}\bar{e}_{-\theta}(-2)\right. \\ &\qquad \left.+(\lambda_0-s+3)\,e_\theta(-1)^{\lambda_0-s+2}\bar{h}_\theta(-1)\right)Q_{i_1}\cdots Q_{i_s}\cdot v_0. \end{aligned}$$

Using another identity

$$[\bar{e}_{-\theta}(1), e_\theta(-1)^n] = n(n-1)e_\theta(-1)^{n-2}\bar{e}_{-\theta}(-1) + ne_\theta(-1)^{n-1}\bar{h}_\theta,$$

we get the desired formula. □

Remark 5.3 *It follows from Theorem 5.2 that* $v_{\lambda_0} \neq 0$. *Moreover the only term which does not involve the odd factors is a non-zero multiple of* $e_\theta(-1)^{\lambda_0+1}$. *Therefore the submodule of the Verma module* $M(\Lambda + h^\vee d)$ *generated by* v_{λ_0} *is again a Verma module.*

Theorem 5.3 *We have the following isomorphism*

$$L(\Lambda + h^\vee d) \cong M(\Lambda + h^\vee d) / \left\langle v_{\lambda_0}, f_i^{\lambda_i+1}1, i = 1, \ldots, l \right\rangle,$$

where $\left\langle v_{\lambda_0}, f_i^{\lambda_i+1}1, i = 1, \ldots, l \right\rangle$ *denotes the submodule of* $M(\Lambda + h^\vee d)$ *generated by the singular vectors* $v_{\lambda_0}, f_i^{\lambda_i+1}1, i = 1, \ldots, l$.

Proof. Since the weights of $v_{\lambda_0}, f_i^{\lambda_i+1}1, i = 1, \ldots, l$ are $\Lambda + h^\vee d - (\lambda_i + 1)\alpha_i, i = 0, 1, \ldots, l$ respectively, we have the following isomorphism of $\hat{\mathfrak{g}}$-modules:

$$\begin{aligned}
&\left\langle v_{\lambda_0}, f_i^{\lambda_i+1}1, i = 1, \ldots, l \right\rangle \\
&= \sum_{i=0}^{l} M(\Lambda + h^\vee d - (\lambda_i + 1)\alpha_i) \\
&= \sum_{i=0}^{l} F \otimes \bar{M}(\Lambda - (\lambda_i + 1)\alpha_i) \\
&= F \otimes \overline{\left\langle f_i^{\lambda_i+1}1, i = 0, 1, \ldots, l \right\rangle}.
\end{aligned}$$

Therefore we have the following isomorphism of $\hat{\hat{\mathfrak{g}}}$-modules:

$$\begin{aligned}
&M(\Lambda + h^\vee d) \Big/ \left\langle v_{\lambda_0}, f_i^{\lambda_i+1}1, i = 1, \ldots, l \right\rangle \\
&= \frac{F \otimes \bar{M}(\Lambda)}{F \otimes \overline{\left\langle f_i^{\lambda_i+1}1, i = 0, 1, \ldots, l \right\rangle}} \\
&= F \otimes \bar{L}(\Lambda) \\
&= L(\Lambda + h^\vee d).
\end{aligned}$$

□

References

[A] A. Astashkevich, *On the structure of Verma modules over Virasoro and Neveu-Schwarz algebras*, preprint (1993).

[BPZ] A. Belavin, A. Polyakov and A. Zamolodchikov, *Infinite conformal symmetries in two-dimensional quantum field theory*, Nucl. Phys. **B241** (1984), 333–380.

[B] R. Borcherds, *Vertex algebras, Kac-Moody algebras, and the Monster*, Proc. Natl. Acad. Sci. USA. **83** (1986), 3068–3071.

[DMZ] C. Dong, G. Mason and Y. Zhu, *Discrete series of the Virasoro algebra and the moonshine module*, preprint.

[FF] B.L. Feigin and D.B. Fuchs, *Verma modules over the Virasoro algebra*, Lect. Notes Math. **1060** (1984) 230–245.

[FHL] I. B. Frenkel, Y. Huang and J. Lepowsky, *On axiomatic approaches to vertex operator algebras and modules*, Mem. Amer. Math. Soc., vol 104, No. 494 (1993).

[FLM] I. B. Frenkel, J. Lepowsky and A. Meurman, *Vertex operator algebras and the Monster*, Academic Press, New York (1988).

[FZ] I. B. Frenkel and Y. Zhu, *Vertex operator algebras associated to representations of affine and Virasoro algebra*, Duke Math. J., vol. 66, No. 1 (1992) 123–168.

[K] V. Kac, Infinite dimensional Lie algebras, third edition, Cambridge University Press (1990).

[KS] Y. Kazama and H. Suzuki, *New $N = 2$ superconformal field theories and superstring compactification*, Nucl. Phys. **B321** (1989), 232–268.

[KT] V. Kac and I. Todorov, *Superconformal current algebras and their unitary representations*, Comm. Math. Phys. **102** (1985), 337–347.

[KW] V. Kac and M. Wakimoto, *Modular invariant representations of infinite-dimensional Lie algebras and superalgebras*, Proc. Natl. Acad. Sci. USA, vol. 85 (1988), 4956–4960.

[KW] V. Kac and M. Wakimoto, *Unitarizable highest weight representations of the Virasoro, Neveu-Schwarz and Ramond algebras*, Lect. notes Phys. **261** (1986), 345–371.

[L] B.H. Lian, *On the classification of simple vertex operator algebras*, preprint (1992).

[T] H. Tsukada, *Vertex operator superalgebra*, Comm. Alg. **18(7)** (1990), 2249–2274.

[W] W. Wang, *Rationality of Virasoro vertex operator algebras*, Duke Math. J., IMRN, vol. 71, No. 1 (1993), 197–211.

[Z] Y. Zhu, *Vertex operator algebras, elliptic functions and modular forms*, Ph.D. dissertation, Yale Univ. (1990).

Department of Mathematics, MIT, Cambridge, MA 02139–4307
email addresses: kac@math.mit.edu, wqwang@math.mit.edu

Contemporary Mathematics
Volume 175, 1994

Topological invariants for 3-manifolds using representations of mapping class groups II: Estimating tunnel number of knots

TOSHITAKE KOHNO

Introduction

Let M be a closed oriented 3-manifold and k a knot in M. The purpose of this paper is to give a lower bound for the tunnel number $t(k)$ in terms of the Jones-Witten invariants for a knot in 3-manifold. The tunnel number is a topological invariant for a knot in 3-manifold defined in the following way. By adjoining arcs $\gamma_1, \cdots, \gamma_t$ to the knot k, we get a situation so that the closure of the complementary space of the regular neighborhood of $k \cup \gamma_1 \cup \cdots \cup \gamma_t$ is a handlebody (see Figure 1). The minimal number t needed in the above description is called the tunnel number and is denoted by $t(k)$.

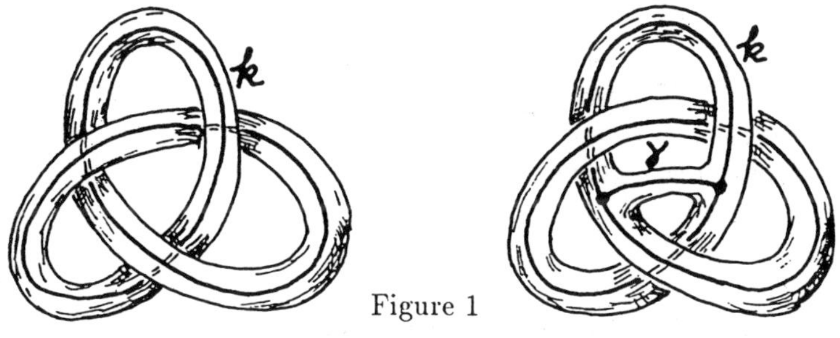

Figure 1

In [W], Witten introduced topological invariants for a link L in M associated with a finite dimensional complex simple Lie algebra $\mathcal{G}$ and a positive integer K called a level. This invariant has the following Dehn surgery description using invariants of colored framed link in S^3, which was studied extensively by Reshetikhin-Turaev [RT] using the notion of modular Hopf algebras and quantized universal enveloping algebras in the sense of Drinfel'd [D1] and Jimbo [J] at roots of unity.

Mathematics Subject Classification. 81T40, 57M25, 20F36.
Partially supported by Grant-in-Aid for Scientific Research on Priority Areas 231 "Infinite Analysis".

We denote by P_+ the set of dominant integral weights and we put

$$P_+(K) = \{\lambda \in P_+; 0 \leq (\lambda, \theta) \leq K\}$$

where $(\cdot,\cdot)$ is the Cartan-Killing form normalized so that $(\theta, \theta) = 2$ for the longest root θ. Let L be an oriented framed link in S^3 with m components. For a color $\lambda : \{1, \cdots, m\} \to P_+(K)$ we can define an invariant of a colored oriented framed link denoted by $J(L, \lambda)$ by means of the monodromy of Wess-Zumino-Witten model at level K. Let M be a closed oriented 3-manifold and L a link in M. We represent the pair (M, L) by the Dehn surgery on a framed link L' in S^3. Namely, we have disjoint links L and L' in S^3 and (M, L) is obtained by the Dehn surgery on the framed link L'. The number of components of L and L' is denoted by m and n respectively. For a color $\lambda : \{1, \cdots, m\} \to P_+(K)$ we define $Z_{\mathcal{G},K}(M, L; \lambda)$ by

$$Z_{\mathcal{G},K}(M, L; \lambda) = C^{sign(L')} \sum_{\mu} S_{0\mu(1)} \cdots S_{0\mu(n)} J(L \cup L', \lambda \cup \mu)$$

where the sum is taken for any color of L', $\mu : \{1, \cdots, n\} \to P_+(K)$. Here we follow a normalization due to Kirby-Melvin [KM] and *sign* stands for the signature of the linking matrix. The numbers C and $S_{0\lambda}$ are given explicitly by

$$C = \exp\left(-\frac{\pi\sqrt{-1}\,K \dim \mathcal{G}}{4(K + h^*)}\right)$$

$$S_{0\lambda} = \frac{1}{(K + h^*)^{l/2}} \left(\frac{vol\,\Lambda^w}{vol\,\Lambda^r}\right)^{1/2} \prod_{\alpha \in \Delta_+} 2 \sin \frac{\pi(\lambda + \rho, \alpha)}{K + h^*}.$$

Here in the above formulae, h^* denotes the dual Coxter number, $vol\,\Lambda^w$ is the volume of the weight lattice, $vol\,\Lambda^r$ is the volume of the coroot lattice, Δ_+ is the set of positive roots, and ρ is the half sum of positive roots.

A main result in this paper is the inequality

$$|Z_{\mathcal{G},K}(M, k; \lambda)| \leq S_{00}^{-t(k)-1}$$

for any $\lambda \in P_+(K)$, which gives a lower estimate for the tunnel number (Theorem 4.3). In particular, for a knot k in S^3 we obtain

$$|J(k, \lambda)| \leq S_{00}^{-t(k)-1}$$

for any $\lambda \in P_+(K)$. Considering the case $\mathcal{G} = sl(2, \mathbf{C})$, this inequality shows that the tunnel number has a lower bound described by special values of the Jones polynomial and its parallel version. More precisely, for $\lambda = 0, 1, \cdots, K$, using γ cable k^γ, we put

$$J(k, \lambda) = \sum_{j=0}^{[\lambda/2]} (-1)^j \binom{\lambda - j}{j} J(k^{\lambda - 2j}, 1)$$

where $J(k,1)$ is a variant of the Jones polynomial at $q = e^{2\pi\sqrt{-1}/(K+2)}$ (see 4.4). Our main result implies, in particular, that

$$|J(k,\lambda)| \leq \left(\sqrt{\frac{2}{K+2}} \sin\frac{\pi}{K+2}\right)^{-t(k)-1}$$

for any $\lambda = 0, 1, \cdots, K$. This case might be treated alternatively using quantum $6j$-symbols, which will be discussed in a separate publication.

Let us note that, considering the case k is empty, we obtain

$$|Z_{\mathcal{G},K}(M)| \leq S_{00}^{-g(M)}$$

which gives a lower estimate for the Heegaard genus of M denoted by $g(M)$. This inequality was also obtained by Walker [Wa] in a general setting of topological quantum field theory in the sense of Atiyah [A] (see also [Koh2] and [G]).

To prove our result an interpretation of the Jones polynomial in terms of the monodromy of the conformal field theory plays an essential role. Let Σ be a closed oriented surface of genus g. As the monodromy of the Wess-Zumino-Witten model we have a projective unitary representation of the mapping class group

$$\rho : \mathcal{M}_g \to GL(\mathcal{H}_\Sigma)$$

on the space of conformal blocks $\mathcal{H}_\Sigma$. In the preceding article [Koh1] we studied Witten's 3-manifold invariants based on this representation of the mapping class group and a Heegaard splitting. This approach using conformal field theory was also pursued independently by Crane [C] and by Cappell-Lee-Miller [CLM] from different points of view (see also an earlier account of Kontsevich [Ko]). We show that $Z_{\mathcal{G},K}(M,k;\lambda)$ is expressed in the form

$$S_{00}^{-g} < v_\lambda^*, \rho(h)v_0 >$$

with $v_0, v_\lambda \in \mathcal{H}_\Sigma$ using the gluing map h appearing in the Heegaard splitting description. Now our inequality is a consequence of the unitarity of the representation ρ.

The paper is organized in the following way. In section 1, we summarize basic facts on braiding and fusing matrices arising from the holonomy of the Knizhnik-Zamolodchikov equation [KZ]. A detailed treatment in the case of $sl(2,\mathbf{C})$ can be found in [TK]. We also refer the readers to [TUY] where the conformal field theory on the universal family of stable curves is formulated. In section 2, we review a method to obtain a linear operator associated with a colored framed tangle using the monodromy of conformal field theory. In section 3, we compare two approaches to Witten's invariants, Dehn surgery and Heegaard splitting (see also [P] and [G] for relations of various approaches). In section 4, we show our main result on a lower bound for the tunnel number. Section 5 focuses the symmetry coming from the Dynkin diagram automorphism. Generalizing the symmetry principle developed in [KM] in the case of $sl(2, \mathbf{C})$, we obtain a practical formula to compute our invariants for $sl(n, \mathbf{C})$. Using this method, we show periodicity of invariants for certain homology 3-spheres in the case of $sl(n, \mathbf{C})$.

1. A brief review of conformal field theory

Let $\mathcal{G}$ be a finite dimensional complex simple Lie algebra and we fix a positive integer K called a level. We denote by P_+ the set of dominant integral weights and we put

$$P_+(K) = \{\lambda \in P_+; 0 \leq (\lambda, \theta) \leq K\}$$

where $(\cdot, \cdot)$ is the Cartan-Killing form normalized so that $(\theta, \theta) = 2$ for the longest root θ. Let $\widehat{\mathcal{G}}$ denote the affine Lie algebra defined as a canonical central extension of the loop algebra $\mathcal{G} \otimes \mathbf{C}[t, t^{-1}]$. It is known by Kac (see [K]) that we can associate to $\lambda \in P_+(K)$ an irreducible representation $\mathcal{H}_\lambda$ of $\widehat{\mathcal{G}}$ called an integrable highest weight module, which contains V_λ, a finite dimensional irreducible $\mathcal{G}$ module of highest weight λ. By means of Sugawara construction we have operators L_n, $n \in \mathbf{Z}$, on $\mathcal{H}_\lambda$ satisfying the relation of the Virasoro Lie algebra with central charge $c = K \dim \mathcal{G}/(K + h^*)$ where h^* is the dual Coxeter number. With respect to the action of L_0 we can decompose $\mathcal{H}_\lambda$ into $\oplus \mathcal{H}_{\lambda,d}$, $d \in \mathbf{Z}_{\geq 0}$ where $\mathcal{H}_{\lambda,d}$ is the eigenspace of L_0 with eigenvalue $\Delta_\lambda + d$. Here Δ_λ is called a conformal weight and is given by

$$\Delta_\lambda = \frac{(\lambda, \lambda + 2\rho)}{2(K + h^*)}$$

with ρ, the half sum of the positive roots of $\mathcal{G}$. We denote by $\widehat{\mathcal{H}_\lambda}$ the direct product $\prod_{d \geq 0} \mathcal{H}_{\lambda,d}$ and we put $\mathcal{H} = \oplus_{\lambda \in P_+(K)} \mathcal{H}_\lambda$ and $\widehat{\mathcal{H}} = \oplus_{\lambda \in P_+(K)} \widehat{\mathcal{H}_\lambda}$.

Let us recall that a primary field of type $\mu \in P_+(K)$ is an operator

$$\Phi(z) : \mathcal{H} \otimes V_\mu \to \widehat{\mathcal{H}}$$

holomorphic with respect to $z \in \mathbf{C} \setminus \{0\}$ satisfying the following conditions.

$$[L_m, \Phi(v, z)] = z^m \{z \frac{\partial}{\partial z} + (m + 1)\Delta_\mu\} \Phi(v, z)$$
$$[X \otimes t^m, \Phi(v, z)] = z^m \Phi(Xv, z) \quad \text{for } X \in \mathcal{G}$$

where we write $\Phi(v, z)(u)$ for $\Phi(z)(u \otimes v)$. A component of the primary field sending $\mathcal{H}_\lambda \otimes V_\mu$ to $\widehat{\mathcal{H}_\nu}$ is called a chiral vertex operator of type $(\lambda, \mu; \nu)$ and the space of all such chiral vertex operators is denoted by $\mathcal{H}_{\lambda\mu}^{\nu}$. This space is naturally embedded in the space of intertwiners $Hom_{\mathcal{G}}(V_\lambda \otimes V_\mu, V_\nu)$.

For $\lambda_1, \cdots, \lambda_n \in P_+(K)$, we set

$$\mathcal{H}_{\lambda_1 \cdots \lambda_n} = \oplus_{\mu_1, \cdots, \mu_{n-1} \in P_+(K)} \mathcal{H}_{0\lambda_1}^{\mu_1} \otimes \mathcal{H}_{\mu_1\lambda_2}^{\mu_2} \otimes \cdots \mathcal{H}_{\mu_{n-1}\lambda_n}^{0}$$

which is considered to be a subspace of $Hom_{\mathcal{G}}(V_{\lambda_1} \otimes \cdots \otimes V_{\lambda_n}, \mathbf{C})$ (see Figure 1.1).

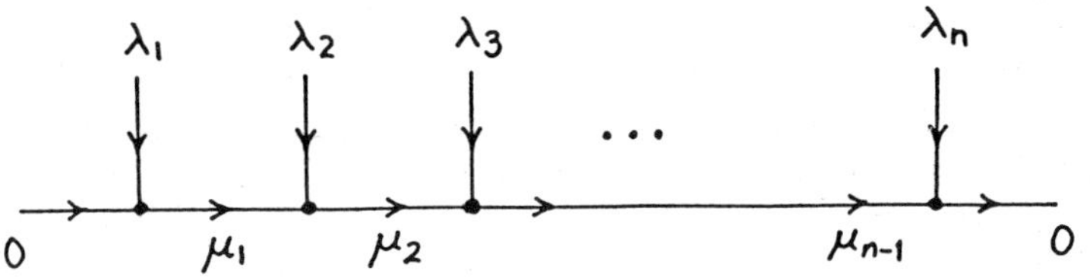

Figure 1.1

Let $\Phi_j(z), 1 \leq j \leq n$, be chiral vertex operators of type $(\mu_{j-1}, \lambda_j; \mu_j)$ where $\mu_0 = \mu_n = 0$. For the highest weight vector $|0> \in \mathcal{H}_0$ and for $<0| \in \mathcal{H}_0^\dagger$ we define the n point function

$$\phi(z) = <0|\Phi_n(z_n) \cdots \Phi_1(z_1)|0>$$

which is known to satisfy the Knizhnik-Zamolodchikov equation

$$\frac{\partial \phi}{\partial z_i} = (K + h^*) \sum_{j \neq i} \frac{\Omega_{ij}}{z_i - z_j} \phi. \tag{1-2}$$

It turns out that the n point function $\phi(z)$ is holomorphic in the region $0 \leq |z_1| \leq \cdots \leq |z_n|$ and is analytically continued to a multi-valued holomorphic function on the space

$$X_n = \{(z_1, \cdots, z_n) \in \mathbf{C}^n; z_i \neq z_j \, \text{for}\, i \neq j\}.$$

Considering the holonomy of the Knizhnik-Zamolodchikov equation in the case of the 4 point functions, we have the following linear isomorphisms called braiding and fusing operators

$$B \begin{bmatrix} \lambda_2 & \lambda_3 \\ \lambda_1 & \lambda_4 \end{bmatrix} : \oplus_\mu \mathcal{H}_{\lambda_1\lambda_2}^{\mu} \otimes \mathcal{H}_{\mu\lambda_3}^{\lambda_4} \to \oplus_{\mu'} \mathcal{H}_{\lambda_1\lambda_3}^{\mu'} \otimes \mathcal{H}_{\mu'\lambda_2}^{\lambda_4} \tag{1-3}$$

$$F \begin{bmatrix} \lambda_2 & \lambda_3 \\ \lambda_1 & \lambda_4 \end{bmatrix} : \oplus_\mu \mathcal{H}_{\lambda_1\lambda_2}^{\mu} \otimes \mathcal{H}_{\mu\lambda_3}^{\lambda_4} \to \oplus_{\mu'} \mathcal{H}_{\lambda_2\lambda_3}^{\mu'} \otimes \mathcal{H}_{\lambda_1\mu'}^{\lambda_4} \tag{1-4}$$

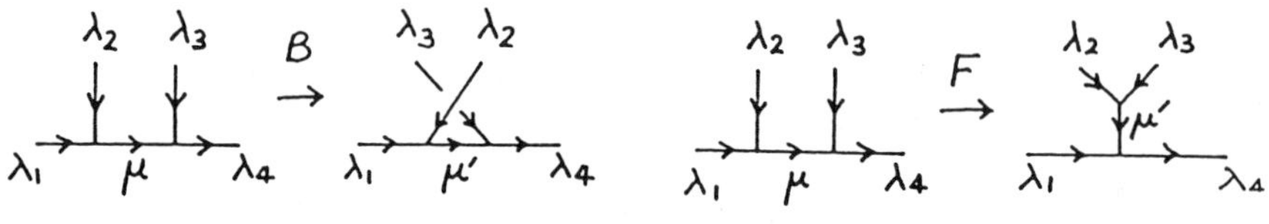

Figure 1.5

which are often indicated schematically as in Figure 1.5. The pentagon and hexagon relations among these operators follow from the integrability of the Knizhnik-Zamolodchikov equation (see also [D2]).

We recall basic results on the modular properties of the characters of integrable highest weight modules. Let $\mathcal{H}_\lambda$, $\lambda \in P_+(K)$, be the integrable highest weight module of $\widehat{\mathcal{G}}$ and we consider its character $\chi_\lambda(\tau) = Tr_{\mathcal{H}_\lambda} q^{L_0 - c/24}$ with $q = e^{2\pi\sqrt{-1}\tau}$, $Im\,\tau > 0$. Using the action of the classical Weyl group W on the weight lattice of $\mathcal{G}$, for λ, $\mu \in P_+(K)$ we set

$$(1\text{-}6)\qquad \begin{aligned} S_{\lambda\mu} &= \alpha_{\mathcal{G},K} \sum_{w \in W} det(w) \exp\left(-\frac{2\pi\sqrt{-1}}{K+h^*}(w(\lambda+\rho), \mu+\rho)\right), \\ T_{\lambda\mu} &= \delta_{\lambda\mu} \exp 2\pi\sqrt{-1}\left(\Delta_\lambda - \frac{c}{24}\right). \end{aligned}$$

with the normalization constant

$$(1\text{-}7)\qquad \alpha_{\mathcal{G},K} = \frac{\sqrt{-1}^{|\Delta_+|}}{(K+h^*)^{l/2}} \left(\frac{vol\,\Lambda^w}{vol\,\Lambda^r}\right)^{1/2}$$

where Δ_+ is the set of positive roots of $\mathcal{G}$, l is the rank of $\mathcal{G}$, $vol\,\Lambda^w$ is the volume of the weight lattice, $vol\,\Lambda^r$ is the volume of the coroot lattice, and c is the central charge of the Virasoro algebra.

A fundamental result in [KP] (see also [KW]) is as follows.

1.8 Theorem. *(Kac-Peterson) The characters χ_λ, $\lambda \in P_+(K)$, behaves with respect to the modular transformations as*

$$\chi_\lambda(-1/\tau) = \sum_{\mu \in P_+(K)} S_{\lambda\mu}\, \chi_\mu(\tau),$$

$$\chi_\lambda(\tau+1) = \exp 2\pi\sqrt{-1}\left(\Delta_\lambda - \frac{c}{24}\right) \chi_\lambda(\tau).$$

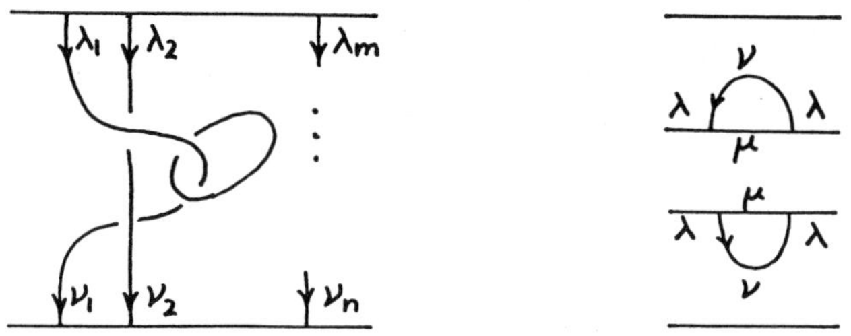

Figure 2.1

The matrices $S = (S_{\lambda\mu})$ and $T = (T_{\lambda\mu})$ are unitary and symmetric and satisfy the relation

$$(ST)^3 = S^2 = (\delta_{\lambda\mu^\dagger}),$$

where $\lambda^\dagger = -w(\lambda)$ with the longest element $w \in W$ and is the highest weight for the dual representation V_λ^*. We make use of the above relation in the following form.

1.9 **Lemma.** *With the central charge c of the Virasoro algebra we put*

$$C = \left(\exp 2\pi\sqrt{-1}\frac{c}{24}\right)^{-3}.$$

Then we have

$$C\sum_{\mu} S_{\lambda\mu}S_{\mu\nu} \exp 2\pi\sqrt{-1}(\Delta_\lambda + \Delta_\mu + \Delta_\nu) = S_{\lambda\nu}.$$

2. Framed tangles and associated operators

Let T be a diagram of an oriented framed tangle, where we suppose that the diagram is good in the sense that the framing is represented as a blackboard framing (see [KM]). By a color of the tangle T we mean a assignment of an element of $P_+(K)$ to each arc in T and we adopt a usual convention that if we change the the orientation of an arc we replace the associated λ by $\lambda^\dagger$.

Let us consider a colored framed tangle shown in Figure 2.1. The fusion operators defined in 1.4 give linear maps

$$F_{\nu 0}[\lambda] : \oplus_\mu \mathcal{H}^\mu_{\lambda\nu} \otimes \mathcal{H}^\lambda_{\mu\nu^\dagger} \to \mathcal{H}^0_{\nu\nu^\dagger} \otimes \mathcal{H}^\lambda_{\lambda 0} \cong \mathbf{C}$$
$$F_{0\nu}[\lambda] : \mathbf{C} \cong \mathcal{H}^0_{\nu\nu^\dagger} \otimes \mathcal{H}^\lambda_{\lambda 0} \to \oplus_\mu \mathcal{H}^\mu_{\lambda\nu} \otimes \mathcal{H}^\lambda_{\mu\nu^\dagger}.$$

By normalizing these, we define annihilation and creation operator by

$$\sqrt{S_{0\nu}/S_{00}}F_{\nu 0}[\lambda] \quad \text{and} \quad \sqrt{S_{0\nu}/S_{00}}F_{0\nu}[\lambda]$$

respectively.

Let T be an oriented framed (m,n) tangle with colors shown as in Figure 2.1. By decomposing T into elementary tangle as in Figure 2.2 and by associating to each elementary tangle the braiding, creation or annihilation operator defined above, we obtain a linear map

$$J(T):\mathcal{H}_{\lambda_1\cdots\lambda_n}\to\mathcal{H}_{\nu_1\cdots\nu_m}.$$

In particular, for the diagram of an oriented framed link L with m components and a color $\lambda:\{1,\cdots,m\}\to P_+(K)$ we have a complex number $J(L,\lambda)$. It has been shown by many authors that it is an invariant of an oriented colored framed link (see for example [MS] and [F]). Let us notice that to prove the invariance of $J(T)$ under the move shown in Figure 2.2, we can make use of the equality $S_{0\lambda}/S_{00}=1/F_\lambda$ known in conformal field theory, where

$$F_\lambda=F_{00}\begin{bmatrix}\lambda & \lambda\\ \lambda & \lambda\end{bmatrix}.$$

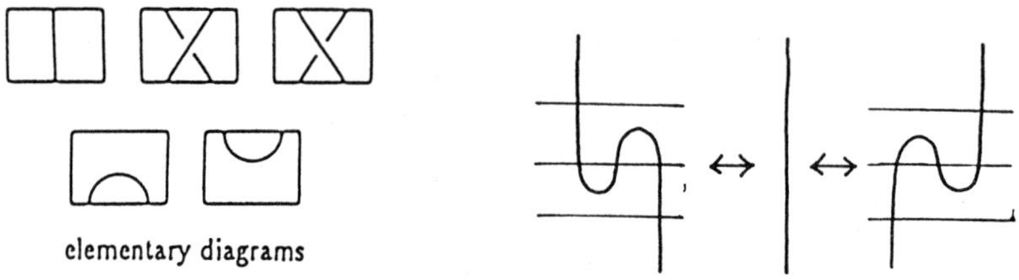

Figure 2.2

We put

$$\mathcal{H}_n=\oplus_{\lambda_1,\cdots,\lambda_n\in P_+(K)}\mathcal{H}_{\lambda_1,\cdots,\lambda_n}.$$

As is explained in [MS] and [Koh1] we obtain a representation of the mapping class group of an n-holed sphere

$$\rho:\mathcal{M}_{0,n}\to GL(\mathcal{H}_n).$$

The Dehn twist with respect to a circle parallel to the i-th hole acts as a scalar multiplication by $\exp(-2\pi\sqrt{-1}\Delta_{\lambda_i})$ on $\mathcal{H}_{\lambda_1\cdots\lambda_n}$. Let us denote by σ_i, $1\le i\le n-1$, the braiding operations interchanging the i-th th and $(i+1)$-st holes in the anti-clockwise direction. The braiding operators $\rho(\sigma_1\sigma_2\cdots\sigma_j)^{j+1}, 1\le j\le n-1$, act as scalar multiplication by

$$\exp\left(2\pi\sqrt{-1}(\Delta_{\mu_{j+1}}-\sum_{k=1}^{j+1}\Delta_{\lambda_k})\right).$$

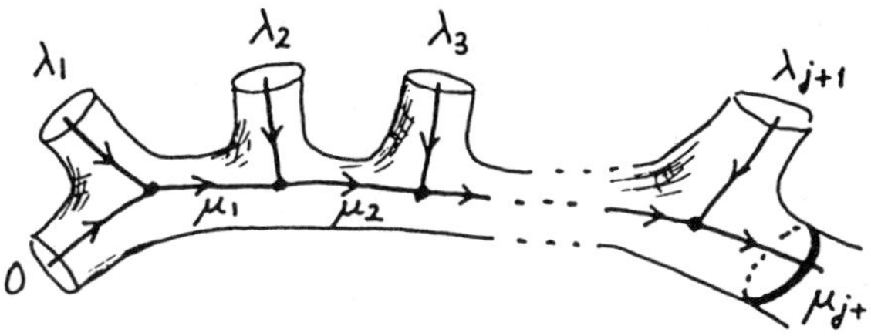

Figure 2.3

on $\mathcal{H}_{0\lambda_1}^{\mu_1} \otimes \mathcal{H}_{\mu_1\lambda_2}^{\mu_2} \otimes \cdots \otimes \mathcal{H}_{\mu_{n-1}\lambda_n}^{0}$ (see Figure 2.3).

Let us recall that the fusion algebra $R_{\mathcal{G},K}$ of the conformal field theory for $\widehat{\mathcal{G}}$ with level K is the algebra over $\mathbf{C}$ generated by ϕ_λ, $\lambda \in P_+(K)$, with relations

$$\phi_\lambda \cdot \phi_\mu = \sum_\nu N_{\lambda\mu}^\nu \phi_\nu$$

where $N_{\lambda\mu}^\nu$ is the dimension of the space of chiral vertex operators $\mathcal{H}_{\lambda\mu}^\nu$. It follows from the isomorhpisms 1.3 and 1.4 that $R_{\mathcal{G},K}$ is a commutative and associative algebra. Using the move depicted in Figure 2.4, it is clear from our construction that we have the following proposition (see also [MoS]).

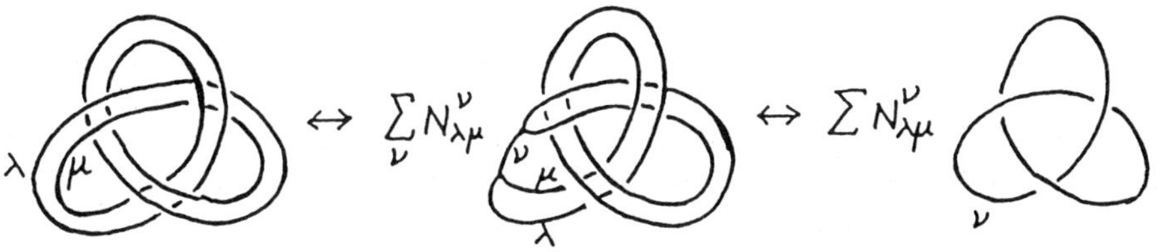

Figure 2.4

2.5 **Proposition**. *Let L be an oriented framed link with m components. The link invariant $J(L,\lambda)$ for $\lambda : \{1,\cdots,m\} \to P_+(K)$ gives a linear map compatible with the above product*

$$(R_{\mathcal{G},K})^{\otimes m} \to \mathbf{C}.$$

By the construction of the creation and annihilation operators we have the following lemma.

2.6 **Lemma**. *Let T_1 and T_2 be oriented framed $(1,1)$-tangle diagrams with colors composed as in Figure 2.7. We consider the link diagram $\widehat{T_1 \circ T_2}$ obtained by closing the composition $T_1 \circ T_2$. Then, we have*

$$J(\widehat{T_1 \circ T_2}) = J(\widehat{T_1})J(\widehat{T_2})\frac{S_{00}}{S_{0\lambda}}.$$

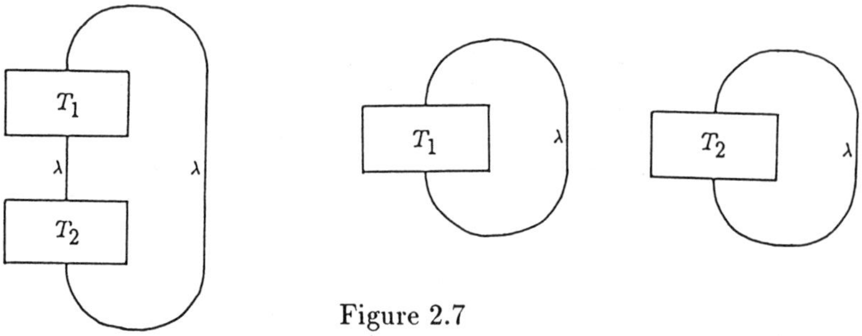

Figure 2.7

2.8 Proposition. *Let T be a colored oriented framed tangle. Then, with respect to the following local modifications of the tangle we have*

$$J \,)O^\lambda = \frac{S_{0\lambda}}{S_{00}} \, J)$$

$$J \, \alpha^\lambda = \exp 2\pi \sqrt{-1} \Delta_\lambda \, J \, (^\lambda,$$

$$J \, \overset{\lambda}{\underset{}{\mathcal{O}}} {}^\mu = \frac{S_{\lambda\mu}}{S_{0\lambda}} \, J \, \{^\lambda.$$

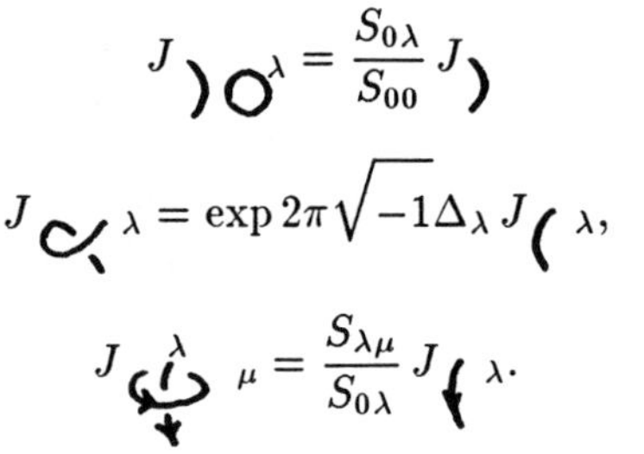

Outline of Proof: The first equality follows from our normalization of the creation and annihilation operators. The second equality corresponds to the fact that our construction gives a representation of the mapping class group of an n-holed sphere so that the Dehn twist with respect to a circle parallel to the boundary is represented by the scalar multiplication by $\exp(-2\pi\sqrt{-1}\Delta_\lambda)$. To show the last equality let us consider the oriented Hopf link H with colors $\lambda, \mu \in P_+(K)$ as shown in Figure 2.10a. We compute the associated invariant $J(H; \lambda, \mu)$. By means of the presentation of the Hopf link depicted in Figure 2.10b it follows from Proposition 2.5 that

$$\exp 2\pi \sqrt{-1}(-\Delta_\lambda - \Delta_\mu) J(H; \lambda, \mu) = \sum_\nu N^\nu_{\lambda\mu} \exp 2\pi \sqrt{-1}(-\Delta_\nu) \frac{S_{0\nu}}{S_{00}}.$$

Applying the Verlinde identity

$$(2\text{-}9) \qquad\qquad N^\nu_{\lambda\mu} = \sum_{\alpha \in P_+(K)} \frac{S_{\lambda\alpha} S_{\mu\alpha} S^*_{\nu\alpha}}{S_{0\alpha}}$$

for the dimension of the space of the chiral vertex operators (see [V], [MS] and [K]), we have

$$J(H; \lambda, \mu) = \sum_{\nu, \alpha} \frac{S_{\lambda\alpha} S_{\mu\alpha} S^*_{\nu\alpha} S_{0\nu}}{S_{0\alpha} S_{00}} \exp 2\pi \sqrt{-1}(\Delta_\lambda + \Delta_\mu - \Delta_\nu).$$

Noticing that $S_{0\lambda}$ is real for $\lambda \in P_+(K)$, we have

$$C^{-1}\sum_{\nu} S_{\nu\alpha}S_{0\nu}\exp 2\pi\sqrt{-1}(-\Delta_\alpha - \Delta_\nu) = S_{0\alpha}$$

by Lemma 1.9. This identity implies

$$J(H;\lambda,\mu) = \frac{S_{\lambda\mu}}{S_{00}}$$

by applying Lemma 1.9 again. The assertion follows immediately combining with Lemma 2.6.

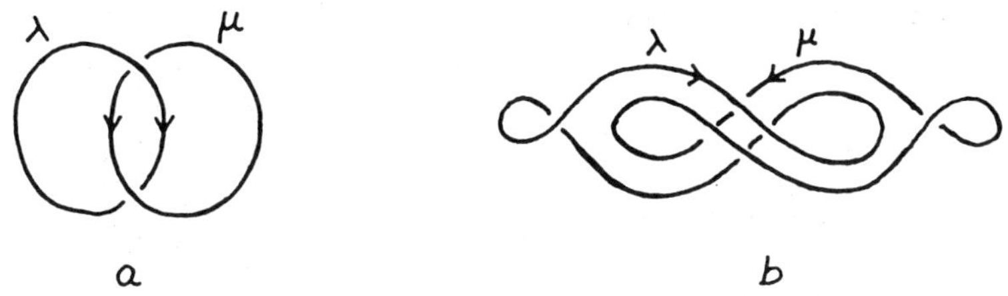

Figure 2.10

3. Projective representations of mapping class groups and 3-manifold invariants

Let M be a closed oriented 3-manifold obtained by the Dehn surgery on a framed link L with m components in S^3. For a complex simple Lie algebra $\mathcal{G}$ and a positive integer K we consider

$$Z_{\mathcal{G},K}(M) = C^{sign(L)}\sum_{\lambda} S_{0\lambda(1)}\cdots S_{0\lambda(m)}\, J(L,\lambda) \tag{3-1}$$

where the sum is for any color $\lambda : \{1,\cdots,m\} \to P_+(K)$, $sign(L)$ denotes the signature of the linking matrix of L and C is as in Lemma 1.9. Let us also recall that $S_{0\lambda}$ is expressed as

$$S_{0\lambda} = \frac{1}{(K+h^*)^{l/2}}\left(\frac{vol\,\Lambda^w}{vol\,\Lambda^r}\right)^{1/2}\prod_{\alpha\in\Delta_+} 2\sin\frac{\pi(\lambda+\rho,\alpha)}{K+h^*}.$$

To define $J(L,\lambda)$ we fix an orientation on the link L, but the above sum does not depend on the orientation on the link since we have the involution † on the set $P_+(K)$. As is shown in the case of $sl(2,\mathbf{C})$ by Reshetikhin-Turaev [RT] and by Kirby-Melvin [KM] using the quantum group at roots of unity, our construction based on the conformal field theory leads us to the following result.

3.2 Theorem. *The above $Z_{\mathcal{G},K}(M)$ is a topological invariant for M and does not depend on the choice of a framed link appearing in the Dehn surgery description. More precisely, if there exists an orientation preserving homeomorphism $M_1 \cong M_2$, we have $Z_{\mathcal{G},K}(M_1) = Z_{\mathcal{G},K}(M_2)$.*

Outline of Proof: The proof is completely analogous to that in [RT]. It is enough to prove that the sum 3.1 is invariant under Kirby moves shown in Figure 3.3 (see [Ki] and [FR]). In the case $n = 0, 1$ this follows directly from Lemma 1.9 and Lemma 2.6. To proceed inductively with respect the number of strands n, we use the fusing operation for the first two strands passing though the circle in Figure 3.3 and apply the consistency for brading and fusing operators. Since this argument is used by many authors, we omit the details.

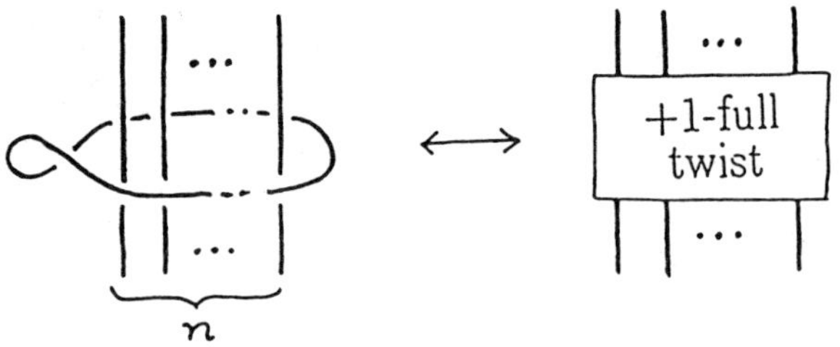

Figure 3.3

Let us notice that in contrast to the Witten's original normalization in [W], we have

$$Z_{\mathcal{G},K}(S^3) = 1, \quad Z_{\mathcal{G},K}(S^1 \times S^2) = S_{00}^{-1}.$$

Let Σ be a closed oriented surface of genus g and we denote by $\mathcal{M}_g$ the mapping class group defined as the group of orientation preserving diffeomorphisms of Σ modulo isotopy. We define the space of conformal blocks for Σ in a combinatorial manner. First, we fix a cut system $\Gamma = \{C_1, \cdots, C_g\}$ illustrated as in Figure 3.5 in the sense of [HT] to get a $2g$ holed sphere and we define $\mathcal{H}_\Gamma$ by

$$\mathcal{H}_\Gamma = \oplus_{\lambda_1, \cdots, \lambda_g \in P_+(K)} \mathcal{H}_{\lambda_1 \lambda_1^\dagger \cdots \lambda_g \lambda_g^\dagger}. \tag{3-4}$$

Let $\alpha_i, 2 \leq i \leq g$, $\beta_i, 1 \leq i \leq g$, and $\delta_i, 1 \leq i \leq g$, be Lickorish generators of the mapping class group $\mathcal{M}_g$, which are Dehn twists along circles shown in Figure 3.5 (see [L]). We are going to construct a projective representation of $\mathcal{M}_g$ on the vector space $\mathcal{H}_\Gamma$ defined above.

To each x_i, $(x = \alpha, \beta, \delta)$ we associate a $(2g, 2g)$ tangle $T(x_i)$ depicted as in Figure 3.5 and for x_i^{-1} we associate its mirror image $T(x_i^{-1})$. Here we put $\delta_1 = \alpha_1$. Let T_0 be a $(2g, 2g)$ tangle shown in Figure 3.5. Given $x \in \mathcal{M}_g$ we express x using Lickorish generators as $x_{i_1}^{\epsilon_1} \cdots x_{i_k}^{\epsilon_k}$, $\epsilon_i = \pm 1$ and we define a tangle $T(x)$ by the composition

$$T(x) = T_0 \circ T(x_{i_1}^{\epsilon_1}) \circ \cdots \circ T(x_{i_k}^{\epsilon_k}).$$

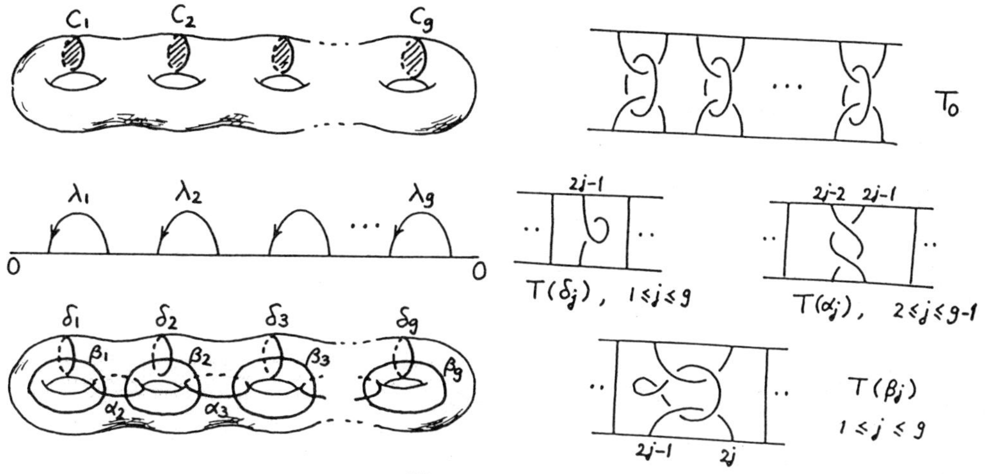

Figure 3.5

Let $T(x)$ be an oriented framed $(2g, 2g)$ tangle defined for $x \in \mathcal{M}_g$ in the previous paragraph by fixing a way of writing x in terms of Lickorish generators. We divide the set of arcs in $T(x)$ into $E^0(x)$, $E^1(x)$ and $L(x)$ defined by

$$
\begin{aligned}
E^0(x) &= \{\gamma; \partial\gamma \subset \mathbf{R}^2 \times \{0\}\}, \\
E^1(x) &= \{\gamma; \partial\gamma \subset \mathbf{R}^2 \times \{1\}\}, \\
L(x) &= \{\gamma; \partial\gamma = \phi\}
\end{aligned}
$$

(see Figure 3.6). We define a linear map

$$\rho(x) : \mathcal{H}_\Gamma \to \mathcal{H}_\Gamma$$

in the following way. For

$$u \in \mathcal{H}_{\lambda_1\lambda_1^\dagger\cdots\lambda_g\lambda_g^\dagger}, \quad v \in \left(\mathcal{H}_{\nu_1\nu_1^\dagger\cdots\nu_g\nu_g^\dagger}\right)^*$$

we associate the colors to the arcs in $E^0(x)$ and $E^1(x)$ as shown in Figure 3.6. As is explained above, each time we associate a color $\mu : \{1, 2, \cdots, m\} \to P_+(K)$ to the components of the oriented framed link $L(x) = L_1 \cup \cdots \cup L_m$ we have a linear map $J(T(x), \mu) : \mathcal{H}_\Gamma \to \mathcal{H}_\Gamma$. We define $\rho(x)$ by

$$
\begin{aligned}
< v, \rho(x)u > &= \sqrt{S_{0\lambda_1}\cdots S_{0\lambda_g}}\sqrt{S_{0\nu_1}\cdots S_{0\nu_g}} \\
&\quad \sum_\mu C^{sign(L(x))} S_{0\mu(1)}\cdots S_{0\mu(m)} J(T(x), \mu)
\end{aligned}
\tag{3-7}
$$

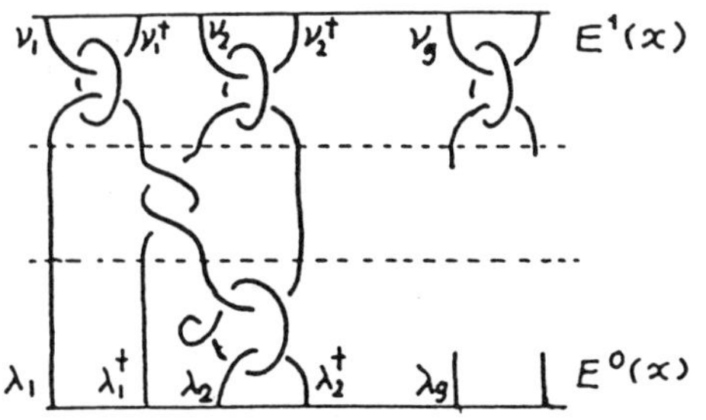

Figure 3.6

where the sum is for all colors of the components of the link $L(x)$ and $< \cdot, \cdot >$ stands for the canonical paring $\mathcal{H}_\Gamma^* \times \mathcal{H}_\Gamma \to \mathbf{C}$.

For $x, y \in \mathcal{M}_g$ we fix a way of writing x and y in terms of Lickorish generators and we define $(2g, 2g)$ tangles $T(x)$, $T(y)$ and $T(xy)$ in the above way. Corresponding to this we have oriented framed links $L(x)$, $L(y)$ and $L(xy)$. We put

$$\sigma(x, y) = sign(L(xy)) - sign(L(x)) - sign(L(y)) \tag{3-8}$$

and we define $\xi(x, y)$ by

$$\xi(x, y) = C^{\sigma(x,y)}.$$

3.9 **Proposition**. *The above construction defines a projective unitary representation*

$$\rho : \mathcal{M}_g \to GL(\mathcal{H}_\Gamma)$$

with a 2-cocycle ξ. Namely, we have

$$\rho(xy) = \xi(x, y)\rho(x)\rho(y)$$

for any $x, y \in \mathcal{M}_g$.

Proof: Let us first notice that for $x \in \mathcal{M}_g$ the Dehn surgery on the framed link $L(x)$ in the tangle diagram $T(x)$ gives a 3-manifold obtained by attaching a handlebody V_g with its second copy by a diffeomorphism $x : \partial V_g \to \partial V_g$. Let us suppose that x and x' are equal in the mapping class group. Writing x and x' in terms of Lickorish generators, we obtain tangles $T(x')$ and $T(x)$. By means of a relative version of Kirby's theorem (see [Ki] and [RT]) we can conclude that the tangle diagram $T(x')$ is obtained from $T(x)$ by a finitely many steps of Kirby moves allowing any arc in the tangle diagram as a strand passing through the circle shown in Figure 3.3. From our construction the operator $\rho(x)$ is invariant under these moves, which shows the well-definedness.

If we compose x and y, by means of the formula 3.6 we see that one gets an extra factor $\xi(x, y)$. Since ρ is a projective representation and a value of the

2-cocycle ξ is a root of unity, to prove the unitarity it suffices to show that each Lickorish generator is represented by a unitary matrix. For α_i and δ_i it follows immediately from the known fact in conformal field theory that under a certain normalization of chiral vertex operators the braiding and fusing operators are represented by unitary matrices. For β_i, the unitarity of $\rho(\beta_i)$ follows from $\rho(\beta_i^{-1}) = \rho(\beta_i)^*$. This completes the proof.

Let M be a closed oriented 3-manifold. It is well-known that M admits a Heegaard splitting $M = V_g \cup_h (-V_g)$, which means that M is obtained from a handlebody of genus g denoted by V_g and its copy by attaching the boundary with $h : \partial V_g \to \partial V_g$.

The following theorem gives a description of the invariant $Z_{\mathcal{G},K}(M)$ in terms of a Heegaard splitting and the representation of the mapping class group ρ defined in the previous section. We fix a normalization of chiral vertex operators such that $\rho(h)$ is represented by a unitary matrix for any $h \in \mathcal{M}_g$. This determines a Hermitian metric on $\mathcal{H}_\Gamma$ invariant under the action of $\mathcal{M}_g$. We take $v_0 \in \mathcal{H}_\Gamma$ from $\mathcal{H}_{00}^0 \otimes \cdots \otimes \mathcal{H}_{00}^0 \cong \mathbf{C}$ so that v_0 corresponds to the tensor product of identity operators. Let us denote by v_0^* its dual element in the dual space $\mathcal{H}_\Gamma^*$.

3.10 **Theorem.** *Let $M = V_g \cup_h (-V_g)$ be a Heegaard splitting with $h \in \mathcal{M}_g$. Then,*

$$S_{00}^{-g} < v_0^*, \rho(h) v_0 >$$

is a topological invariant for M and we have

$$Z_{\mathcal{G},K}(M) = S_{00}^{-g} < v_0^*, \rho(h) v_0 > .$$

Proof: The topological invariance of $S_{00}^{-g} < v_0^*, \rho(h) v_0 >$ might be seen directly from Reidemeister-Singer's stabilization theorem (see [Cr]). By performing an elementary stabilization to get $\tilde{h} \in \mathcal{M}_{g+1}$, we have

$$< v_0^*, \rho(\tilde{h}) v_0 >= S_{00} < v_0^*, \rho(h) v_0 > .$$

Hence it is enough to show that v_0 is invariant by any element of the mapping class group extended to a diffeomorphism of the handlebody. A system of generators of such subgroup of $\mathcal{M}_g$ was obtained by Suzuki [S] and we can check that v_0 is invariant by Suzuki's generators (see [Kohl] for details). As is noticed in the proof of Proposition 3.9, by the Dehn surgery on the framed link $L(h)$ we obtain our 3-manifold M. Applying the formula 3.6, we obtain the equality.

4. Estimating Heegaard genus and tunnel number

Let M be a closed oriented 3-manifold and let k be a knot in M. We denote by $E(k)$ the exterior of k. Namely, $E(k)$ is the closure of the complementary set of the tubular neighborhood $N(k)$ denoted by $cl(M - N(k))$. We represent $E(k)$ as

$$E(k) = cl(V - N(k)) \cup_h (-V)$$

with a gluing map $h : \partial V \to \partial V$. Here V is a handlebody and the knot k is embedded in a simple position as shown in Figure 4.1, which means that there exists a separating disk D with $\partial D \subset \partial V$ so that k is embedded as a core of the solid torus obtained by cutting the handlebody along D. Let g be the genus of ∂V. The minimal number $g-1$ such that the knot exterior $E(k)$ admits the above splitting is called the tunnel number of k and is denoted by $t(k)$.

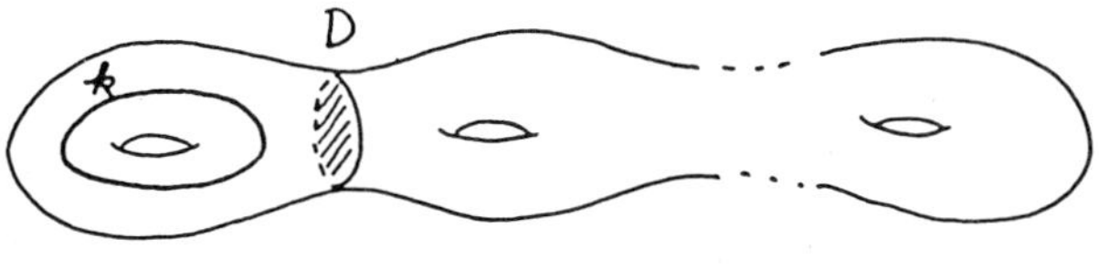

Figure 4.1

Let M be a closed oriented 3-manifold and L a link in M. We represent the pair (M, L) by the Dehn surgery on a framed link L' in S^3. Namely, we have disjoint links L and L' in S^3 and (M, L) is obtained by the Dehn surgery on a framed link L'. Let $\mathcal{G}$ be a finite dimensional complex simple Lie algebra and K a positive integer. The number of components of L and L' is denoted by m and n respectively. For a color $\lambda : \{1, \cdots, m\} \to P_+(K)$ we define $Z_{\mathcal{G},K}(M, L; \lambda)$ by

$$Z_{\mathcal{G},K}(M, L; \lambda) = C^{sign(L')} \sum_{\mu} S_{0\mu(1)} \cdots S_{0\mu(n)} J(L \cup L', \lambda \cup \mu) \tag{4-2}$$

where the sum is taken for any color of L', $\mu : \{1, \cdots n\} \to P_+(K)$. As in [RT] one can show that $Z_{\mathcal{G},K}(M, L; \lambda)$ is a topological invariant of the colored link L in M up to a root of unity coming from the choice of framing of L.

4.3 Theorem. *Let M be a closed oriented 3-manifold and let k be a knot in M. Let $\mathcal{G}$ be a finite dimensional complex simple Lie algebra and K a positive integer. For any $\lambda \in P_+(K)$ we have*

$$|Z_{\mathcal{G},K}(M, k; \lambda)| \leq S_{00}^{-t(k)-1}$$

which gives a lower bound for the tunnel number $t(k)$. Here S_{00} is given by

$$S_{00} = \frac{1}{(K+h^*)^{l/2}} \left(\frac{vol\,\Lambda^w}{vol\,\Lambda^r}\right)^{1/2} \prod_{\alpha\in\Delta_+} 2\sin\frac{\pi(\rho,\alpha)}{K+h^*}.$$

In particular, for a knot k in S^3 we have

$$|J(k,\lambda)| \leq S_{00}^{-t(k)-1}$$

for any $\lambda \in P_+(K)$.

Proof: Let us suppose that the knot exterior $E(k)$ admits a splitting $E(k) = cl(V - N(k)) \cup_h (-V)$ with a handlebody V of genus g. We take v_λ from

$$\mathcal{H}_{0\lambda}^{\lambda} \otimes \mathcal{H}_{\lambda\lambda^\dagger}^{0} \otimes \mathcal{H}_{00}^{0} \otimes \cdots \otimes \mathcal{H}_{00}^{0} \cong \mathbf{C}$$

so that v_λ has norm one with respect to the Hermitian metric on $\mathcal{H}_\Gamma$ invariant under the action of $\mathcal{M}_g$ via ρ. It is clear from the construction in the previous section that we have

$$|Z_{\mathcal{G},K}(M,k;\lambda)| = S_{00}^{-g}| < v_\lambda^*, \rho(h)v_0 > |.$$

Since $\rho(h)$ is unitary, we have

$$|Z_{\mathcal{G},K}(M,k;\lambda)| \leq S_{00}^{-g}.$$

Let us recall that S_{00} is a positive real number and the matrix S is unitary, therefore we have $S_{00} \leq 1$. Hence the above inequality implies our assertion. This completes the proof.

Let k be a framed knot in S^3 and let us consider the case $\mathcal{G} = sl(2,\mathbf{C})$. The invariant $J(k,\lambda)$ can be expressed in terms of special values of the Jones polynomial and its parallel version. We parametrize highest weights $\lambda \in P_+(K)$ by integers $\lambda = 0, 1, \cdots, K$. Namely, the representation corresponding to λ has dimension $\lambda+1$. Let L be an oriented framed link in S^3 and we denote by V_L the original Jones polynomial defined in [Jo]. We put $r = K+2$. As is shown in [KM] using $U_q(sl(2,\mathbf{C}))$, for a framed link L with color $\lambda = 1$ we have

$$J(L,1) = \left(2\cos\frac{\pi}{r}\right) t^{3L\cdot L} V_L(e^{-\frac{2\pi\sqrt{-1}}{r}}) \tag{4-4}$$

with $t = e^{\frac{\pi\sqrt{-1}}{2r}}$ for any orientation on L where $L \cdot L$ denotes the total linking number. This assertion follows easily in our setting of conformal field theory by looking at the eigenvalues of the braiding matrix, which gives the skein relation. Let k be a framed knot in S^3 and γ a non-negative integer. We define the cable k^γ by replacing k with γ parallel pushoffs defined by means of the framing. Using this notation we have

$$J(k,\lambda)=\sum_{j=0}^{[\lambda/2]}(-1)^j\binom{\lambda-j}{j}J(k^{\lambda-2j},1). \tag{4-5}$$

Here we put $J(k^0,1)=1$. Let us recall that

$$S_{00}=\sqrt{\frac{2}{r}}\sin\frac{\pi}{r}$$

in our situation. Considering, in particular, the case $\lambda=1$, we are led to the following corollary.

4.6 **Corollary.** *Let k be a knot in S^3 and $t(k)$ its tunnel number. For a positive integer $r\geq 3$ we have the following lower estimate for $t(k)$ in terms of a special value of the Jones polynomial.*

$$|V_k(e^{\frac{2\pi\sqrt{-1}}{r}})|\leq\left(2\cos\frac{\pi}{r}\right)^{-1}\left(\sqrt{\frac{2}{r}}\sin\frac{\pi}{r}\right)^{-t(k)-1}$$

Let M be a closed oriented 3-manifold. We define the Heegaard genus $g(M)$ to be the minimal number g such that M admits a Heegaard splitting of genus g. Considering the case the knot k is empty in Theorem 4.3, we obtain the following lower estimate for the Heegaard genus.

4.7 **Corollary.** *We have*

$$|Z_{\mathcal{G},K}(M)|\leq S_{00}^{-g(M)}.$$

4.8 *Remark.* This inequality was also obtained by Walker [Wa] in a general framework of topological quantum field theory (see also [Koh2] and [G]). Let k be a knot in S^3. By computing the Witten invariant for the 2-fold branched covering of S^3 branched along k, the above inequality can be used to give a lower bound for the bridge index of the knot k.

5. Symmetry in the case of $sl(n, \mathbf{C})$

In this section we consider the case $\mathcal{G} = sl(n, \mathbf{C})$. Witten's invariants associated with classical Lie algebras have been treated in [TW] and [W] using the algebras developed in [GW], and also in [M] by means of the skein theory. In [KT] we generalized the symmetry principle due to [KM] based on the universal R constructed in [T]. Here we focus this symmetry from a point of view of conformal field theory.

The set $P_+(K)$ admits a natural action of the cyclic group $\mathbf{Z}_n$ derived from the Dynkin diagram automorphism of the corresponding affine Lie algebra in the following way. The set of weights $P_+(K)$ is expressed as

$$P_+(K) = \{\sum_{i=1}^{n-1} a_i\Lambda_i \,;\, a_i \in \mathbf{Z},\, a_i \geq 0,\, \sum_{i=1}^{n-1} a_i \leq K\}$$

using the fundamental weights Λ_i, $1 \leq i \leq n-1$. Let $\widehat{\Lambda}_i, 0 \leq i \leq n-1$, denote the fundamental weights of the affine Lie algebra $\widehat{sl(n, \mathbf{C})}$. The corresponding set of dominant integral weights at level K is

$$\widehat{P}_+(K) = \{\sum_{i=0}^{n-1} a_i\widehat{\Lambda}_i \,;\, a_i \in \mathbf{Z},\, a_i \geq 0,\, \sum_{i=0}^{n-1} a_i = K\}.$$

We have a natural identification $j : \widehat{P}_+(K) \to P_+(K)$ defined by $j(\lambda) = \sum_{i=1}^{n-1} a_i\Lambda_i$ for $\lambda = \sum_{i=0}^{n-1} a_i\widehat{\Lambda}_i$. The cyclic group $\mathbf{Z}_n$ acts on the set $\widehat{P}_+(K)$ by

$$\sigma(\widehat{\Lambda}_i) = \widehat{\Lambda}_{i+1}$$

where the suffix is taken modulo n. By means of the above identification this induces a $\mathbf{Z}_n$ action on $P_+(K)$. The action of σ on the set $P_+(K)$ is a rotation of angle $2\pi/n$ with respect to the barycenter of the alcove. In particular, in the case of $sl(2, \mathbf{C})$ this is a reflection defined by $\sigma(\lambda) = K - \lambda$. As a fundamental domain of this cyclic group action, we have

$$D(n, K) = \left\{ \sum_{i=1}^{n-1} a_i\Lambda_i \in P_+(K) \,;\, K - \sum_{j=1}^{n-1} a_j \geq a_i,\, 1 \leq i \leq n-1 \right\}.$$

Let us notice that the above action has a fixed point

$$Fix(\sigma) = \frac{K}{n}\sum_{i=1}^{n-1} \Lambda_i$$

if $K \equiv 0 \bmod n$ and is fixed point free otherwise.

Let us describe the behavior of the conformal weight $\Delta_\lambda = \frac{(\lambda,\lambda+2\rho)}{2(K+n)}$ and the modular transformation S matrix under the above action of the cyclic group $\mathbf{Z}_n$. For $\lambda = \sum_{i=1}^{n-1} a_i\Lambda_i$, we put $|\lambda| = \sum_{i=1}^{n-1} ia_i$, which is equal to the number of nodes in the associated Young diagram. By a direct computation we have the following.

5.1 Lemma. *Under the action σ, the conformal weight and the matrix S changes as*

$$\Delta_{\sigma(\lambda)} - \Delta_\lambda = \frac{1}{n}\left(\frac{(n-1)K}{2} - |\lambda|\right)$$

$$S_{\sigma(\lambda)\mu} = \exp\left(\frac{2\pi\sqrt{-1}|\mu|}{n}\right) S_{\lambda\mu}.$$

Let L be an oriented framed link with components L_i, $1 \le i \le m$. We denote by $L_i \cdot L_j$ the linking number of the i-th and j-th components.

5.2 Lemma. *Let k be an oriented framed knot in S^3 with color $\lambda \in P_+(K)$. Then, with respect to the action of σ we have*

$$\frac{J(k,\sigma(\lambda))}{J(k,\lambda)} = \exp\left(\frac{\pi\sqrt{-1}}{n}((n-1)K - 2|\lambda|)N\cdot N\right).$$

Proof. We set

$$I(N,\lambda) = \frac{J(N,\sigma(\lambda))}{J(N,\lambda)}.$$

Let us recall that on the space of chiral vertex operators $\mathcal{H}^{\mu}_{\lambda\lambda}$ the braiding operation interchanging two holes corresponding to the weight λ has eigenvalue

$$\rho^{\mu}_{\lambda\lambda} = \epsilon \exp \pi\sqrt{-1}(\Delta_\mu - 2\Delta_\lambda)$$

where ϵ is the parity and is ± 1. We have a natural isomorphism $\mathcal{H}^{\mu}_{\lambda\lambda} \cong \mathcal{H}^{\sigma^2(\mu)}_{\sigma(\lambda)\sigma(\lambda)}$ and we have

$$\rho^{\sigma^2(\mu)}_{\sigma(\lambda)\sigma(\lambda)} = \exp\left(\frac{\pi\sqrt{-1}}{n}((n-1)K - 2|\lambda|)\right)\rho^{\mu}_{\lambda\lambda}$$

by applying Lemma 5.1 and by examining the sign (see also [NRS] and [KN]). It follows that this braiding operator gives the above phase factor to $I(N,\lambda)$.

Using Lemma 5.1 again, we observe that the creation and annihilation operators do not contribute to $I(N, \lambda)$. This completes the proof.

Let $L = L_1 \cup \cdots \cup L_m$ be an oriented framed link in S^3 with m components with color $\mu : \{1, \cdots, m\} \longrightarrow P_+(K)$. Let us consider an oriented framed link $L \cup N$ adding one more component N. We assign a color $\lambda \in P_+(K)$ to the component N.

Using the argument in the proof of Lemma 5.2 and applying Lemma 2.6, we obtain the following.

5.3 Proposition (SYMMETRY PRINCIPLE). *If we change the color of the component N by the action of σ, then we have*

$$\frac{J(L \cup N, \mu \cup \sigma(\lambda))}{J(L \cup N, \mu \cup \lambda)}$$
$$= \exp\left(\frac{\pi\sqrt{-1}}{n}\left(((n-1)K - 2|\lambda|)N \cdot N - 2\sum_{j=1}^{m} |\mu(j)| L_j \cdot N\right)\right).$$

Let L be an oriented framed link in S^3 with m components $L_i, 1 \leq i \leq m$. Let us recall that L determines a smooth 4-manifold W_L obtained by adding 2-handles to the 4-ball B^4 along the components L_i. Each L_i determines an element of $H_2(W_L; \mathbf{Z})$, which is also denoted by L_i. Let $\lambda : \{1, \cdots, m\} \to P_+(K)$ be a color of the link L. By means of the above identification, $x \in H_2(W_L; \mathbf{Z}_n)$ may be written as $\sum_{i=1}^{m} x_i L_i$ with $x_i \in \mathbf{Z}_n$. We define a new color λ^x by

$$\lambda^x(i) = \sigma^{x_i}(\lambda(i)), \quad 1 \leq i \leq m$$

using the $\mathbf{Z}_n$ action on $P_+(K)$. Let us now describe $J(L, \lambda^x)/J(L, \lambda)$. We define a map $\phi_\lambda : H_2(W_L; \mathbf{Z}_n) \to \mathbf{Z}_{2n}$ by

$$\phi_\lambda\left(\sum_{i=1}^{m} x_i L_i\right)$$
$$= K\sum_{i=1}^{m} L_i \cdot L_i x_i(n - x_i) - 2K\sum_{i<j} L_i \cdot L_j x_i x_j - 2\sum_{i,j} |\lambda(j)| L_i \cdot L_j x_i.$$

Now applying symmetry principle inductively, we have the following.

5.4 Theorem. *Using the above notation, we have*

$$\frac{J(L,\lambda^x)}{J(L,\lambda)} = \exp\left(\frac{\pi\sqrt{-1}}{n}\phi_\lambda(x)\right).$$

5.5 *Remark.* (i)Let us notice that if $K \equiv 0 \bmod n$, then ϕ_λ is $\mathbf{Z}$ linear with

$$\phi_\lambda(L_i) = (n-1)K\, L_i \cdot L_i - 2\sum_j |\lambda(j)| L_i \cdot L_j.$$

(ii) In view of the estimate of the tunnel number discussed in the previous section, it would be practical to observe the equality

$$|J(L,\lambda^x)| = |J(L,\lambda)|.$$

In particular, in the case $\mathcal{G} = sl(2,\mathbf{C})$ we have $|J(k,\lambda)| = |J(k,K-\lambda)|$.

Let $\lambda : \{1,\cdots,m\} \to P_+(K)$ be a color of L. We denote by T_λ a sublink of L consisting of the components with color unequal to the fixed point $Fix(\sigma)$. We consider the orbit of λ denoted by $orb(\lambda) = \{\lambda^x\,;\, x \in H_2(W_L;\mathbf{Z}_n)\}$. As a subspace of $H_2(W_L;\mathbf{Z}_n)$ we define U_λ by $U_\lambda = \oplus_{\lambda(i)\neq Fix(\sigma)} \mathbf{Z}_n L_i$. We put

$$m_{L,\lambda} = \sum_{\lambda^x \in orb(\lambda)} \exp\left(\frac{\pi\sqrt{-1}}{n}\phi_\lambda(x)\right) = \sum_{x\in U_\lambda} \exp\left(\frac{\pi\sqrt{-1}}{n}\phi_\lambda(x)\right).$$

We denote by $Z_{n,K}(M)$ the 3-manifold invariant $Z_{\mathcal{G},K}(M)$ in the case $\mathcal{G} = sl(n,\mathbf{C})$. Considering the orbit decomposition of the sum 3.1 with respect to the $\mathbf{Z}_n$ action, we obtain the following.

5.6 Proposition. *Let M be a closed oriented 3-manifold obtained by the Dehn surgery on a framed link L with m components. Then, the 3-manifold invariant $Z_{n,K}(M)$ may be written as*

$$C^{sign(L)} \sum_\lambda m_{L,\lambda}\, S_{0\lambda(1)} \cdots S_{0\lambda(k)}\, J(L,\lambda),$$

where the sum is for any color λ satisfying $Im(\lambda) \subset D(n,K)$.

Let us now consider the case $K \equiv 0 \bmod n$. We have the following result.

5.7 Theorem. *If $K \equiv 0 \bmod n$, then we have*

$$Z_{n,K}(M) = C^{sign(L)} \sum_{\lambda} n^{|T_\lambda|} S_{0\lambda(1)} \cdots S_{0\lambda(m)} J(L, \lambda),$$

where the sum is for any color $\lambda : \{1, \cdots, m\} \to D(n, K)$ satisfying the condition

$$\phi_\lambda(L_i) = (n-1)K\, L_i \cdot L_i - 2\sum_{j=1}^{m} |\lambda(j)|\, L_i \cdot L_j \equiv 0 \bmod 2n, \quad 1 \le i \le m.$$

Proof. Let us note that U_λ is a free $\mathbf{Z}_n$ module of rank $|T_\lambda|$ and that $\phi_\lambda|U_\lambda : U_\lambda \to \mathbf{Z}_{2n}$ is a $\mathbf{Z}$ linear map. The coefficient $m_{L,\lambda}$ is written as a Monsky sum $\sum_{x \in U_\lambda} \exp\left(\frac{\pi\sqrt{-1}}{n}\phi_\lambda(x)\right)$, which is equal to $n^{|T_\lambda|}$ if $\phi_\lambda|U_\lambda$ is identically 0 and is equal to 0 otherwise. Now the condition $\phi_\lambda|U_\lambda = 0$ is replaced by $\phi_\lambda = 0$ on $H_2(W_L; \mathbf{Z}_n)$. Indeed, if $\phi_\lambda(L_i) \neq 0$ for some i such that $\lambda(i) = Fix(\sigma)$, then $J(L, \lambda) = 0$ by the symmetry principle. This completes the proof.

Let us denote by $N_{1/a}$ the homology 3-sphere obtained by $1/a$-Dehn surgery on a knot N in S^3. Then, we have the following Corollary.

5.8 Corollary. *If $K \equiv 0 \bmod 2n$, then the following periodicity holds for any integer a.*

$$Z_{n,K}(N_{1/a}) = Z_{n,K}(N_{1/(a+n+K)}).$$

Proof. The 3-manifold $N_{1/a}$ is obtained from the Dehn surgery on a 2 component framed link $L = L_1 \cup L_2$, where $L_1 = N$ with 0 framing and L_2 is its unknotted meridian with $-a$ framing. Noticing that the signature of the linking matrix is 0, it follows from Lemma 2.6 and Theorem 5.7 that

$$Z_{n,K}(N_{1/a}) = \sum_{\lambda} n^{|T_\lambda|} S_{\lambda(1)\lambda(2)} S_{0\lambda(2)} \exp(-2\pi\sqrt{-1}\, a\Delta_{\lambda(2)}) J(N, \lambda(1)),$$

where the sum is for any $\lambda : \{1, 2\} \to D(n, K)$ such that $\phi_\lambda(L_i) \equiv 0 \bmod 2n$ for $i = 1, 2$. Hence it is enough to consider λ satisfying $|\lambda(i)| \equiv 0 \bmod n$. Let us observe that if a weight λ satisfies $|\lambda| \equiv 0 \bmod n$, then $(\lambda, \lambda + 2\rho)$ is an even integer and it follows that $\left(\exp(2\pi\sqrt{-1}\Delta_\lambda)\right)^{K+n} = 1$. Combining with this, we obtain the assertion.

REFERENCES

[A] M. F. Atiyah, *Topological quantum field theories*, Publ. Math. IHES **68** (1988), 175–186.

[CLM] S. Cappell, R. Lee and E. Y. Miller, *Invariants of 3-manifolds from conformal field theory*, preprint, 1990.

[Cr] R. Craggs, *A new proof of the Reidemeister-Singer theorem on stable equivalence of Heegaard splitting*, Proc. Amer. Math. Soc. **57** (1976), 143–147.

[C] L. Crane, *2d Physics and 3d topology*, Commun. Math. Phys. **135** (1991), 615–640.

[D1] V. G. Drinfel'd, *Quantum groups*, Proc. Int. Cong. Math. Berkeley (1987), 798–820.

[D2] V. G. Drinfel'd, *Quasi-Hopf algebras*, Algebra and Analysis **1-6** (1989), 114–148.

[FR] R. Fenn and C. Rourke, *On Kirby's calculus of links*, Topology **18** (1979), 1–15.

[F] J. Fröhlich, *Two-dimensional conformal field theory and three-manifold topology*, Int. J. Mod. Phys. **4-20** (1989), 5321–5399.

[G] S. Garoufalidis, *Relations among 3-manifold invariants*, Thesis, The University of Chicago, 1992.

[GW] F. M. Goodman and H. Wenzl, *Littlewood-Richardson coefficients for Hecke algebras at roots of unity*, Adv. in Math. **82** (1990), 244–265.

[HT] A. Hatcher and W. Thurston, *A presentation for the mapping class group of a closed orientable surface*, Topology **19** (1980), 221–237.

[J] M. Jimbo, *A q-difference analogue of $U(\mathcal{G})$ and the Yang-Baxter equation*, Letters in Math. Phys. **10** (1985), 63–69.

[Jo] V. F. R. Jones, *Hecke algebra representations of braid groups and link polynomials*, Ann. of Math. **126** (1987), 355–388.

[K] V. G. Kac, "Infinite dimensional Lie algebras," third edition.

[KP] V. G. Kac and D. H. Peterson, *Infinite dimensional Lie algebras, theta functions and modular forms*, Adv. in Math. **53** (1984), 125–264.

[KW] V. G. Kac and M. Wakimoto, *Modular and conformal invariance in the representation theory of affine Lie algebras*, Adv. in Math. **40** (1988), 156–236.

[Ki] R. Kirby, *A calculus for framed links*, Invent. Math. **45** (1978), 35–56.

[KM] R. Kirby and P. Melvin, *The 3-manifold invariants of Witten and Reshetikhin-Turaev for $sl(2, \mathbf{C})$*, Invent. Math. **105** (1991), 473–545.

[KZ] V. G. Knizhnik and A. B. Zamolodchikov, *Current algebra and Wess-Zumino models in two dimensions*, Nucl. Phys. **B247** (1984), 83–103.

[Koh1] T. Kohno, *Topological invariants for 3-manifolds using representations of mapping class groups I*, Topology **31** (1992), 203–230.

[Koh2] T. Kohno, *Three-manifold invariants derived from conformal field theory and projective representations of modular groups*, Int. J. Mod. Phys. B **6** (1992), 1795–1805.

[KT] T. Kohno and T. Takata, *Symmetry of Witten's 3-manifold invariants for $sl(n, \mathbf{C})$*, Journal of Knot Theory and Its Ramifications **2-2** (1993), 149–169.

[Ko] M. Kontsevich, *Rational conformal field theory and invariants of 3-dimenensional manifolds*, preprint.

[KN] A. Kuniba and T. Nakanishi, *Level-rank duality in fusion RSOS models*, in Modern Quantum Field Theory, eds. A. Das et al., World Scientific, 1991, 344–374.

[L] W. B. R. Lickorish, *A finite generators for the homeotopy group of a 2-manifold*, Proc. Camb. Phil. Soc. **60** (1964), 769–778 (ibid. 62 (1966) 679-681).

[MS] G. Moore and N. Seiberg, *Classical and quantum conformal field theory*, Commun. Math. Phys. **123** (1989), 177–254.

[MoS] H. R. Morton and P. M. Strickland, *Jones polynomial invariants for knots and satellites*, Math. Proc. Camb. Phil. Soc. **109** (1991), 83–103.

[M] H. R. Morton, *Invariants of links and 3-manifolds from skein theory and from quantum groups*, preprint, 1992.

[NRS] S. G. Naculich, H. A. Riggs and H. J. Schnitzer, *Simple current symmetries, rank-level duality, and linear skein relations for Chern-Simons graphs*, preprint, 1992.

[P] S. Piunikhin, *Reshetikhin-Turaev and Kontsevich-Kohno-Crane 3-manifold invariants coincide*, preprint.

[RT] N. Y. Reshetikhin and V. G. Turaev, *Invariants of 3-manifolds via link polynomials and quantum groups*, Invent. Math. **103** (1991), 547–597.

[S] S. Suzuki, *On homeomorphisms of a 3-dimensional handlebody*, Can. J. Math. **29(1)** (1977), 111–124.

[T] T. Takata, *Universal R at roots of unity for $sl(N+1, \mathbf{C})$*, preprint, 1992.

[TK] A. Tsuchiya and Y. Kanie, *Vertex operators in conformal field theory on P^1 and monodromy representations of braid groups*, Advanced Studies in Pure Math. **16** (1988), 297–372.

[TUY] A. Tsuchiya, K. Ueno and Y. Yamada, *Conformal field theory on universal family of stable curves with gauge symmetries*, Advanced Studies in Pure Math. **19** (1989), 459–566.

[TW] V. G. Turaev and H. Wenzl, *Quantum invariants of 3-manifolds associated with classical simple Lie algebras*, preprint, 1991.

[V] E. Verlinde, *Fusion rules and modular transformations in 2-D conformal field theory*, Nucl. Phys. **B300** (1988), 360–376.

[Wa] K. Walker, *On Witten's 3-manifold invariants (and talk given at KNOTS 90, Osaka)*, preprint.

[We] H. Wenzl, *Braids and invariants of 3-manifolds*, preprint, 1992.

[W] E. Witten, *Quantum field theory and the Jones polynomial*, Commun. Math. Phys. **121** (1989), 351-399.

Department of Mathematical Sciences, University of Tokyo, Komaba, Tokyo 153, Japan
E-mail address: t-kohno@tansei.cc.u-tokyo.ac.jp

Contemporary Mathematics
Volume **175**, 1994

Poisson Lie Groups, Quantum Duality Principle, and the Quantum Double

M.A.Semenov-Tian-Shansky

Abstract

The Heisenberg double of a Hopf algebra may be regarded as a quantum analogue of the cotangent bundle of a Lie group. Quantum duality theory describes relations between a Hopf algebra, its dual, and their Heisenberg double in a way which extends both the theory of coadjoint orbits and the classical Fourier transform. We also describe the twisted Heisenberg double which is relevant for the study of nontrivial deformations of the quantized universal enveloping algebras.

1 Introduction

The standard way to describe quantum deformations of simple finite dimensional or affine Lie algebras is by means of generators and relations (generalizing the classical Chevalley - Serre relations) [7, 11]. A dual approach, due to Faddeev, Reshetikhin, and Takhtajan [8], [9], is to construct quantum universal enveloping algebras as deformations of coordinate rings of functions on Lie groups (regarded as affine algebraic groups). Of course, construction of a quantum deformation of the algebra $Fun(G)$ was one of the first results of the quantum group theory and is, in fact, a direct generalization of the Baxter commutation relation $RT_1T_2 = T_2T_1R$. A nontrivial fact, first observed by Faddeev, Reshetikhin, and Takhtajan, is that the dual algebra $Fun_q(G)^*$ may also be regarded as a deformation of a function algebra on a Lie group (namely, on the dual group G^*, see below)[1] . More generally, the FRT construction is related to the *quantum duality principle* which we are now going to state.

*1991 Mathematics Subject Classification: Primary 16A15, 16A24, 16A49 Secondary 58F05, 58F06

†This paper is in final form and no version of it will be submitted for publication elsewhere

‡The author is grateful to the Organizing Committee of the Mt.Holyoke Conference and to Université de Bourgogne, Dijon, France, for financial subsistence

[1]In a disguised form this fact was also mentioned in Drinfeld's report [7]('Equivalence of the category of QFSH-agebras and the category of QUE-algebras')

Observe, first of all, that in the semiclassical approximation quantum deformations of function algebras are determined by Poisson brackets on Lie groups. The class of Poisson brackets related to deformations of algebras $Fun(G)$ in the category of Hopf algebras is described by the following well-known axiom [6].

Definition 1.1 *A Poisson bracket on a Lie group G defines on G the structure of a Poisson Lie group if multiplication*

$$m : G \times G \to G$$

is a Poisson map

A Poisson bracket satisfying this axiom is degenerate and, in particular, is identically zero at the unit element of the group. Linearizing it at this point defines the structure of a Lie algebra in the space $T_e^*G \simeq \boldsymbol{g}^*$. The pair $(\boldsymbol{g}, \boldsymbol{g}^*)$ is called the *tangent Lie bialgebra* of G. Lie brackets in $\boldsymbol{g}$ and $\boldsymbol{g}^*$ satisfy the following compatibility condition:

Let $\varphi : \boldsymbol{g} \to \boldsymbol{g} \wedge \boldsymbol{g}$ be the dual of the commutator map $[,]_* : \boldsymbol{g}^* \wedge \boldsymbol{g}^* \to \boldsymbol{g}^*$. Then φ is a 1-cocycle on $\boldsymbol{g}$ (with respect to the natural action of $\boldsymbol{g}$ on $\boldsymbol{g} \wedge \boldsymbol{g}$).

Let c_{ij}^k, f_c^{ab} be the structure constants of $\boldsymbol{g}, \boldsymbol{g}^*$ with respect to the dual bases $\{e_i\}, \{e^i\}$ in $\boldsymbol{g}, \boldsymbol{g}^*$. The compatibility condition means that

$$c_{ab}^s f_s^{ik} - c_{as}^i f_b^{sk} + c_{as}^k f_b^{si} - c_{bs}^k f_a^{si} + c_{bs}^i f_a^{sk} = 0.$$

This condition is symmetric with respect to exchange of c and f. Thus if $(\boldsymbol{g}, \boldsymbol{g}^*)$ is a Lie bialgebra, then $(\boldsymbol{g}^*, \boldsymbol{g})$ is also a Lie bialgebra. Let G^* be a (connected simply connected) Lie group which corresponds to $\boldsymbol{g}^*$. Since the correspondence between Poisson Lie groups and Lie bialgebras is functorial, G^* is also a Poisson Lie group (called the dual of G).

Passing to quantization, we may (at least if there are no obstructions, see [7]) construct two Hopf algebras $Fun_q(G)$, $Fun_q(G^*)$ which correspond to the Poisson—Hopf algebras $Fun(G), Fun(G^*)$.The *quantum duality principle* then asserts that these algebras are dual to each other as Hopf algebras. More precisely, let h be the deformation parameter (for simplicity we chose h to be the same for both algebras). There exists a nondegenerate bilinear pairing

$$Fun_q(G) \otimes Fun_q(G^*) \to \mathbf{C}[[h]]$$

which sets the algebras $Fun_q(G), Fun_q(G^*)$ into duality as Hopf algebras. Hence, in particular, we have, up to an appropriate completion,

$$Fun_q(G^*) \simeq U_q(\boldsymbol{g}).$$

In the dual way, we have also

$$Fun_q(G) \simeq U_q(\boldsymbol{g}^*).$$

As a simple example, let us consider *trivial Lie bialgebra*. Let $\boldsymbol{g}$ be an arbitrary Lie algebra, $\boldsymbol{g}^*$ its dual equipped with the zero Lie bracket. In this case the Poisson bracket on G is trivial. The dual group of G is the additive group of the space $\boldsymbol{g}^*$ equipped with the Lie—Poisson bracket [5], [20] of $\boldsymbol{g}$. In the present case the algebra $Fun(G)$ does not deform at all (since the germ of a deformation defined by the Poisson bracket is identically zero). The deformation of $Fun(\boldsymbol{g}^*)$ may be identified with the universal enveloping algebra $U(\boldsymbol{g})$ [5], a function $\psi \in Fun(\boldsymbol{g}^*)$ being regarded as the symbol of a left invariant (pseudo)differential operator on G. The pairing $Fun(G) \otimes Fun(\boldsymbol{g}^*) \to \mathbf{C}[[h]]$ is given by

$$< \varphi, \psi > \;=\; \int_{G\times \boldsymbol{g}^*} \varphi(x)\psi(p) \exp i/h < p, \log x > dxdp \qquad (1.1)$$

The integration measure $dxdp$ in (1.1) is of course the Liouville measure on $G \times \boldsymbol{g}^* \simeq T^*G$. The emergence of T^*G in this context is not accidental. The pairing (1.1) canonically generates an action

$$Fun_q(\boldsymbol{g}^*) \;\otimes\; Fun(G) \to Fun(G) : \psi \otimes \varphi \to < id \otimes \psi, \Delta\varphi >$$

which is the usual action of $U(\boldsymbol{g})$ on $Fun(G)$ by left-invariant derivations. Let $\mathcal{H}$ be the associative algebra generated by $U(\boldsymbol{g})$ and $Fun(G)$ regarded as differential operators and multiplication operators in $Fun(G)$, respectively. Then $\mathcal{H}$ is a quantum deformation of the Poisson algebra of functions on T^*G with the canonical Poisson bracket. The algebra $\mathcal{H}$ arises along with its irreducible (Schroedinger) representation. As we shall see below, $\mathcal{H}$ is a special case of the *Heisenberg double.*

Thus even for trivial Lie bialgebras the quantum duality principle is quite meaningful: it includes, e.g., the usual Fourier duality. In general case, quantum duality principle may also be regarded as a generalization of the Fourier transform (cf. Section 4).

For semisimple Lie algebras quantum deformations of Poisson algebras $Fun(G)$, $Fun(G^*)$ may easily be constructed if we know the corresponding quantum R-matrices. This allows also to check the duality principle. The need to know the quantum R-matrices in advance is a certain drawback of this approach, as compared to the standard definition of Drinfeld and Jimbo. On the other hand, the FRT construction alllows to define, along with the well-known algebras, such as $Fun_q(G)$ and $Fun_q(G^*) \simeq U_q(\boldsymbol{g})$, a large family of their relatives. The two simplest algebras in this family are the *Heisenberg double* and the *Drinfeld double.* Both are naturally described as quantizations of some geometrically defined Poisson algebras. (An elementary algebraic definition in these cases is also possible.[2]) We have seen already, on an elementary example, that the

[2]The geometric construction of the Heisenberg double was proposed almost simultaneously by the author and by Alekseev and Faddeev [2].The author has also pointed out its direct algebraic definition. Recently, an algebraic construction of a family of operator algebras which includes the Heisenberg double and the Drinfeld double, was proposed by S.P.Novikov [16]

Heisenberg double is the quantum analogue of the algebra $Fun(T^*G)$. Recall that connections between the canonical Poisson bracket on T^*G and the Lie—Poisson bracket on $\boldsymbol{g}^*$ form the basis of the standard geometric construction of irreducible representations of G (the so called 'orbits method'). Mutual relations between the algebras $Fun_q(G), Fun_q(G^*)$ and their Heisenberg double form the precise analogue of the orbits method (the author hopes to describe this subject more thoroughly in a separate paper).

In many cases the Heisenberg double admits non-trivial deformations associated with outer automorphisms of the underlying algebras. In this way we are led to the notion of the *twisted double*. This allows notably to define the twisted algebra $U_q(\boldsymbol{g})_\tau^{\otimes N}$ which may be regarded as a lattice version of a current algebra, the twisting carefully reproducing the effects of central extension [3].[3] The author does not know any direct elementary definition of the twisted double or of the related algebras, so here the geometric approach seems indispensable.

For general semisimple Lie algebras the logics of our construction is as follows:

1. We use the Drinfeld—Jimbo construction to define quantum deformations of the universal enveloping algebras

2. The universal R-matrix theory [12, 13] and the representation theory for $U_q(\boldsymbol{g})$ yield explicit formulae for quantum R-matrices and for the FRT generators of the algebras $Fun_q(G), Fun_q(G^*)$ in terms of the Drinfeld—Jimbo generators.

3. Finally, we may describe all related algebras (the Heisenberg double, the twisted double and their subalgebras).

In the present article we shall leave aside the theory of universal R-matrices and admit (in the spirit of [8]) that all necessary R-matrices are known in advance.

Acknowledgement. The author had numerous fruitful discussions on the matters discussed in this paper with A.Alekseev, L.D.Faddeev and N.Reshetikhin.

2 Quasiclassical Duality Theory

Let $(\boldsymbol{g}, \boldsymbol{g}^*)$ be Lie bialgebra. Put $\boldsymbol{d} = \boldsymbol{g} \oplus \boldsymbol{g}^*$. There is a unique structure of a Lie algebra on $\boldsymbol{d}$ such that

(i) $\boldsymbol{g}, \boldsymbol{g}^* \subset \boldsymbol{d}$ are Lie subalgebras.

(ii) The canonical bilinear form on $\boldsymbol{d}$ given by

$$< (X_1, f_1),\ (X_2, f_2) > \ = \ f_1(X_2) \ + \ f_2(X_1) \tag{2.1}$$

[3]Recently Fock and Rosly introduced a still more general class of algebras which are associated with arbitrary graphs [10].

is $ad - \boldsymbol{d}$-invariant.

Let $P_{\boldsymbol{g}}, P_{\boldsymbol{g}^*}$ be the projection operators onto $\boldsymbol{g}, \boldsymbol{g}^* \subset \boldsymbol{d}$ parallel to the complementary subalgebra. The linear operator

$$r_{\boldsymbol{d}} = P_{\boldsymbol{g}} - P_{\boldsymbol{g}^*} \tag{2.2}$$

is skew-symmetric with respect to (2.1) and may be identified with an element of $\wedge^2 \boldsymbol{d}$. The formula

$$[X, Y]_* = \frac{1}{2}(\, [r_{\boldsymbol{d}}X,\, Y] + [X,\, r_{\boldsymbol{d}}Y]\,) \tag{2.3}$$

defines in the space $\boldsymbol{d}^* \simeq \boldsymbol{d}$ a Lie bracket which makes $(\boldsymbol{d}, \boldsymbol{d}^*)$ a Lie bialgebra. As a linear space again $\boldsymbol{d}^* = \boldsymbol{g} \oplus \boldsymbol{g}^*$ but this time $\boldsymbol{g}, \boldsymbol{g}^*$ are complementary ideals in $\boldsymbol{d}^*$; moreover, $\boldsymbol{d}^*/\boldsymbol{g}^*$ is isomorphic to $\boldsymbol{g}$ and $\boldsymbol{d}^*/\boldsymbol{g}$ is *anti-isomorphic* to $\boldsymbol{g}^*$. The Lie bialgebra $(\boldsymbol{d}, \boldsymbol{d}^*)$ is called the double of $(\boldsymbol{g}, \boldsymbol{g}^*)$. Clearly, the dual Lie bialgebras $(\boldsymbol{g}, \boldsymbol{g}^*)$ and $(\boldsymbol{g}^*, \boldsymbol{g})$ have a common double.

Let D be the (connected simply connected) Lie group with the Lie algebra $\boldsymbol{d}$. One can define on D several important Poisson structures. The two simplest ones are defined as follows.

For $\varphi \in C^\infty(D)$ let $X_\varphi, X'_\varphi \in \boldsymbol{d}$ be its left and right gradients defined by the formulae

$$\begin{aligned} < X_\varphi(x), \xi > &= \frac{d}{dt}_{t=0} \varphi(e^{t\xi}x), \\ < X'_\varphi(x), \xi > &= \frac{d}{dt}_{t=0} \varphi(xe^{t\xi}), \quad \xi \in \boldsymbol{d}. \end{aligned} \tag{2.4}$$

Put

$$\{\varphi, \psi\}_\pm = \frac{1}{2} < r_{\boldsymbol{d}}X_\varphi, X_\psi > \pm \frac{1}{2} < r_{\boldsymbol{d}}X'_\varphi, X'_\psi > . \tag{2.5}$$

The bracket $\{,\}_-$ equips D with the structure of a Poisson Lie group; its tangent Lie bialgebra is precisely $(\boldsymbol{d}, \boldsymbol{d}^*)$. The bracket $\{,\}_+$ is non-degenerate (at least on an open dense subset in D) and defines (almost everywhere) on D a symplectic structure. If $(\boldsymbol{g}, \boldsymbol{g}^*)$ is a trivial Lie bialgebra, the group $D = G \times \boldsymbol{g}^*$ is the semi-direct product of G and the additive group of $\boldsymbol{g}^*$. Thus D may be identified with T^*G. It is easy to check that $\{,\}_+$ coincides in this case with the canonical Poisson bracket on T^*G. The bracket $\{,\}_-$ is highly degenerate: it is the direct product of the Lie—Poisson bracket on $\boldsymbol{g}^*$ and the trivial bracket on G.

In general, $\{,\}_+$ is also an analogue of the canonical Poisson bracket on the cotangent bundle. In order to describe its relations with the Poisson brackets on G and on G^*, let us first recall some simple facts on Poisson reduction [19, 14].

Let M be a Poisson manifold. An action $H \times M \to M$ is called *admissible* if the space of H-invariant functions is a Lie subalgebra of the Poisson algebra $Fun(M)$. Admissible actions may *not* preserve Poisson brackets on M. If

$H \times M \to M$ is an admissible action and the quotient space M/H is smooth, we may identify the algebras $Fun(M/H)$ and $Fun(M)^H$. Hence there exists a Poisson structure on M/H such that the canonical projection $\pi : M \to M/H$ is a Poisson map. The space M/H is called the *reduced Poisson manifold.* Even if M is symplectic, the reduced Poisson bracket on M/H is usually degenerate. The difficult part of reduction is the description of its symplectic leaves. To solve this problem we need the notion of *dual pairs* [20].

Assume that there are *two* transformation groups H, H' acting on M. Admisssible actions $H \times M \to M, H' \times M \to M$ are said to be *dual* to each other if the subalgebras of invariants $Fun(M)^H$, $Fun(M)^{H'}$ are the centralizers of each other in the Poisson algebra $Fun(M)$. Assume that M is symplectic and H, H' are dual transformation groups of M. In that case symplectic leaves in M/H are the connected components of the sets $\pi({\pi'}^{-1}(x)), x \in M/H'$; in a similar way, symplectic leaves in M/H' are the connected components of the sets $\pi'(\pi^{-1}(y)),\ y \in M/H$.

Theorem 2.1 *Let G, G^* be the dual Poisson—Lie groups, D_+ their double equipped with the Poisson bracket $\{,\}_+$ (2.5). Then*

(i) Natural actions of $G(G^*)$ on D by left ang right translations are admissible.

(ii) The actions of $G(G^*)$ on D_+ by left and right translations form a dual pair.

Elements in D which admit a unique factorization

$$x = g \cdot g^* = \tilde{g}^* \cdot \tilde{g}\,, \quad g, \tilde{g} \in G\,, \quad g^*, \tilde{g}^* \in G^*,$$

form an open dence subset in D. Thus G may be identified with an open dense subset in D/G^*, or in $G^*\backslash D$, and G^* with an open dense subset in D/G, or in $G\backslash D$.

Theorem 2.2 *(i) $G \subset D/G^*$ is a Poisson submanifold; the induced Poisson structure on G is anti-isomorphic to the original one.*

(ii) In a similar way, G^* is a Poisson submanifold in D/G; the induced Poisson bracket on G^* coincides with the original one.[4]

For trivial Lie bialgebras $D = T^*G, D/G \simeq \boldsymbol{g}^*$, and Theorem 2.2 amounts to the well known connection between the canonical Poisson bracket on T^*G and the Lie—Poisson bracket on $\boldsymbol{g}^*$. Theorem 2.1 then asserts that the Hamiltonians of left and right translations on T^*G are in involution with respect to the

[4]Sign difference in (i), (ii) is due to the minus sign in (2.2).

canonical Poisson bracket.This result plays the key role in the classical 'orbits method'.

We shall be mainly concerned with a special class of Lie bialgebras, the so called *factorizable Lie bialgebras.* Let $\boldsymbol{g}$ be a semisimple Lie algebra equipped with a fixed nondegenerate invariant inner product. We use it to fix an isomorphism of the dual space $\boldsymbol{g}^*$ with $\boldsymbol{g}$. The bialgebra structure on $\boldsymbol{g}$ is defined by the cobracket

$$\phi(X) = -\frac{1}{2}[r, X \otimes 1 + 1 \otimes X],$$

where $r \in \wedge^2 \boldsymbol{g}$ is a classical r-matrix. Since we identified $\boldsymbol{g}^*$ with $\boldsymbol{g}$, we may regard r as a skew symmetric linear operator in $\boldsymbol{g}$. Assume that r satisfies the *modified classical Yang—Baxter identity*

$$[rX, rY] = r(\,[rX,\ Y] + [X,\ rY]\,) - [X,Y]. \tag{2.6}$$

The Lie bracket on $\boldsymbol{g}^* \simeq \boldsymbol{g}$ which corresponds to (2.6) is given by

$$[X,Y]_* = \frac{1}{2}([rX,Y] + [X,rY]) \tag{2.7}$$

and by virtue of (2.6) satisfies the Jacobi identity. Put $r_\pm = \frac{1}{2}(r \pm id)$. Then (2.6) implies that $r_\pm$ regarded as a mapping from $\boldsymbol{g}^*$ into $\boldsymbol{g}$ is a Lie algebra homomorphism. Put $\boldsymbol{d} = \boldsymbol{g} \oplus \boldsymbol{g}$ (direct sum of two copies).The mapping

$$\boldsymbol{g}^* \to \boldsymbol{d} \quad : X \mapsto (X_+,\ X_-), \quad X_\pm = r_\pm X,$$

is a Lie algebra embedding. Thus we may identify $\boldsymbol{g}^*$ with a Lie subalgebra in $\boldsymbol{d}$. Let $\boldsymbol{g}^\delta \subset \boldsymbol{d}$ be the diagonal subalgebra. Equip $\boldsymbol{d}$ with the inner product

$$\ll (X,X'),(Y,Y') \gg = < X,Y > - < X',Y' > . \tag{2.8}$$

Proposition 2.1 *(i) As a linear space*

$$\boldsymbol{d} = \boldsymbol{g}^\delta \oplus \boldsymbol{g}^*.$$

(ii) Let $P_{\boldsymbol{g}}$, $P_{\boldsymbol{g}^*}$ be the projection operators onto $\boldsymbol{g}^\delta$, $\boldsymbol{g}^* \subset \boldsymbol{d}$ parallel to the complementary subalgebra,$r_{\boldsymbol{d}} = P_{\boldsymbol{g}} - P_{\boldsymbol{g}^*}$. Then $r_{\boldsymbol{g}}$ is skew with respect to the inner product (2.8) and satisfies (2.6). Hence it defines on $\boldsymbol{d}$ the structure of a Lie bialgebra.

(iii) $(\boldsymbol{d}, \boldsymbol{d}^*)$ is canonically isomorphic to the double of $(\boldsymbol{g}, \boldsymbol{g}^*)$.

The explicit expression for $r_{\boldsymbol{d}} \in End(\boldsymbol{g} \oplus \boldsymbol{g})$ in terms of the original r-matrix is given by

$$r_{\boldsymbol{d}} = \begin{pmatrix} r & -2r_+ \\ 2r_- & -r \end{pmatrix}. \tag{2.9}$$

A Lie bialgebra $(\boldsymbol{g}, \boldsymbol{g}^*)$ with the properties as above is called a *factorizable Lie bialgebra*. Thus the double of a factorizable Lie bialgebra is isomorphic (as a Lie algebra) to the square of $\boldsymbol{g}$.

Now let G be a linear algebraic group with the Lie algebra $\boldsymbol{g}$. Put $D = G \times G$. Embedding $\boldsymbol{g}^* \hookrightarrow \boldsymbol{d}$ may be extended to a homomorphism $G^* \hookrightarrow D$; we shall identify G^* with the corresponding subgroup in D. Almost all elements $(x, y) \in D$ admit a representation

$$(x,y) = (L^+, L^-) \cdot (T,T)^{-1}, \tag{2.10}$$

where $(L^+, L^-) \in G^*, \quad (T,T) \in G^\delta \subset D$.

Let (ρ, V) be an exact matrix representation of G. The algebra $Fun(G)$ is generated by matrix coefficients $\rho(x)_{ij}, \ \rho(y)_{ij}$. Matrices $L^\pm, T$ may be regarded as (almost everywhere regular) functions of x, y. Hence the matrix coefficients $\rho(L^\pm)_{ij}, \ \ \rho(T)_{ij}$ give another system of generators of the (suitably enlarged) algebra $Fun(D)$. Functions $\rho(L^\pm_{ij}, \rho(T)_{ij}$ are rational functions on D with singularities at those points $(x,y) \in G$ for which factorization (2.10) does not exist.

It is convenient to define the Poisson structure on D in terms of the generators of $Fun(D)$. We use the standard tensor notation to suppress matrix indices. Thus we write $T_1 = T \otimes id, \ \ T_2 = id \otimes T$, etc. The Poisson bracket $\{T_1^V, T_2^W\}$ is, by definition, a matrix in $End(V \otimes W)$ whose matrix coefficients are the Poisson brackets $\{\rho_V(T)_{ij}, \rho_W(T)_{kl}\}$. The superscripts V, W will sometimes be omitted. We shall need explicit formulae for the Poisson brackets of two systems of generators of $Fun(D)$. We have

$$\begin{aligned}
\{x_1, x_2\}_\pm &= \frac{1}{2}\left(r_{VW}\ x_1 x_2 \ \pm\ x_1 x_2\ r_{VW}\ \right), \\
\{y_1, y_2\}_\pm &= \frac{1}{2}\left(r_{VW}\ y_1 y_2 \ \pm\ y_1 y_2\ r_{VW}\ \right), \\
\{y_1, x_2\}_\pm &= r^+_{VW}\ y_1 y_2 \ \pm\ y_1 x_2\ r^+_{VW},
\end{aligned} \tag{2.11}$$

$$\begin{aligned}
\{T_1, T_2\}_\pm &= \frac{1}{2}\left[r_{VW}, T_1 T_2\right], \\
\{L_1^\pm, L_2^\pm\}_\pm &= \frac{1}{2}\left[r_{VW}, L_1^\pm L_2^\pm\right], \\
\{L_1^+, L_2^-\}_\pm &= \left[r^+_{VW}, L_1^+ L_2^-\right], \\
\{L_1^\pm, T_2\}_+ &= L_1^\pm T_2\, r^\pm_{VW}, \\
\{L_1^\pm, T_2\}_- &= 0.
\end{aligned} \tag{2.12}$$

Here $r_{VW} = (\rho_V \otimes \rho_W) r \in End(V \otimes W)$ and r is regarded as an element of $\wedge^2 \boldsymbol{g}$. Note that Poisson brackets $\{,\}_\pm$ in terms of the generators $L^\pm, T$ differ only by the expression for $\{L_1^{\pm}, T_2\}$. Recall that D_- is a Poisson Lie group. Formulae (2.12) show that as a Poisson manifold (though of course not as a group) D_- is a direct product of its subgroups G, G^*. The brackets $\{,\}_+$ are non-degenerate.

Since D_- is a Poisson Lie group, the diagonal map

$$\Delta : Fun(D_-) \to Fun(D_- \times D_-)$$

is a homomorphism of Poisson algebras. This map may easily be described in terms of the generators x, y:

$$\Delta x \;=\; x \dot{\otimes} x, \quad \Delta\, y \;=\; y \dot{\otimes} y,$$

or, in a more accurate notation,

$$\Delta(\rho(x)_{ij}) \;=\; \sum_k \rho(x)_{ik} \otimes \rho(x)_{kj},$$

and similarly for $\Delta(\rho(y))$. It is easy to check that multiplication in D induces a Poisson mapping $D_- \times D_+ \to D_+$. Hence the diagonal map also defines on $Fun(D_+)$ the structure of a left $Fun(D_-)$-comodule. In the obvious notation we may write

$$C(x_+) \;=\; x_- \dot{\otimes} x_+, \quad C(y_+) \;=\; y_- \dot{\otimes} y_+.$$

Another group which is also important for the duality theory is the *dual double* D^*. By definition, D^* is the Poisson—Lie group which corresponds to the Lie bialgebra $(\boldsymbol{d}^*, \boldsymbol{d})$. Since $\boldsymbol{d}^* \simeq \boldsymbol{g} \oplus \boldsymbol{g}^*$ is a direct sum of Lie algebras, $D^* \;=\; G \times G^*$ as a Lie group. However, the Poisson structure on D^* does *not* split. The algebra $Fun(D^*)$ is generated by matrix coefficients $\rho(L^{\pm}, \rho(T)$. Let us denote these generators by ${}^*L^{\pm}$, ${}^*T$ in order to distinguish them from the generators (2.10). We have

$$\begin{aligned}
\{{}^*T_1\,,\,{}^*T_2\} &= \frac{1}{2}\,[r\,,\,{}^*T_1\,{}^*T_2],\\
\{{}^*L_1^{\pm}\,,\,{}^*L_2^{\pm}\} &= \frac{1}{2}\,[r\,,\,{}^*L_1^{\pm}\,{}^*L_2^{\pm}],\\
\{{}^*L_1^{+}\,,\,{}^*L_2^{-}\} &= [r\,,\,{}^*L^{+}\,{}^*L_2^{-}\},\\
\{{}^*L_1^{\pm}\,,\,{}^*T_2\} &= [r^{\pm}\,,\,{}^*L_1^{\pm}\,{}^*T_2].
\end{aligned} \tag{2.13}$$

The coproduct in $Fun(D^*)$ is given by

$$\Delta^*T \;=\; {}^*T \dot{\otimes} {}^*T, \quad \Delta\,{}^*L^{\pm} ={}^* L^{\pm}\dot{\otimes}{}^*L^{\pm}.$$

Let us return to the study of the bracket $\{,\}_+$ on D and discuss the relations between various systems of generators of $Fun(D_+)$. Observe first of all that due to factorization formula (2.10) we may regard $L^{\pm}$ as generators of the algebra $Fun(D/G^{\delta}$ and T as generators of the algebra $Fun(G^{*}\backslash D)$ .One can also factorize (x, y) in the opposite order:

$$(x, y) = (\hat{T}, \hat{T}) \cdot (\hat{L}^{+}, \hat{L}^{-})^{-1} \tag{2.14}$$

Clearly, the matrix coefficients of $\hat{T}$ generate the algebra $Fun(G^{\delta}\backslash D)$. As we know, canonical projections

$$D/G \xleftarrow{\pi} D \xrightarrow{\prime\pi} G\backslash D, \qquad D/G^{*} \xleftarrow{p} D \xrightarrow{p'} G^{*}\backslash D \tag{2.15}$$

form dual pairs. Hence the generators $\hat{L}^{\pm}, \hat{T}$ satisfy the Poisson bracket relations (2.12) and, moreover,

$$\{L_1^{\pm}, \hat{L}_2^{\pm}\} = \{T_1, \hat{T}_2\} = 0.$$

Decompositions (2.10, 2.14) are analogous to the definition of left and right momenta for a rigid body [1], or to chiral decompositions in Conformal Field Theory (the last analogy is discussed in [2]).

The quotient spaces D/G^{δ}, $G^{\delta}\backslash D$ may be canonically identified with the group G itself. The projection maps $\pi, \hat{\pi}$ are given by

$$\pi(x, y) = xy^{-1}, \qquad \hat{\pi}(x, y) = y^{-1}x.$$

This allows to introduce another set of generators for the algebras $Fun(D/G)$, $Fun(G\backslash D)$. Put $L_V = \rho_V(xy^{-1}), \hat{L}_V = \rho_V(y^{-1}x)$ (here, as above, (ρ, V) is an exact linear representation of G). Clearly, we have

$$L_V = L_V^{+}(L_V^{-})^{-1}, \hat{L}_V = (\hat{L}_V^{+})^{-1}\hat{L}_V^{-}.$$

Note that matrix coefficients of $L_V, \hat{L}_V$ are everywhere regular functions on D. The Poisson brackets for these generators are given by

$$\{L_1, L_2\} = L_1 r_+ L_2 + L_2 r_- L_1 - \frac{1}{2} L_1 L_2 r - \frac{1}{2} r L_1 L_2, \tag{2.16}$$

$$\{L_1, T_2\} = L_1 T_2 r_- - T_2 r_+ L_1. \tag{2.17}$$

According to the general theory, symplectic leaves in $D/G \simeq G$ are the connected components of the sets $\pi(\pi'^{-1}(x)), x \in G\backslash D$. In the present case $\pi(x, y) = xy^{-1}$, $\pi'(x, y) = y^{-1}x$. Thus symplectic leaves in G are simply the conjugacy classes; moreover,the action of G on itself by conjugations is a Poisson mapping

$$G \times G \to G : (x, L) \mapsto xLx^{-1}. \tag{2.18}$$

Here the bracket $\{L_1, L_2\}$ is given by (2.16), and the bracket for the matrix coefficients of x is given by the standard formula

$$\{x_1 \, , \, x_2\} \; = \; \frac{1}{2} \, [r \, , \, x_1 x_2]. \tag{2.19}$$

By duality, we get a morphism

$$Fun(D/G) \to Fun(G) \otimes Fun(D/G) : \quad L \mapsto x_1 L_2 x_1^{-1}. \tag{2.20}$$

The action (2.18) and the dual coaction (2.20) is an example of the so called *dressing transformations* [19].

Let us finally note that the Casimir functions which form the center of the Poisson algebra $Fun(D/G)$ are precisely central functions on G. The generators of the ring of Casimir functions are given by

$$C_k \; = \; \mathrm{tr} \; L^k \; = \; \mathrm{tr} \; \hat{L}^k. \tag{2.21}$$

3 Quantization

Let A be a factorizable quasitriangular Hopf algebra [7, 17]. By definition, a quasitriangular Hopf algebra is a Hopf algebra with a distinguished element $R \in A \otimes A$ satisfying the following properties:

(i)

$$\Delta'(x) \; = \; R\Delta(x)R^{-1} \tag{3.1}$$

for any $x \in A$. (Here Δ' denotes the opposite coproduct in A.)

(ii)

$$(\Delta \otimes id)R \; = \; R_{13}R_{23}, \qquad (id \otimes \Delta)R \; = \; R_{13}R_{12}, \tag{3.2}$$

(iii)

$$(\epsilon \otimes id)R \; = \; (id \otimes \epsilon)R \; = \; 1.$$

Under these assumptions R is invertible and $R^{-1} \; = \; (S \otimes id)R \; = \; (id \otimes S)R$, where S is the antipode of A.

Let σ be the permutation map in $A \otimes A$. Put $R_+ \; = \; R$, $R_- \; = \; \sigma(R^{-1})$. Let A^* be the dual Hopf algebra of A, and A^0 the same algebra with the opposite coproduct. Let $R^\pm : A^0 \to A$ be the maps

$$R^\pm : \; f \; \mapsto < f \; \otimes \; id \; , \; R_\pm > \tag{3.3}$$

Then by (3.1) $R^\pm$ are Hopf algebra homomorphisms. Let us consider the combined mapping

$$A^0 \overset{(R^+ \otimes R^-)\Delta^0}{\longrightarrow} A \otimes A \overset{m(id \otimes S^{-1})}{\longrightarrow} A. \tag{3.4}$$

A quasitriangular Hopf algebra is called *factorizable* if the composition map is a linear space isomorphism (for infinite-dimensional spaces we require that it is an isomorphism of an open dense subspace in A^0 onto an open dense subspace in A).

Choose a linear basis $\{e_i\}$ in A and let $\{e^i\}$ be the dual basis in A^0. Let

$$T = e^i \otimes e_i \in A^0 \otimes A \tag{3.5}$$

be the canonical element. We shall write, using the standard tensor notation,

$$\begin{aligned} T_1T_2 &= e^ie^j \otimes e_i \otimes e_j \in A^* \otimes A \otimes A, \\ T_2T_1 &= e^je^i \otimes e_i \otimes e_j \in A^* \otimes A \otimes A. \end{aligned}$$

From (3.1) we have

$$T_2T_1 = R\, T_1T_2\, R^{-1}. \tag{3.6}$$

Put

$$L^{\pm} = (R^{\pm} \otimes id)\, T \in A \otimes A. \tag{3.7}$$

We have

$$\begin{aligned} L_2^{\pm}L_1^{\pm} &= R_+\, L_1^{\pm}L_2^{\pm}\, R_+^{-1}, \\ L_2^{-}L_1^{+} &= R_+L_1^{+}L_2^{-}R_+^{-1}. \end{aligned} \tag{3.8}$$

Put $L = L_+^{-1}\, L_-$.Then we get from (3.8)

$$R_+^{-1}L_2R_+\, L_1 = L_1\, R_-^{-1}L_2R_-. \tag{3.9}$$

Now let ρ_V, ρ_W be repesentations of A in linear spaces V, W. Put

$$\begin{aligned} T^V &= (id \otimes \rho_V)\, T \in A^0 \otimes \mathrm{End}V, \\ L^{\pm V} &= (id \otimes \rho_V)\, L_{\pm} \in A \otimes \mathrm{End}V. \end{aligned}$$

Then relations (3.6, 3.8, 3.9) yield

$$\begin{aligned} T_2^WT_1^V &= R^{VW}T_1^VT_2^WR^{VW\ -1}, \\ L_2^{\pm\ W}L_1^{\pm\ V} &= R^{VW}L_1^{\pm\ V}L_2^{\pm\ W}(R^{VW})^{-1}, \\ L_2^{-\ W}L_1^{+\ V} &= R^{VW}L_1^{+\ V}L_2^{-\ W}(R^{VW})^{-1}, \\ (R_+^{VW})^{-1}L_2^WR_+^{VW}L_1^V &= L_1^V(R_-^{VW})^{-1}L_2^WR_-^{VW}. \end{aligned} \tag{3.10}$$

Assume now that A is quasi-classical. This means that A is a free module over the ring $\mathbf{C}[[h]]$, where h is a deformation parameter, and, moreover, A/hA is isomorphic (as a Hopf algebra) to the universal enveloping algebra of a Lie algebra $\mathfrak{g}$. Formula (3.1) then implies that

$$R_{\pm} = 1 + hr_{\pm} + o(h),$$

where $r_{\pm} \in U(\boldsymbol{g})^{\otimes 2}$. It is easy to see that in fact $r_{\pm} \in \boldsymbol{g} \otimes \boldsymbol{g}$ where $\boldsymbol{g}$ is identified with the subalgebra of primitive elements in $U(\boldsymbol{g})$. Formulae (3.2) imply that $r_{\pm}$ satisfy the classical Yang—Baxter identity. The factorization map (3.4) induces an isomorphism of linear spaces $\boldsymbol{g}^* \to \boldsymbol{g}$, i.e. an invariant inner product on $\boldsymbol{g}$. Thus $\boldsymbol{g}$ has the structure of a factorizable Lie bialgebra. Let us define the associated Poisson Lie groups G, G^* and Poisson algebras $Fun(G)$, $Fun(G^*)$, as in Section 2. We shall assume that G, G^* are classical algebraic linear groups and that an exact representation (ρ, V) of G agrees with the representation ρ_V of A. The algebras $Fun(G), Fun(G^*)$ are generated by the matrix coefficients of matrices $\rho_V(T), \rho_V(L^{\pm})$,respectively. There are obvious relations which express the symmetry conditions for $T \in G$ and describe the image of G^* in G under the mappings $r_{\pm}$. (For instance, if $G = GL(n)$ given with its standard matrix representation, and r is the standard r-matrix on $gl(n)$ associated with the Gauss decomposition , then the symmetry conditions are void and $r_{\pm}(G)$ are the opposite Borel subgroups in $GL(n)$.) All these relations may be explicitly quantized, i.e. they admit canonical deformations which make them compatible with the quantum commutation relations [8]. In the sequel, we shall not write down these relations explicitly. The reader may assume that $\boldsymbol{g} = gl(n)$, however, an extension to other classical Lie algebras is always straightforward.

Proposition 3.1 *Associative algebras $Fun_q(G), Fun_q(G^*)$ generated by the matrix coefficients of T^V, $L^{\pm\ V}$ (or L^V) and relations (3.10) (along, if necessary, with the quantum symmetry relations) are quantizations of the Poisson algebras $Fun(G), Fun(G^*)$,respectively.*

The generators $T^V, L^{\pm\ V}$ are expressed through the canonical element (3.5) and the universal R-matrix. This defines the homomorphisms

$$Fun_q(G^*) \to A, \quad Fun_q(G) \to A^*. \tag{3.11}$$

For $A = U_q(G)$ we may use explicit formulae for the universal R-matrix [12, 13] to express the generators $L^{\pm\ V}, T^V$ in terms of the Drinfeld—Jimbo generators. (For $\boldsymbol{g} = sl_2$ these formulae are elementary and are given in [8])

Remark. One can check, using the results of [8] that the mappings (3.11) are actually isomorphisms (i.e. there are no extra relations in the algebras)

The construction below makes sense for any Hopf algebra. Let us consider an action $A \otimes A^* \to A^*$ by 'left derivations'

$$x \otimes f \mapsto D_x f = < x \otimes id , \Delta f > . \tag{3.12}$$

Obviously, we have

$$< x , f > = \epsilon(D_x f), \tag{3.13}$$

i.e the canonical pairing between A and A^* is given by the 'value of the derivative at the unit element'. Let us consider an asssociative algebra $\mathcal{H}$ of operators in A^* generated by derivations $D_x, x \in A$, and by multiplication operators

$$m_f : \phi \mapsto f\phi, f \in A^*. \tag{3.14}$$

Definition 3.1 *The algebra $\mathcal{H}$ is called the Heisenberg double of A.*

*Example.*If $A = S(V)$ is the symmetric algebra of a linear space V, $\mathcal{H}$ is the enveloping algebra of the Heisenberg algebra generated by $V \oplus V^*$.

Assume again that A is factorizable. We shall write down the commutation relations for $\mathcal{H}(A)$ in the form that allows a comparison with the Poisson algebra $Fun(D_+)$. Let us introduce the operator-valued matrices

$$\begin{aligned} D_{L^\pm} &= (D \otimes id)\ L^\pm \in \text{End } A^* \otimes A, \\ m_T &= (id \otimes m)\ T \in A \otimes \text{End } A^*. \end{aligned}$$

Obviously,

$$D : A \to \text{End } A^* \ , \ m : A^* \to \text{End } A^*$$

are homomorphisms of algebras, so it is sufficient to compute the commutation relations between the matrices $D_{L^\pm}$, or D_L and m_T.

Proposition 3.2 *We have*

$$\begin{aligned} D_{L_1^\pm} \circ T_2 &= T_2 \circ D_{L_1^\pm} R_\pm, \\ D_{L_1} \circ T_2 &= T_2 \circ R_+ D_{L_1} R_-^{-1}, \\ < L_1^\pm, T_2 > &= R_\pm. \end{aligned} \tag{3.15}$$

In order to interprete these relations let us consider an associative algebra generated by the matrix coefficients of the matrices $T^V, L^{\pm\ V}$ (or L^V) and relations (3.10) and the supplementary relations

$$\begin{aligned} L_1^{\pm\ V} T_2^W &= T_2^W L_1^{\pm\ V} R_\pm^{VW}, \\ L_1^V T_2^W &= T_2^W R_+^{VW} L_1^V (R_-^{VW})^{-1}. \end{aligned} \tag{3.16}$$

(We always tacitly assume that L, T satisfy also the symmetry relations which hold for classical groups.)

Proposition 3.3 *Assume that A is a quasi-classical factorizable Hopf algebra, as in 3.1.Then*

(i) The algebra (3.10, 3.16) is a quantization of the Poisson algebra $Fun(D_+)$.

(ii) Realization of L, T by derivation and multiplication operators in A^* gives an exact representation of this algebra in EndA^*.

Alternatively, we can use another set of generators of the Heisenberg double for which the commutation relations have a more symmetric form.

Proposition 3.4 *Put* $X = L_+T^{-1}$, $Y = L_-T^{-1}$. *Then the following relations hold:*

$$\begin{aligned} R_+X_1X_2 &= X_2X_1R_-^{-1}, \\ R_+Y_1Y_2 &= Y_2Y_1R_-^{-1}, \\ R_+X_1Y_2 &= Y_2X_1R_+^{-1}. \end{aligned} \tag{3.17}$$

The algebra generated by the matrix coefficients of X, Y *and relations (3.17) is a quantization of the Poisson algebra (2.11).*

Along with the Heisenberg double one can also define the *Drinfeld double.* Recall that the double of a Lie bialgebra is a factorizable Lie bialgebra, and the associated classical r-matrices are essentially the projection operators. In complete analogy, the Drinfeld double $\mathcal{D}(A)$ of a Hopf algebra A is a canonically defined factorizable Hopf algebra with the underlying linear space $A \otimes A^*$.

Theorem (Drinfeld, [7]) 1 *There exists a unique structure of a Hopf algebra on* $\mathcal{D}(A) = A \otimes A^*$ *such that*

(i) $A, A^* \subset \mathcal{D}$ *are subalgebras in* $\mathcal{D}$.

(ii) As a coalgebra, $\mathcal{D} = A \otimes A^0$.

(iii) The universal R-matrix is the image of the canonical element $T \in A \otimes A^*$ *(cf. (3.5)) under the natural embedding* $A \otimes A^* \hookrightarrow \mathcal{D} \otimes \mathcal{D}$.

The Hopf algebra $\mathcal{D}$ *is factorizable ([17]).*

It is interesting to observe that $\mathcal{D}(A)$ also admits an operator realization. Let us first define the (right) adjoint action of an arbitrary Hopf algebra on itself:

$$\mathrm{Ad} : A \otimes A \to A : x \otimes y \mapsto Sy^{(1)}xy^{(2)}, \tag{3.18}$$

where $\Delta y = y^{(1)} \otimes y^{(2)}$ is the coproduct in A. The *coadjoint action* $\mathrm{Ad}^* : A^* \otimes A \to A^*$ is defined by

$$< \mathrm{Ad}^*(f \otimes x) \, , \, y > = < f \, , \, \mathrm{Ad}(x \otimes y) > . \tag{3.19}$$

From the definition one easily gets that

$$\mathrm{Ad}^*(f \otimes x) = < Sf^{(1)}f^{(3)}, x > f^{(2)}, \tag{3.20}$$

where $\Delta^{(2)}f = f^{(1)} \otimes f^{(2)} \otimes f^{(3)}$. Let us consider the algebra $\mathcal{D}$ of operators acting on A^* generated by multiplication operators $m_f, f \in A^*$ and the operators $\mathrm{Ad}^*x, x \in A$.

Proposition 3.5 *As an associative algebra, $\mathcal{D}$ is isomorphic to the Drinfeld double of A.*

Let us assume now that A is a quasiclassical factorizable Hopf algebra. We shall describe its Drinfeld double as a quantization of a Poisson algebra. The quantum duality principle suggests that $\mathcal{D}\,(A) \simeq Fun_q(D^*)$ where D^* is the dual double of G, G^*. We shall see now that this is indeed the case. Let $T \in A^* \otimes A$ be the canonical element (see (3.5)); put $L^\pm = (R^\pm \otimes id)T$. We fix a representation (ρ, V) and set

$$\begin{aligned} \hat{\mathcal{L}}^{\pm\ V} &= (Ad^* \otimes \rho_V)L^\pm, \\ \hat{\mathcal{T}}^V &= (m \otimes \rho_V)T. \end{aligned}$$

(Here $m : A^* \longrightarrow \mathrm{End}A^*$ is the representation of A^* by multiplication operators, as in (3.14).)

Proposition 3.6 *(i) The operator matrices $\hat{\mathcal{L}}^{\pm\ V}, \hat{\mathcal{L}}^V$ satisfy*

$$\hat{\mathcal{L}}_2^{\pm\ V}\hat{\mathcal{T}}_1^V = R_\pm^{VV}\hat{\mathcal{T}}_1^V\hat{\mathcal{L}}_2^{\pm\ V}(R_\pm^{VV})^{-1}. \tag{3.21}$$

(ii) The associative algebra generated by the matrix coefficients of $\hat{\mathcal{L}}^{\pm\ V}$,$\hat{\mathcal{T}}^V$ and relations (3.21, 3.10) is a quantization of the Poisson algebra $Fun(D^*)$ with the Poisson bracket relations (2.13)

(iii) The coproduct in $\mathcal{D}\,(A)$ is given by

$$\begin{aligned} \Delta\hat{\mathcal{T}} &= \hat{\mathcal{T}}_2 \dot{\otimes} \hat{\mathcal{T}}_1, \\ \Delta\hat{\mathcal{L}}^\pm &= \hat{\mathcal{L}}_1^\pm \dot{\otimes} \hat{\mathcal{L}}_2^\pm. \end{aligned}$$

In a similar way, the algebra $Fun_q(D_-)$ may be identified with the *dual* of the Drinfeld double.

Proposition 3.7 *(i) The associative algebra $Fun_q(D_-)$ generated by matrix coefficients of the matrices $\mathcal{L}^\pm$, $\mathcal{T}$, relations (3.6,3.8) and the supplementary relation*

$$\mathcal{T}_1\mathcal{L}_2^\pm = \mathcal{L}_2^\pm\mathcal{T}_1 \tag{3.22}$$

is isomorphic to the dual of the Drinfeld double $\mathcal{D}\,(A)^*$.

(ii) the algebra $Fun_q(D_-)$ is a quantization of the Poisson algebra $Fun(D_-)$ with relations (2.12).

In order to describe the coproduct in $Fun_q(D_-)$ it is convenient to introduce another system of generators suggested by (2.11). Put

$$\check{X} = \mathcal{L}^+\mathcal{T}, \qquad \check{Y} = \mathcal{L}^-\mathcal{T}. \tag{3.23}$$

Proposition 3.8 *(i) The matrix coefficients of $\check{X}, \check{Y}$ satisfy the following commutation relations*

$$\begin{aligned} \check{X}_2\check{X}_1 &= R_+\check{X}_1\check{X}_2R_+^{-1}, \\ \check{Y}_2\check{Y}_1 &= R_+\check{Y}_1\check{Y}_2R_+^{-1}, \\ \check{Y}_2\check{X}_1 &= R_+\check{X}_1\check{Y}_2R_+^{-1}. \end{aligned} \tag{3.24}$$

(ii) Formulae

$$\Delta\check{X} = \check{X}\dot{\otimes}\check{X}, \quad \Delta\check{Y} = \check{Y}\dot{\otimes}\check{Y}$$

define on $Fun_q(D_-)$ the coalgebra structure which is in agreement with that of $\mathcal{D}(A)^*$.

(iii) Let X, Y be the matrices generating $H(A)$ and satisfying relations (3.17). Formulae

$$C(X) = \check{X}\dot{\otimes}\check{X}, \; C(Y) = \check{Y}\dot{\otimes}\check{Y}$$

define on $\mathcal{H}(A)$ the structure of a left $Fun_q(D_-)$-comodule.

Let us now return to the study of the Heisenberg double. We shall describe the quantum analogue of the Poisson reduction and the dual pair described in (2.15). We have just mentioned that $\mathcal{H}(A)$ has a natural structure of a left $\mathcal{D}(A)^*$-comodule. In a similar way, one can define on $\mathcal{H}(A)$ the structure of a right $\mathcal{D}(A)^*$-comodule. We may specialize these formulae to get on $\mathcal{H}(A)$ the structure of a left and right $Fun_q(G)$-comodule. Namely, put

$$\begin{aligned} C_L(X) &= T\dot{\otimes}X, \; C_L(Y) = T\dot{\otimes}Y, \\ C_R(X) &= X\dot{\otimes}T^{-1}, \; C_R(Y) = Y\dot{\otimes}T^{-1}, \end{aligned} \tag{3.25}$$

where T is the generator matrix for $Fun_q(G)$, $T^{-1} = (S\otimes id)T$, and S is the antipode of $Fun_q(G)$. The necessary formal properties of C_L, C_R are easily verified.

Let now $Fun_q(G\backslash D)$, $Fun_q(D/G) \subset Fun_q(D_+)$ be the subalgebras of left (right) coinvariants. By definition,

$$\begin{aligned} \phi \in Fun_q(g\backslash D) &\Longleftrightarrow C_L((\phi) \in 1 \otimes Fun_q(D_+), \\ \psi \in Fun_q(D/G) &\Longleftrightarrow C_R(\psi) \in Fun_q(D_+) \otimes 1. \end{aligned}$$

Proposition 3.9 *(i) The algebra $Fun_q(G\backslash D)$ is generated by the matrix coefficients of $L' = X^{-1}Y$; in a similar way, the algebra $Fun_q(D/G)$ is generated by $L = XY^{-1}$.*

(ii) The matrices L, L' satisfy the commutation relations (3.9); thus

$$Fun_q(D/G) \simeq Fun_q(G\backslash D) \simeq Fun_q(G^*)$$

(iii) Moreover, $L_1L_2' = L_2'L_1$*, i.e. the subalgebras* $Fun_q(D/G)$*,* $Fun_q(G\backslash D)$ *centralize each other in* $Fun_q(D_+) \simeq \mathcal{H}(A)$*.*

Clearly, the algebra $Fun_q(D/G)$ inherits the structure of a left $Fun_q(G)$-comodule. It is given by

$$C(L) = T_1L_2T_1^{-1}. \tag{3.26}$$

Formula (3.26) gives in the present case a quantum analogue of 'dressing transformations'. The next assertion is similar to the description of the center of the Poisson algebra $Fun(D/G)$; in the same time, it provides a quantum version of the well known Gelfand theorem describing the center of a universal enveloping algebra.

Proposition 3.10 *The center of* $Fun_q(D/G)$ *coincides with the subalgebra of coinvariants of the coaction (3.26).*

We shall study quantum dressing transformations more thoroughly in a separate paper.

4 The Heisenberg Double and the Quantum Fourier Transform

So far we have studied the commutation relations for $L^{\pm\ V}, T^V$ fixing the representation (ρ, V) of our Hopf algebra. If, instead, we consider arbitrary irreducible representations of A and A^* and their tensor products, we may relate the quantum duality principle with the intuitively attractive quantum Fourier transform.[5]

Let A be a Hopf algebra, A^* its dual. Let $\hat{A} = \mathrm{Spec}(A)$ be a set of its irreducible representations. We do not fix a $*$-structure on A and hence do not assume that rerpresentations $\pi \in \hat{A}$ are unitary. For $\pi \in \hat{A}$ let V_π be the corresponding A-module. Let $Fun(\hat{A})$ be the space of functions on $\hat{A}$ such that $\phi_\lambda \in \mathrm{End}V_\lambda$; $Fun(\hat{A})$ is an algebra with respect to pointwise multiplication. It is natural to identify the dual space $Fun(\hat{A})^*$ with the space of *matrix-valued measures* on $\hat{A}$. For $\phi \in Fun(\hat{A})$, $\sigma = d\sigma(\lambda) \in Mes(\hat{A})$ we put [6]

$$<\sigma, \phi> = \int_{\hat{A}} \mathrm{tr}_{V_\lambda}(\phi(\lambda)d\sigma(\lambda)).$$

We shall assume that the set $\hat{A}$ is closed with respect to the tensor product. More precisely, we suppose that for any $\lambda_1, \lambda_2 \in \hat{A}$

$$V_{\lambda_1} \otimes V_{\lambda_2} \simeq \int \bigoplus_{\lambda\in\hat{A}} V_\lambda \otimes W_\lambda,$$

[5]A different version of the quantum Fourier transform is dicussed in [15].

[6]In our exposition we shall ignore all convergence questions, so the reader may assume that all representations are finite-dimensional.

where W_λ is the multiplicity space.

Let us introduce the Clebsch—Gordan coefficients by the formula

$$\lambda_1 \otimes \lambda_2(x) = \int_{\hat{A}} \mathrm{tr}_{V_\lambda}(\lambda(x) dC(\lambda \mid \lambda_1, \lambda_2)).$$

By definition, $dC(\lambda \mid \lambda_1, \lambda_2)$ is a function of variables λ_1, λ_2 and a measure in variable λ with values in $\mathrm{End}(V_{\lambda_1} \otimes V_{\lambda_2} \otimes V_\lambda)$ which commutes with all intertwiners $I \in \mathrm{End}_A(V_{\lambda_1} \otimes V_{\lambda_2} \otimes V_\lambda)$.

In a similar way we define the set $\hat{A}^* = \mathrm{Spec}(A^*)$ and the spaces $Fun(\hat{A}^*)$, $Mes(\hat{A}^*)$. Let us denote by $\mathcal{R}$ the canonical element in $A^* \otimes A$ [7] and put

$$\begin{aligned} T_\rho &= (id \otimes \rho)\mathcal{R}, \quad \rho \in \hat{A}, \\ L_\lambda &= (\lambda \otimes id)\mathcal{R}, \quad \lambda \in \hat{A}^*. \end{aligned} \tag{4.1}$$

Define the *spectral representation* $T: \ Mes(\hat{A}) \to A$ by

$$T(\phi) = \int_{\hat{A}} \mathrm{tr}_{V_\rho}(T_\rho d\phi(\rho)). \tag{4.2}$$

In a similar way, the spectral representation $L: \ Mes(\hat{A}^*) \to A$ is defined by

$$L(\psi) = \int_{\hat{A}^*} \mathrm{tr}(L_\lambda d\psi(\lambda)). \tag{4.3}$$

We define the convolutions of measures

$$\begin{aligned} *: \ & Mes(\hat{A}) \otimes Mes(\hat{A}) \to Mes(\hat{A}), \\ : \ & Mes(\hat{A}^*) \otimes Mes(\hat{A}^*) \to Mes(\hat{A}^*), \end{aligned} \tag{4.4}$$

demanding that

$$\begin{aligned} T(\phi_1 * \phi_2) &= T(\phi_1)T(\phi_2), \\ L(\psi_1 * \psi_2) &= L(\psi_1)L(\psi_2). \end{aligned} \tag{4.5}$$

Proposition 4.1 *(i) The convolution of measures on $\hat{A}$ is given by*

$$\phi_1 * \phi_2(\rho) = \int_{\hat{A} \times \hat{A}} tr_{V_{\rho_1} \otimes V_{\rho_2}} C(\rho \mid \rho_1, \rho_2) d\phi_1(\rho_1) d\phi_2(\rho_2),$$

where $C(\rho \mid \rho_1, \rho_2)$ are the Clebsch—Gordan coefficients of A.

(ii) In a similar way, the convolution of measures on $\hat{A}^*$ is given by

$$\psi_1 * \psi_2(\lambda) = \int_{\hat{A}^* \times \hat{A}^*} tr_{W_{\lambda_1} \otimes W_{\lambda_2}} C^*(\lambda \mid \lambda_1, \lambda_2) d\psi_1(\lambda_1) d\psi_2(\lambda_2),$$

where $C^*(\lambda \mid \lambda_1, \lambda_2)$ are the Clebsch—Gordan coefficients of A^*.

[7]We are changing slightly the notation introduced in Section 3. Indeed, $\mathcal{R}$ coincides with the universal R-matrix for $\mathcal{D}(A)$.

In terms of the spectral representations the coupling $A^* \otimes A \to \mathbf{C}$ takes the form

$$< \phi \mid \psi > = \int \mathrm{tr}_{V_\rho \otimes W_\lambda}(d\phi(\rho) R_{\rho\lambda} d\psi(\lambda)), \tag{4.6}$$

where

$$R_{\rho\lambda} = (\rho \otimes \lambda)\mathcal{R}. \tag{4.7}$$

Let us define the Fourier transform

$$\phi : \ Mes(\hat{A}^*) \to Fun(\hat{A})$$

by

$$\Phi\phi(\lambda) = \int \mathrm{tr}_{V_\lambda}(d\phi(\rho) R_{\rho\lambda}). \tag{4.8}$$

Proposition 4.2 *We have*

$$\Phi(\phi_1 * \phi_2) = \Phi\phi_1 \Phi\phi_2.$$

In a similar way we define the conjugate Fourier transform

$$\hat{\Phi} : \ Mes(\hat{A}) \to Fun(\hat{A}^*).$$

Example. Let $A = S(V)$ be the symmetric algebra of a linear space V. Then $A^* \simeq S(V^*)$. Choose a basis e_i in V and let e^i be the dual basis in V^*. Then the canonical element in $A^* \otimes A$ is given by

$$\mathcal{R} = \exp(\sum \mathrm{e}^{\mathrm{i}} \otimes \mathrm{e}_{\mathrm{i}}).$$

Irreducible representations of A and A^* are 1-dimensional and we have $\hat{A} = V^*$, $\hat{A}^* = V$. The convolutions (4.5) become ordinary convolutions of measures on a linear space, and the Fourier transform is the ordinary Fourier—Laplace transform on V.

Let us define an action

$$Mes(\hat{A}^*) \otimes Mes(\hat{A}) \to Mes(\hat{A})$$

by the formula

$$\phi \otimes \psi \to \Phi\phi \cdot \psi. \tag{4.9}$$

Proposition 4.3 *The algebra of operators acting on $Mes(\hat{A})$ which is generated by convolution operators (4.4) and multiplication operators (4.9) is isomorphic to the Heisenberg double $\mathcal{H}(A)$.*

5 The Twisted Double, and Deformations of Classical and Quantum Algebras

Let us return to the study of Poisson algebras of functions and consider certain types of their deformations. We begin with a simple example: deformations of the Lie—Poisson bracket on the dual space of a Lie algebra.Let $\boldsymbol{g}$ be a Lie algebra such that $H^2(\boldsymbol{g}) \neq 0$. Fix a nontrivial cocycle $\omega \in C^2(\boldsymbol{g})$, and let $\hat{\boldsymbol{g}}_\omega = \boldsymbol{g} \oplus \mathbf{R}$ be the associated central extension of $\boldsymbol{g}$. The dual space $\hat{\boldsymbol{g}}_\omega^*$ is naturally isomorphic to $\boldsymbol{g}^* \oplus \mathbf{R}$. Let e be an affine coordinate on $\mathbf{R}$. Clearly, e lies in the center of the Poisson algebra $Fun(\hat{\boldsymbol{g}}_\omega^*) \simeq Fun(\boldsymbol{g}) \otimes Fun(\mathbf{R})$. Fixing the value of e we get a 1-parameter family of Poisson brackets on $\boldsymbol{g}$ which may be regarded as a deformation of the original Lie—Poisson bracket. Its 'universal' deformation is parametrized by the elements of $H^2(\boldsymbol{g})$.

Let us consider similar deformations for general Poisson Lie groups. We shall again begin with the deformations associated with central extensions (cf. [18]).

Let $(\boldsymbol{g}, \boldsymbol{g}^*)$ be a factorizable Lie bialgebra, $\partial \in \mathrm{Der}(\boldsymbol{g}, \boldsymbol{g}^*)$ its derivation. By definition, this means that ∂ is a derivation of $\boldsymbol{g}$ which is skew with respect to the inner product on $\boldsymbol{g}$ and commutes with $r \in \mathrm{End}\boldsymbol{g}$. Formula

$$\omega(X,Y) = (X , \partial Y) \tag{5.1}$$

defines a 2-cocycle on $\boldsymbol{g}$. Thus we get an embedding $\mathrm{Der}(\boldsymbol{g}, \boldsymbol{g}^*) \hookrightarrow C^2(\boldsymbol{g})$.

Example. In typical applications $\boldsymbol{g} = L\boldsymbol{a}$ is a loop algebra equipped with a standard r-matrix, $\partial = \partial_x$ is the derivative in loop parameter. In this case it is easy to see that the class group $[\mathrm{Der}(\boldsymbol{g}, \boldsymbol{g}^*)]$ (i.e. the quotient group of all derivations modulo inner derivations) is isomorphic to $H^2(L\boldsymbol{a})$. For a simple Lie algebra $\boldsymbol{a}$ the group $H^2(L\boldsymbol{a}) \simeq \mathbf{R}$ is generated by the cocycle (5.1).

Let $\hat{\boldsymbol{g}} = \boldsymbol{g} \oplus \mathbf{R}$ be the central extension of $\boldsymbol{g}$ associated with the cocycle (5.1). Then $\hat{\boldsymbol{g}}$ has a canonical structure of a Lie bialgebra; the dual Lie algebra $\hat{\boldsymbol{g}^*}$ is the semi-direct product

$$\hat{\boldsymbol{g}}^* = \boldsymbol{g}^* \uplus \mathbf{R}\partial.$$

More precisely, let us define the commutator in $\hat{\boldsymbol{g}}^*$ by

$$[f + \alpha\partial , g + \beta\partial] = [f , g]_* + \alpha\partial r(g) - \beta\partial r(f). \tag{5.2}$$

Proposition 5.1 *The pair $(\hat{\boldsymbol{g}}, \hat{\boldsymbol{g}}^*)$ is a Lie bialgebra.*

It is easy to describe the double of $(\hat{\boldsymbol{g}}, \hat{\boldsymbol{g}}^*)$. We shall do it under the simplifying assumption

$$\partial - \partial \cdot r^2 = 0. \tag{5.3}$$

(This condition holds for standard r-matrices on loop algebras.)

Put

$$\boldsymbol{d} = \boldsymbol{g} \oplus \boldsymbol{g}, \quad \hat{\boldsymbol{d}} = \boldsymbol{d} \oplus \mathbf{R} \cdot \mathbf{c} \oplus \mathbf{R} \cdot \partial.$$

We define the inner product on $\hat{\boldsymbol{d}}$ by

$$\ll (X_1, Y_1, \alpha_1, \beta_1)\ ,\ (X_2, Y_2, \alpha_2, \beta_2) \gg \ = \\ < X_1, X_2 > - < Y_1, Y_2 > + \alpha_1\beta_2 + \alpha_2\beta_1.$$

Extend the derivation ∂ to $\boldsymbol{d}$ by $\hat{\partial}(X,Y) \ = \ (\partial X, -\partial Y)$ and define the cocycle $\omega_{\boldsymbol{d}}$ which gives the 'central ' component of the Lie bracket on $\hat{\boldsymbol{d}}$ by

$$\omega_{\boldsymbol{d}}(a,b) \ = \ll a, \hat{\partial} b \gg . \tag{5.4}$$

Proposition 5.2 *Assume that condition (5.3) holds. Then $\hat{\boldsymbol{d}}$ is isomorphic to the double of $(\hat{\boldsymbol{g}}, \hat{\boldsymbol{g}}^*)$; embeddings $\boldsymbol{g}, \boldsymbol{g}^* \hookrightarrow \hat{\boldsymbol{d}}$ are given by*

$$(X,\ \alpha) \ \mapsto \ (X,\ X,\ \alpha c)\ \ ,\ \ (f,\ \beta) \ \mapsto \ (r_+ f,\ r_- f,\ \beta\ \partial).$$

Let Γ be the group of automorphisms of $(\boldsymbol{g}, \boldsymbol{g}^*)$ generated by ∂. The Lie group which corresponds to $\hat{\boldsymbol{g}}^*$ is the semi-direct product $G^* \bowtie \Gamma$. According to the quantum duality principle, the quantized universal enveloping algebra $U_q(\hat{\boldsymbol{g}})$ may be identified with $Fun_q(\hat{G}^*)$. As a first step towards the study of this algebra let us consider the Poisson algebra $Fun(\hat{G}^*)$.

Proposition 5.3 *Let e be an affine coordinate on $\Gamma \simeq \mathbf{R}^\times$. Then e lies in the center of the Poisson algebra $Fun(\hat{G}^*)$*

The function e has the meaning of *central charge.* Proposition (5.3) means that the bracket on $\hat{G}^*$ depends on e as a parameter, i.e. we get a 1-parameter family $\{,\}_e$ of Posson brackets on G^*. After quantization we get a 1-parameter family of algebras $Fun_q(G^*)_e \simeq U_q(\hat{\boldsymbol{g}})/(e = \text{const})$ which may be regarded as quotient algebras of $U_q(\hat{\boldsymbol{g}})$ obtained by 'setting $e = \text{const}$'.

The algebra $Fun(G^*)$ may be obtained from a larger algebra $Fun(D_+)$ by reduction. In a similar way, the algebras $Fun(G^*)$ may be obtained by reduction from the algebra of functions on the *twisted double* of (G, G^*).

Remarkably, the definition of the twisted double and the Poisson structure on $G^*_- \ = \ G^* \times \{e\} \subset \hat{G}^*$ depends not on the derivation ∂ itself, but only on the automorphism $\exp(e\partial) \in \Gamma$. This allows to define the twisted double and the twisted bracket on G^* in a more general setting.

Put $D \ = \ G \times G$. Let $\tau \in \text{Aut}(G)$ be an automorphism of G. We assume that the corresponding automorphism of $\boldsymbol{g}$ preserves the inner product in $\boldsymbol{g}$ and commutes with r .Put

$$\hat{\tau} \ = \ \begin{pmatrix} 1 & 0 \\ 0 & \tau \end{pmatrix} \in \text{Aut}(G \times G),$$

$${}^\tau r_{\boldsymbol{d}} \ = \ \hat{\tau} r_{\boldsymbol{d}} \hat{\tau}^{-1} \ = \ \begin{pmatrix} r & -2r_+ \circ \tau^{-1} \\ 2\tau \circ r_- & -r \end{pmatrix}, \tag{5.5}$$

and define the twisted Poisson bracket on D by

$$2\{\phi,\psi\}_\tau = \ll r_{\boldsymbol{d}}X, Y \gg + \ll^\tau r_{\boldsymbol{d}}X', Y' \gg . \tag{5.6}$$

(Here X, Y, X', Y' are the left and right gradients of ϕ, ψ, respecxtively, defined by (2.4).) Note that the Jacobi identity for (5.6) follows from the classical Yang—Baxter identity (2.6) for $r_{\boldsymbol{d}}$.

The group D equipped with the Poisson bracket (5.6) is called the twisted double and is denoted by D_τ. Assume, as in Section 2, that G is a classical matrix group with exact representation (ρ, V). The algebra $Fun(D_\tau)$ is generated by the matrix coefficients of $\rho(x), \rho(y)$. The Poisson bracket relations for the generators are given by

$$\begin{aligned}
\{x_1, x_2\}_\tau &= \frac{1}{2}\,(r^{VV} x_1 x_2 + x_1 x_2 r^{VV}),\\
\{y_1, y_2\}_\tau &= \frac{1}{2}\,(r^{VV} y_1 y_2 + y_1 y_2 r^{VV}),\\
\{y_1, x_2\}_\tau &= -r_+^{VV} y_1 x_2 - y_1 x_2^\tau r_+^{VV},
\end{aligned} \tag{5.7}$$

where

$$\begin{aligned}
{}^\tau r_+ &= (id \otimes \tau) r_+ \in \boldsymbol{g} \otimes \boldsymbol{g},\\
r^{VV} &= \rho \otimes \rho(r) \in \mathrm{End}(V \otimes V).
\end{aligned}$$

Let G^* be the dual group of G . As in Section 2, we identify it with a subgroup in D.

Proposition 5.4 *(i) The actions of G on D_τ by left and right translations defined by*

$$\begin{aligned}
G \times D \to D: \quad g(x,y) &= (gx, gy),\\
D \times G \to D: \quad (x,y)g &= (xg, yg^\tau)
\end{aligned}$$

are admissible and form a dual pair.

(ii) In a similar way, the actions of G^* on D_τ defined by

$$\begin{aligned}
G^* \times D \to D: \quad h(x,y) &= (h_+, h_- y),\\
D \times G^* \to D: \quad (x,y)h &= (xh_+, yh_-^\tau)
\end{aligned}$$

are also admissible and form a dual pair.

The quotient spaces $D/G, G\backslash D$ may be identified with G , the projection maps p, p' being given by

$$p : D \to G : (x,y) \mapsto y^{-1}x, \quad p' : D \to G : (x,y) \mapsto x(y^{-1})^{\tau}.$$

(We write $\bar{\tau} = \tau^{-1}$ for shortness.) The reduced Poisson bracket on G is given by

$$\{L_1, L_2\}_\tau = L_1 r_+^\tau L_2 + L_2 r_-^\tau L_1 - \frac{1}{2}(r\, L_1 L_2 + L_1 L_2\, r), \tag{5.8}$$

where $r_+^\tau = (id \otimes \tau) r_+$, $r_-^\tau = (\tau \otimes id) r_-$ (note that $(\tau \otimes \tau) r = r$, due to the condition imposed on τ).

Another description of the reduced spaces D/G, D/G^* results from a factorization problem on D. Observe that almost all elements $(x, y) \in D \times D$ may be represented in the form

$$\begin{aligned} (x, y) &= (L_+, L_-)(T, T^\tau)^{-1}, \\ (L_+, L_-) &\in G^* \subset D,\ T \in G. \end{aligned} \tag{5.9}$$

Thus we may identify (up to a set of positive codimension) $D/G^* \simeq G$, $G \backslash D \simeq G^*$ and compute the reduced brackets for the generators of the algebras $Fun(G)$, $Fun(G^*)$. The reduced bracket on $D/G^* \simeq G$ is the usual *Sklyanin bracket*

$$\{T_1, T_2\} = \frac{1}{2}[r\, ,\ T_1 T_2]$$

and does not depend on twisting. The brackets on $G^* \simeq G \backslash D$ are given by

$$\begin{aligned} \{L_1^\pm\, ,\ L_2^\pm\}_\tau &= \frac{1}{2}\,[r\, ,\ L_1^\pm L_2^\pm], \\ \{L_1^+\, ,\ L_2^-\}_\tau &= r_+ L_1^+ L_2^- - L_1^+ L_2^- r_+^\tau. \end{aligned} \tag{5.10}$$

The identification of the two models for the quotient space $G \backslash D$ is given by the *twisted factorization*

$$L = (L^+)^{-1} (L^-)^\tau.$$

Finally, using the expansion (5.9) we may express the bracket on D_τ in terms of the generators $L^\pm, T$. It is sufficient to compute only the brackets $\{L_1^\pm, T_2\}$. The computation gives

$$\{L_1^\pm\, ,\ T_2\} = L_1^\pm T_2 r_\pm,$$

i.e. the same formula as in the non-twisted case.

Assume now that $\tau = \exp(e\partial)$ where ∂ is a derivation of $(\mathfrak{g}, \mathfrak{g}^*)$ which satisfies (5.3). Then the bracket (5.9) coincides with the bracket on $Fun(G^*)_e$ associated with the central extension of $\mathfrak{g}$.

Remarkably, in formulae (5.6, 5.7, 5.8, 5.10) derivations are replaced by finite automorphisms. This is a manifestation of a general principle: for quantum groups (and even for Poisson Lie groups) differential operators are replaced by difference operators. In the present case we see that deformations of Poisson structures on Lie groups are parametrized by the group $\mathrm{Out}(\mathfrak{g}, \mathfrak{g}^*)$ of outer automorphisms (which may be considerably larger than the infinitezimal group $[\mathrm{Der}(\boldsymbol{g}, \boldsymbol{g}^*)] \simeq H^2(\boldsymbol{g})$).

A typical example when the group $\mathrm{Out}(\boldsymbol{g}, \boldsymbol{g}^*)$ is non-trivial (although $H^2(\boldsymbol{g}) = 0$) is connected with lattice systems. Let us describe this example in more detail.

Let $(\boldsymbol{g}, \boldsymbol{g}^*)$ be a factorizable Lie bialgebra. Put $\mathcal{G} = \oplus^N \boldsymbol{g}$, $\mathcal{G}^* = \oplus^N \boldsymbol{g}^*$. It is convenient to consider elements of $(\mathcal{G}, \mathcal{G}^*)$ as functions on $\mathbf{Z}/N\mathbf{Z}$ with values in $\boldsymbol{g}, \boldsymbol{g}^*$. Let τ be the automorphism of $\mathcal{G}$ induced by the cyclic permutation on $\mathbf{Z}/N\mathbf{Z}$. Clearly, τ is an automorphism of $(\mathcal{G}, \mathcal{G}^*)$. In the 'continuous limit' the periodic lattice $\mathbf{Z}/N\mathbf{Z}$ is replaced by a circle and the automorphism τ by the derivation ∂_x giving rise to the central extension of the loop algebra $L\boldsymbol{g}$. The twisted algebras $Fun(G^N)_\tau$ mimick, in the finite-dimensional setting, the effects of the central extension of $L\boldsymbol{g}$.

The algebra $Fun(G^N)$ is generated by the matrix coefficients of the matrices $\rho_V(L^s)$, $s \in \mathbf{Z}/N\mathbf{Z}$. Specializing formula (5.8), we get the following Poisson bracket relations

$$
\begin{aligned}
\{L_1^k\,,\, L_2^k\} &= -\frac{1}{2}\, r L_1^k L_2^k - \frac{1}{2}\, L_1^k L_2^k r,\\
\{L_1^k\,,\, L_2^{k+1}\} &= L_1^k r_+ L_2^{k+1},\\
\{L_1^k\,,\, L_2^l\} &= 0 \quad \text{for } \mid k-l \mid \geq 2.
\end{aligned}
\tag{5.11}
$$

The bracket (5.11) is not *ultralocal*, i.e. does not decompose into direct product of Poisson brackets on different factors. The properties of this bracket are described by the following theorem.

Theorem 5.1 *(i) Let*

$$M : G^N \to G : (L^1, ..., L^N) \mapsto \overleftarrow{\prod} L^i$$

be the monodromy map. Equip the target group G with the Poisson bracket (2.16). Then M is a Poisson mapping.

(ii) Suppose that N is odd. Then the algebra of Casimir functions on G^N is generated by

$$c_k(L^1, ..., L^N) = \mathrm{tr\ M^k},\ \mathrm{k} = 1, 2, ... \tag{5.12}$$

and symplectic leaves in G^N coincide with the orbits of lattice gauge transformations

$$G^N \times G^N \to G^N : (g, L) \mapsto (g_1 L^1 g_2^{-1}, ..., g_N L^N g_1^{-1}). \tag{5.13}$$

Remark. The property of the monodromy map described in theorem (5.1) plays an important role in the theory of integrable systems. In a more general way, suppose that $r \in \text{End}(\boldsymbol{g} \oplus \boldsymbol{g})$ is an arbitrary solution of the modified classical Yang—Baxter equation (2.6) for the square of $\boldsymbol{g}$,

$$r = \begin{pmatrix} A & B \\ B^t & D \end{pmatrix}, \quad A = -A^t, D = -D^t.$$

Define the Poisson bracket on G by

$$\{L_1\,,\,L_2\}_r \;=\; AL_1L_2 - L_1L_2D + L_1BL_2 - L_2B^tL_1. \tag{5.14}$$

This bracket has an obvious twisted lattice counterpart; the corresponding Poisson bracket relations are given by

$$\begin{aligned} \{L_1^k\,,\,L_2^k\}_{r,\tau} &= AL_1^kL_2^k - L_1^kL_2^kD, \\ \{L_1^k\,,\,L_2^{k+1}\}_{r,\tau} &= L_1^kBL_2^{k+1}, \\ \{L_1^k\,,\,L_2^l\}_{r,\tau} &= 0 \text{ for } \mid k-l \mid \geq 2. \end{aligned} \tag{5.15}$$

Proposition 5.5 *Equip G^N with the bracket (5.15) and the target group with the bracket (5.14). Then $M : G^N \to G$ is a Poisson map iff r satisfies the constraint*

$$A - B^t = D - B.$$

Functions (5.12) are in involution with respect to (5.15).

Note that all solutions of the modified classical Yang—Baxter equation on $\boldsymbol{g} \oplus \boldsymbol{g}$ may be completely classified (cf.[4]) ;the r-matrix (2.9) is a special case of such solution.

Let us now turn to the definition of the twisted quantum double. We shall depart from the definition of the Heisenberg double $\mathcal{H}(A)$ of a factorizable Hopf algebra in terms of the generators X^V, Y^V and relations (3.17). Assume that τ is an automorphism of End(V) such that $(\tau \otimes \tau)R^{VV} = R^{VV}$. Put

$$R^\tau \;=\; (id \otimes \tau)R. \tag{5.16}$$

We define $\mathcal{H}_\tau(A)$ as a free algebra generated by the matrix coefficients of $X, Y \in \mathcal{H}_\tau \otimes \text{End}V$ satisfying the following relations

$$\begin{aligned} X_2X_1 &= R_+X_1X_2R_-, \\ Y_2Y_1 &= R_+Y_1Y_2R_-, \\ Y_2X_1 &= R_+X_1Y_2R_+^\tau. \end{aligned} \tag{5.17}$$

[As in Section 3, we assume that X, Y also satisfy the symmetry relations which characterize classical groups.]

Clearly, the algebra $\mathcal{H}_\tau(A)$ is a quantization of the Poisson algebra $Fun(D_\tau)$.

Let us denote by $Fun_q(G)^\tau$, $Fun_q(G^*)^\tau$ the algebras generated by the matrix coefficients of $T^V, L^{\pm\, V}$ and relations (3.10) in which the R-matrix R^{VV} is replaced by $(R^{VV})^\tau$. The following construction is the quantum analogue of the reduction procedure described in proposition (5.4).

Theorem 5.2 *(i) Formulae*

$$\begin{aligned} C_L(X) &= T \dot{\otimes} X, \quad C_L(Y) &= T \dot{\otimes} Y, \\ C_R(X) &= X \dot{\otimes} T^{-1}, \quad C_R(Y) &= Y \dot{\otimes} T^{-1} \end{aligned} \tag{5.18}$$

defines on $Fun_q(D)_\tau$ the structure of a left $Fun_q(G)$- and a right $Fun_q(G)^\tau$-comodule.

(ii) The subalgebra of left coinvariants (i.e. of elements $f \in Fun_q(D)_\tau$ satisfying $C_L f \in 1 \otimes Fun_q(D)_\tau$) is generated by the matrix coefficients of $L = X^{-1}Y$. The commutation relations for L are given by

$$L_2 R_+ L_1 (R_+^\tau)^{-1} = R_-^\tau L_1 R_-^{-1} L_2. \tag{5.19}$$

(iii) In a similar way, the subalgebra or right coinvariants is generated by the matrix coefficients of $M = X(Y^\tau)^{-1}$; the commutation relations for M are given by

$$R_+^{-1} M_2 R_+^\tau = M_1 (R_-^\tau)^{-1} M_2 R_-. \tag{5.20}$$

(iv) Moreover, $L_1 M_2 = M_2 L_1$, i.e. the subalgebras of left and right coinvariants centralize each other in $Fun_q(D)_\tau$.

One can also define another system of generators for $Fun_q(D)_\tau$.

Proposition 5.6 *The algebra generated by the matrix coefficients of $L^\pm, T$ satisfying the relations*

$$\begin{aligned} L_2^\pm L_1^\pm &= R_+ L_1^\pm L_2^\pm R_+^{-1}, \\ L_2^- L_1^+ &= R_+ L_1^+ L_2^- (R_+^\tau)^{-1}, \\ T_2 T_1 &= R_+ T_1 T_2 R_+^{-1}, \\ L_1^\pm T_2 &= T_2 L_1^\pm R_\pm \end{aligned} \tag{5.21}$$

is isomorphic to $Fun_q(D)_\tau$. The correspondence between the two sets of generators is given by $X = L^+ T^{-1}$, $Y = L^-(T^\tau)^{-1}$ where $T^\tau = (id \otimes \tau)T$.

The main example of twisting is again related to lattice systems. Let us consider again the group G^N consisting of functions on a periodic lattice $\mathbf{Z}/N\mathbf{Z}$ with values in G. The quantum algebra $Fun_q(G^N)$ may be described by generators $T^i, i \in \mathbf{Z}/N\mathbf{Z}$ and relations

$$T_2^i T_1^i = R T_1^i T_2^i R^{-1}, T_1^i T_2^j = T_2^j T_1^i \quad \text{for } i \neq j. \tag{5.22}$$

Put

$$R^{ij} = \begin{cases} R, i = j, \\ I, i \neq j \end{cases} \tag{5.23}$$

We may regard R^{ij} as a linear operator acting in $\overset{N}{\otimes} V$. Let τ be the cyclic permutation, $\tau(i) = (i+1)$ mod N. Then $R^{\tau i, \tau j} = R^{ij}$, and we may use τ to twist our quantum algebras. We get the following algebras associated with the lattice $\mathbf{Z}/N\mathbf{Z}$:

(i) The twisted double generated by the matrix coefficients of X^i, Y^i and relations

$$\begin{aligned} X_2^j X_1^i &= R_+^{ij} X_1^i X_2^j R_-^{ij}, \\ Y_2^j Y_1^i &= R_+^{ij} Y_1^i Y_2^j R_-^{ij}, \\ Y_2^j X_1^i &= R_+^{ij} X_1^i Y_2^j R_+^{i,j+1}. \end{aligned} \tag{5.24}$$

(ii) The algebra $Fun_q(D/G)_\tau \simeq G_\tau^*$ with generators L^i and relations

$$L_2^j R_+^{ij} L_1^i (R_+^{i,j+1})^{-1} = R_-^{i-1,j} L_1^i R_-^{ij} L_2^j. \tag{5.25}$$

Theorem [3] 1 *(i) Let*

$$M = \overleftarrow{\prod} L^i$$

be the monodromy matrix. Then

$$M_2 R_+ M_1 R_+^{-1} = R_- M_1 R_-^{-1} M_2. \tag{5.26}$$

Thus the monodromy map gives rise to an embedding

$$U_q(\mathfrak{g}) \simeq Fun_q(G^*) \hookrightarrow Fun_q(D/G)_\tau$$

(ii) Suppose that N is odd. Then the centers of the algebras $Fun_q(D/G)_\tau$ and $Fun_q(G^*) \subset Fun_q(D/G)_\tau$ coincide.

In view of proposition (3.10) the last assertion may be stated in the following form: *The center of the algebra $Fun_q(D/G)_\tau$ coincides with the subalgebra of gauge invariants.*

The algebra $Fun_q(D/G)_\tau$ may be regarded as a nontrivial deformation of $U_q(\mathfrak{g})^{\otimes N}$. One can show that in the 'continuous limit' this algebra gives the algebra $U(L\mathfrak{g})_e$, i.e. the quotient algebra obtained from the universal enveloping algebra of a Kac—Moody algebra by 'setting the central charge equal to constant'(cf. [3]).

References

[1] V.I.Arnol'd, *Mathematical methods of classical mechanics.* Graduate Texts in Math., v.60, Springer-Verlag, Berlin, New York, 1989 (2nd edition).

[2] A.Yu Alekseev, L.D.Faddeev, *Quantum T^*G as a toy model for conformal field theory*, Commun.Math.Phys.

[3] A.Yu.Alekseev, L.D.Faddeev, M.A.Semenov-Tian-Shansky, *Hidden quantum groups inside Kac–Moody algebras*, Commun. Math.Phys. **149** (1992), 335–346.

[4] A.A.Belavin, V.G.Drinfeld, *Triangle equations and simple Lie algebras*,Soviet Sci. Rev., Sect. C **4**(1984), 93–165. Harwood Academic Publ., New York.

[5] F.A.Berezin, *Some remarks on the associative envelope of a Lie algebra*, Funkts. Anal. Prilozh. **1**(1967) no 2, 1–14.

[6] V.G.Drinfel'd, *Hamiltonian structures on Lie groups, Lie bialgebras and the geometric meaning of the classical Yang–Baxter equation*,Sov.Math.Doklady **27** (1983),68–71.

[7] V.G.Drinfel'd, *Quantum groups*, Proc. Internat. Congr. Math., Berkeley, California, 1986, Amer.Math.Soc., Providence, 1987, pp.798–820.

[8] L.D.Faddeev, N..Yu.Reshetikhin, and L.A.Takhtajan, *Quantum groups*, Algebraic Analysis, vol. 1 (M.Kashiwara, and T.Kawai, eds.), Academic Press, Boston, 1989, pp. 129–139.

[9] N.Yu.Reshetikhin, L.A.Takhtajan, and L.D.Faddeev, *Quantization of Lie groups and Lie algebras*, Leningrad Math. J. **1**(1990), 193–225 .

[10] V.V.Fock, A.A.Rosly, *Poisson structure on moduli of flat connections on Riemann surfaces and r-matrix*, Preprint ITEP 92–72, Moscow, 1992, 20p.

[11] M.Jimbo, *A q-difference analogue of $U(\mathfrak{g})$ and the Yang–Baxter formula*, Let.Math.Phys. **10**(1985), 63–69.

[12] A.N.Kirillov, N.Yu Reshetikhin, *q-Weyl group and a multiplicative formula for universal R-matrices*, Commun. Math.Phys. **134**(1990), 421–431.

[13] S.Z.Levendorski, and Ya.S.Soibelman, *Some applications of quantum Weyl group. 1.The multiplicative formula for universal R-matrix for simple Lie algebras*, J. Geom. Phys. **7**(1990), .

[14] J.H.Lu, *Momentum mappings and Poisson reduction*, Symplectic Geometry, Groupoids, and Integrable Systems (P.Dazord and A.Weinstein, eds.). MSRI Publications **20**(1991), pp 209–226.

[15] V.Lyubashenko, S.Majid, *Braided groups and quantum Fourier transform*, Preprint DATMP/91-26. Cambridge, UK, July 1991, 24p.

[16] S.P.Novikov,*private communication.*

[17] N.Yu.Reshetikhin, M.A.Semenov-Tian-Shansky, *Factorization problems for quantum groups . In: Geometry and Physics. Essays in honour of of I.M.Gelfand (S.Gindikin and I.M.Singer, eds).* North Holland Publishers, 1991, pp 533–550.

[18] N.Yu.Reshetikhin, M.A.Semenov-Tian-Shansky, *Central extensions of quantum current groups*, Lett.Math.Phys. **19**(1990), 133–142.

[19] M.A.Semenov-Tian-Shansky, *Dressing transformations and Poisson group actions*, Publ.Math.RIMS **21**(1985),1237–1260.

[20] A.Weinstein, *Local structure of Poisson manifolds*, J.Diff. Geom. **18**(1983), 523–558.

Physique Mathématique,
Université de Bourgogne
B.P. 138, 21004 Dijon France,
and
V.A.Steklov Matematical Institute,
St.Petersburg, 191011 Russia
Current e-mail address: semenov@satie.u-bourgogne.fr

Contemporary Mathematics
Volume **175**, 1994

Local 4-point functions and the Knizhnik-Zamolodchikov equation

YASSEN S. STANEV and IVAN T. TODOROV

ABSTRACT. We classify all polynomial solutions of the Knizhnik-Zamolodchikov (KZ) equation for the Möbius invariant 4-point amplitude of a primary field of the su_2 conformal current algebra. The exposition is based on (and reviews) recent work of Louis Michel and the authors [1] on $\mathcal{D}_3$ invariant rational 4-point functions and the local extensions of su_2 current algebras.

1. Introduction

The KZ equation [2] for a conformally invariant 4-point amplitude is a system of first order differential equations for a vector valued function $f = f(\eta)$ which can be brought to the form

$$(1.1) \qquad \left(pd - \frac{C_{12}}{\eta} + \frac{C_{23}}{1-\eta} \right) f = 0 \quad , \quad d = \frac{d}{d\eta} \quad .$$

1991 Mathematics Subject Classification. Primary 81T40, 35C05, 20C12.
Supported in part by the Bulgarian National Foundation for Scientific Research under contract F11 and by the American Mathematical Society.
This paper is in final form and no version of it will be submitted for publication elsewhere.

0271-4132/94 $1.00 + $.25 per page

Here C_{12} and C_{23} are (η-independent) hermitean operators acting in a finite dimensional vector space. For unitary representations of the underlying Kac-Moody algebra $\hat{\mathcal{G}}$ the coefficient p takes integer values exceeding the dual Coxeter number $\tilde{h}$ of the simple Lie algebra $\mathcal{G}$ ($p = k+\tilde{h}$ where $k=0,1,2,\ldots$ is the level), while C_{ij} is related to the Casimir operator of $\mathcal{G}$ in the tensor product of representations i and j (its form will be spelled out for the case of su_2 in Sec.2). The variable η is the Möbius (SU(1,1)-) invariant cross ratio

$$(1.2) \qquad \eta = \frac{z_{12}z_{34}}{z_{13}z_{24}} \;, \quad \left(1-\eta = \frac{z_{14}z_{23}}{z_{13}z_{24}} \right) \;, \quad z_{ij} = z_i - z_j$$

where z_i are the (complex) world sheet arguments of the primary conformal field (or "chiral vertex operators" [3]). In a real (Lorentz signature) picture the "physical" values of z belong to the unit circle $|z| = 1$ which appears as a compactified light ray of a 2-dimensional Minkowski space. The three singular points $\eta = 0,1,\infty$ of Eq. (1.1) all correspond to coinciding world sheet arguments.

The vacuum sector Hilbert space $\mathcal{H}_0$ is defined as an irreducible lowest weight $\hat{\mathcal{G}}$ module corresponding to a fixed level k and $\mathcal{G}$ weight zero. A $\hat{\mathcal{G}}$ primary field $V=V_\lambda$ maps $\mathcal{H}_0$ into another irreducible $\hat{\mathcal{G}}$ module $\mathcal{H}_\lambda$ of the same level and weight λ. V is characterized by λ and by its conformal dimension $\Delta=\Delta_\lambda$ (for a review - see [4]). In particular, there is a (unitary) representation $U=U(g)$ of a group SU(1,1) of Möbius transformations

$$(1.3a) \qquad g = \begin{pmatrix} a & b \\ \bar{b} & \bar{a} \end{pmatrix} \; : \; z \to gz = \frac{az+b}{\bar{b}z+\bar{a}} \;, \quad (a\bar{a} - b\bar{b} = 1)$$

acting in the state space of the theory such that

(1.3b) $$U(g)V(z)U(g)^{-1} = \left(\frac{dgz}{dz}\right)^{\Delta} V(gz), \quad \frac{dgz}{dz} = (\bar{b}z + \bar{a})^{-2}$$

and leaving the vacuum vector $|0\rangle \in \mathcal{H}_0$ invariant,

(1.4) $$U(g)|0\rangle = |0\rangle.$$

If $\mathfrak{g}$ is the Lie algebra of a compact Lie group G, then $\mathcal{H}_\lambda$ admits a finite dimensional "minimal energy" subspace $\mathcal{H}_\lambda^{(0)}$ that is an irreducible G - module such that

(1.5a) $$U\begin{pmatrix} a & 0 \\ 0 & \bar{a} \end{pmatrix} \mathcal{H}_\lambda^{(0)} = \left(\frac{a}{\bar{a}}\right)^{\Delta} \mathcal{H}_\lambda^{(0)} \qquad \left(\frac{a}{\bar{a}} = a^2 \right).$$

Furthermore, V can be labeled by vectors v of this subspace in such a way that the state $V(z,v)|0\rangle$ tends to v for $z \to 0$ (cf.[5]):

(1.5b) $$\lim_{z\to 0} V(z,v)|0\rangle = v .$$

The conformally invariant 2-point function is unique up to normalization. We fix the normalization of V by introducing a norm preserving antilinear involution $v \to \bar{v}$ in $\mathcal{H}_\lambda^{(0)}$ such that

(1.6) $$\langle 0|V(z_1,\bar{v}_1)\, V(z_2,v_2)|0\rangle = z_{12}^{-2\Delta} \langle v_1|v_2\rangle$$

where $\langle\ |\ \rangle$ stands for the inner product in $\mathcal{H}_\lambda$. The n-point function of V is a homogeneous function of z_{ij} of degree $-n\Delta$.

There is a "naturality property" of 2-dimensional conformal field theories noted by Moore and Seiberg [6]: If the chiral algebras of observables $\mathcal{A}$ and $\bar{\mathcal{A}}$ are maximally extended, then in the 2-dimensional theory the left movers and the right movers are paired by an automorphism of the fusion rule algebra. Hence, the classification of 2-D rational conformal field theories can be viewed as a two step process. First we classify all extended chiral algebras $\mathcal{A}$, then look for all automorpisms of the fusion rules. It follows from the **ADE** classification of the $su_k(2)$ modular invariant partition functions [7] that the $\mathbf{D}_{even}$ series and the $\mathbf{E}_6$, $\mathbf{E}_8$ exceptional models correspond to an extended observable algebra with diagonal pairing, while $\mathbf{D}_{odd}$ and $\mathbf{E}_7$ correspond to non-trivial automorphisms.

In this paper we present an alternative classification of the extended algebras of observables in the $su_k(2)$ WZNW models which does not use modular invariance. It is based instead on the classification of all $\mathcal{D}_3$ invariant polynomial solutions of the KZ equation. We prove that there are no other extensions of the observable algebra apart from the $\mathbf{D}_{even}$ and $\mathbf{E}_{even}$ modular invariant cases.

The problem is stated as follows. We look for such primary fields V of $\mathfrak{g}$ which are not only local with respect to the currents - a property that all primary fields share- but also with respect to themselves. A necessary condition for this is that V has an integer conformal dimension

$$\Delta(\Lambda) = \frac{1}{2p} C_2(\Lambda) \quad (p = k+\tilde{h}) ,$$

where $C_2(\Lambda)$ is the (second order) Casimir operator for the irreducible representation of weight Λ of the simple compact Lie algebra $\mathfrak{g}$. There are, however, many primary fields of integer dimensions (for $\mathfrak{g} = su_2$ and $p=k+2=p_1^{n_1} \dots p_\nu^{n_\nu}$, where p_i are primes, their number is $2^{\nu-1}$) which are, in general, not local. If V transforms under a selfconjugate representation of $\mathfrak{g}$ then a much more restrictive condition is reflected in the requirement that the 4-point function

$$(1.7)\qquad \begin{aligned} W_4 &= \langle 0|V(z_1)\,V(z_2)\,V(z_3)\,V(z_4)|0\rangle = \\ &= \left(\frac{z_{13}z_{24}}{z_{12}z_{34}z_{23}z_{14}}\right)^{2\Delta} f(\eta) \end{aligned}$$

(where we are skipping the vector arguments v_i) is a rational function of z_{ij}, or more precisely that $f(\eta)$ is a polynomial of degree $\nu=4\Delta$. In fact, every irreducible representation of $\mathcal{g} = su_2$ is selfconjugate. The results of Sec.3. imply that in this case the above requirement is also sufficient for the locality of V. For higher rank groups the representations of integer dimension are typically not equivalent to their conjugate. (For instance, for $(\widehat{su}_3)_{k=3n}$, the representation of weight $\Lambda=(k,0)$ leads to a dimension $\Delta(3n,0)=n$, but it is not equivalent to its conjugate $(0,k)$.) In such a case one has to consider 4-point functions involving two fields of type V, and two, of type V^*. There are three different 4-point functions of this type obtained from one another by locality transformation (i.e., by permuting the arguments).

For the 4-point function (1.7) locality implies that the permuted expectation value satisfies the same equation (1.1) as the original one. Indeed, the KZ operator

$$(1.8)\qquad \mathbf{K} = p\eta(1-\eta)d + \eta C_{23} - (1-\eta)C_{12}$$

is covariant under permutations :

$$(1.9)\qquad \mu(g,\eta)\,\mathbf{K}_g = \mathbf{K} \quad (g \in S_4).$$

Here μ is a scalar multiplier, $\mathbf{K}_g$ is obtained from $\mathbf{K}$ by permuting the world sheet variables z_i and the internal symmetry indices of V (which are suppressed in the above notation).

Conversely, as pointed out in Sec.4 , the condition of permutation symmetry severly restricts the operator $\mathbf{K}$: in the case of SU_2 covariant fields it yields a 1-parameter family of operators. We also discuss the reduction of the resulting equation to a scalar higher order ordinary differential equation corresponding to a singular vector in a minimal conformal model (following the line of ref.[8]).

In Sec.3 we review and complement the result of [1] providing a complete list of polynamial solutions of Eq.(1.1). Thus we classify the (local) bose, fermi and $\mathbb{Z}_4$ - parafermionic extensions of the su_2 current algebras .

2. Homogeneous polynomial realization of the 4-point function of an $\widehat{su}_2$ primary field D_3 - invariance of the KZ equation

For $G=SU_2$ we shall choose for each weight λ (=0,1,...) a "coherent state" vector $v = e_\lambda(\zeta)$ that is a polynomial of degree λ in the complex variable ζ . The field (or coherent state operator) $V(z,\zeta) = V(z; e_\lambda(\zeta))$ obeys, by definition a tarnsformation law reminiscent to (1.3) (1.4) :

$$\begin{pmatrix} \alpha & \beta \\ -\bar\beta & \bar\alpha \end{pmatrix} : V(z,\zeta) \rightarrow (\bar\alpha - \bar\beta\zeta)^{\lambda} V(z,\frac{\alpha\zeta + \beta}{\bar\alpha - \bar\beta\zeta}) \tag{2.1}$$

(for $\alpha\bar\alpha + \beta\bar\beta = 1$). The general 2-point invariant, replacing the inner product in (1.6) is a multiple of

$$< \overline{e_\lambda(\zeta_1)} | e_\lambda(\zeta_2) > = \zeta_{12}^{\lambda} \equiv (\zeta_1 - \zeta_2)^{\lambda} . \tag{2.2}$$

The 4-point function (1.7) is a vector in the λ+1 dimensional subspace of SU_2 invariants in the tensor product of four irreducible representations of weight λ. It appears in the coherent state realization as a homogeneous polynomial of degree λ

in the pair of variables

$$\vec{\xi} = (\xi_1, \xi_2) , \quad \xi_1 = \zeta_{12}\zeta_{34} , \quad \xi_2 = \zeta_{14}\zeta_{23} . \tag{2.3}$$

Similarly, it is a homogeneous function of the pair of world sheet variables

$$\vec{\eta} = (\eta_1, \eta_2) , \quad \eta_1 = z_{12}z_{34} , \quad \eta_2 = z_{14}z_{23} \quad (\eta_1 + \eta_2 = z_{13}z_{24}). \tag{2.4}$$

We shall set

$$W_4 = [\eta_1 \eta_2 (\eta_1 + \eta_2)]^{-2\Delta} \quad H(\vec{\xi}, \vec{\eta}) \tag{2.5}$$

with

$$\eta_1 \eta_2 (\eta_1 + \eta_2) = z_{12} z_{23} z_{34} z_{14} z_{13} z_{24} \quad ; \tag{2.6}$$

H is a homogeneous polynomial of degree λ in $\vec{\xi}$ and a homogeneous function of degree $\nu = 4\Delta$ in $\vec{\eta}$. We shall also write

$$H(\vec{\xi}, \vec{\eta}) = \eta_3^{\nu} \, \xi_3^{\lambda} \, f(\xi, \eta) \, , \quad \nu = 4\Delta \, , \tag{2.7}$$

where

$$\xi_3 = \xi_1 + \xi_2 \, , \quad \xi = \frac{\xi_1}{\xi_3} \, , \quad \eta_3 = \eta_1 + \eta_2 \, , \quad \eta = \frac{\eta_1}{\eta_3} \, . \tag{2.8}$$

The SU_2 current

$$J(z,\zeta) = J_-(z) + 2\zeta \, J_3(z) - \zeta^2 \, J_+(z) \tag{2.9}$$

satisfies the commutation relation

$$\begin{aligned} [J(z_1,\zeta_1), J(z_2,\zeta_2)] &= - \, \delta(z_{12})(\zeta_{12}^2 \partial_{\zeta_2} + 2\zeta_{12}) \, J(z_2,\zeta_2) \, + \\ &+ k \, \zeta_{12}^2 \, \delta'(z_{12}) \quad , \end{aligned} \tag{2.10}$$

the δ function on the circle being normalized by

$$\oint_{|z_2|=|z_1|} \delta(z_{12}) \, f(z_2) \, \frac{dz_2}{2\pi i} = f(z_1) \, . \tag{2.11}$$

The primary field V_λ of su_2 weight λ is characterized by the Ward identities

$$\langle 0|J(z_1,\zeta_1)V(z_2,\zeta_2) = -\frac{1}{z_{12}}(\zeta_{12}^2\partial_{\zeta_2} + \lambda\zeta_{12}) \langle 0|V(z_2,\zeta_2) , \tag{2.12a}$$

$$V(z_1,\zeta_1)J(z_2,\zeta_2)|0\rangle = \frac{1}{z_{12}}(\zeta_{12}^2\partial_{\zeta_1} - \lambda\zeta_{12}) V(z_1,\zeta_1)|0\rangle \tag{2.12b}$$

and the KZ equation

$$(k+2)\partial_z V = :\frac{\lambda}{2} (\partial_\zeta J) V - J \partial_\zeta V: . \tag{2.13}$$

As a consequence the 4-point function W_4 satisfies the differential equation

$$\{ (k+2)\partial_{z_2} + \frac{\Omega_{12}}{z_{12}} - \frac{\Omega_{23}}{z_{23}} - \frac{\Omega_{24}}{z_{24}} \} W_4 = 0 . \tag{2.14}$$

This yields the following system of two equations for the homogeneous function H in (2.5) :

$$\begin{aligned} p\eta_1 \frac{\partial}{\partial\eta_1} H &= \{ (1-\eta)\Omega_{12} - \eta\Omega_{23} + \frac{1}{2} \lambda(\lambda+2) \} H , \\ p\eta_2 \frac{\partial}{\partial\eta_2} H &= \{ \eta\Omega_{23} - (1-\eta)\Omega_{12} + \frac{1}{2} \lambda(\lambda+2) \} H , \end{aligned} \tag{2.15}$$

(for a derivation using the same conventions - see Sec.2 of [1]); here

$$p = k + 2 \quad (k = 0,1,2,\dots), \quad \Omega_{ij} = 2 \vec{I}_i \vec{I}_j , \tag{2.16}$$

where $\vec{I}_i$ is the 3-vector of isospin generators acting on ζ_i ; we have

$$\Omega_{12} = \frac{1}{2}\lambda^2 - \lambda\xi_1 (2\frac{\partial}{\partial\xi_1} - \frac{\partial}{\partial\xi_2}) + \xi_1^2 (\frac{\partial^2}{\partial\xi_1^2} - \frac{\partial^2}{\partial\xi_1\partial\xi_2}) , \tag{2.17a}$$

$$\Omega_{23} = \frac{1}{2}\lambda^2 - \lambda\xi_2 (2\frac{\partial}{\partial\xi_2} - \frac{\partial}{\partial\xi_1}) + \xi_2^2 (\frac{\partial^2}{\partial\xi_2^2} - \frac{\partial^2}{\partial\xi_1\partial\xi_2}) . \tag{2.17b}$$

The sum of the two equations reproduces the Euler homogeneity condition for p,ν,λ related by

$$p\nu \ (= 4(k+2)\Delta) = \lambda(\lambda+2) . \tag{2.18}$$

The nontrivial KZ equation is embodied by their difference

$$(2.19)\qquad \mathbf{K}\,H = 0 \quad , \ \mathbf{K} = p\eta_3\left(\eta_1\frac{\partial}{\partial\eta_1} - \eta_2\frac{\partial}{\partial\eta_2}\right) - 2(\eta_2\Omega_{12} - \eta_1\Omega_{23}) \ .$$

We now proceed to displaying the covariance properties of H and $\mathbf{K}$ under permutation.

The permutation group S_4 involves 6 transpositions s_{ij} $(1\le i<j\le 4)$ that permute the pairs (z_i,ζ_i) and (z_j,ζ_j) only 3 of which, $s_{i\,i+1}$ $(i = 1,2,3)$, are independent, due to the relations

$$(2.20a)\qquad s_{12}s_{23}s_{12} = s_{23}s_{12}s_{23} = s_{13} \ , \ s_{23}s_{34}s_{23} = s_{34}s_{23}s_{34} = s_{24}$$

$$(2.20b)\qquad s_{12}s_{23}s_{34}s_{23}s_{12} = s_{34}s_{23}s_{12}s_{23}s_{34} = s_{14} \ ;$$

each transposition is by definition, an involution : $s_{ij}^2 = 1$.
S_4 has a 4-element normal abelian subgroup, $\mathbb{Z}_2\times\mathbb{Z}_2$ generated by

$$(2.21)\qquad s_{12}s_{34} = s_{34}s_{12} \ \text{and} \ s_{23}s_{14} = s_{14}s_{23} \ ,$$

which acts trivially on $\vec{\xi}$ and $\vec{\eta}$. The 6 element factor group

$$(2.22)\qquad \mathcal{D}_3 = S_4 \,/\, (\mathbb{Z}_2\times\mathbb{Z}_2) \ ,$$

the dihedral group (the symmetry group of an equilateral triangle) acts faithfully in the 2-space (x_1,x_2) where x_i stands for either ξ_i or η_i . It can be viewed as a subgroup of the rotation group in 3-space (x_1,x_2,x_3) that leaves the plane

$$(2.23)\qquad x_3 = x_1 + x_2$$

invariant and is realized by orthogonal 3×3 matrices with elements 0 and ± 1 . $\mathcal{D}_3$ is generated by any pair of reflections among s_1 ($\simeq s_{12} \simeq s_{34}$ mod $\mathbb{Z}_2\times\mathbb{Z}_2$) , s_2 ($\simeq s_{23} \simeq s_{14}$ mod $\mathbb{Z}_2\times\mathbb{Z}_2$) and s ($\simeq s_{13} \simeq s_{24}$ mod $\mathbb{Z}_2\times\mathbb{Z}_2$) :

$$(2.24a)\qquad s_1\begin{pmatrix} x_1 \\ x_2 \\ x_3 \end{pmatrix} = \begin{pmatrix} -1 & 0 & 0 \\ 0 & 0 & 1 \\ 0 & 1 & 0 \end{pmatrix}\begin{pmatrix} x_1 \\ x_2 \\ x_3 \end{pmatrix} \quad \text{or} \quad s_1\begin{pmatrix} x_1 \\ x_2 \end{pmatrix} = \begin{pmatrix} -1 & 0 \\ 1 & 1 \end{pmatrix}\begin{pmatrix} x_1 \\ x_2 \end{pmatrix} ,$$

(2.24b) $$s_2\begin{pmatrix} x_1 \\ x_2 \\ x_3 \end{pmatrix} = \begin{pmatrix} 0 & 0 & 1 \\ 0 & -1 & 0 \\ 1 & 0 & 0 \end{pmatrix}\begin{pmatrix} x_1 \\ x_2 \\ x_3 \end{pmatrix} \quad \text{or} \quad s_2\begin{pmatrix} x_1 \\ x_2 \end{pmatrix} = \begin{pmatrix} 1 & 1 \\ 0 & -1 \end{pmatrix}\begin{pmatrix} x_1 \\ x_2 \end{pmatrix} ,$$

$s = s_1 s_2 s_1 = s_2 s_1 s_2$,

(2.24c) $$s\begin{pmatrix} x_1 \\ x_2 \\ x_3 \end{pmatrix} = \begin{pmatrix} 0 & -1 & 0 \\ -1 & 0 & 0 \\ 0 & 0 & -1 \end{pmatrix}\begin{pmatrix} x_1 \\ x_2 \\ x_3 \end{pmatrix} \quad \text{or} \quad s\begin{pmatrix} x_1 \\ x_2 \end{pmatrix} = \begin{pmatrix} 0 & -1 \\ -1 & 0 \end{pmatrix}\begin{pmatrix} x_1 \\ x_2 \end{pmatrix} .$$

$\mathcal{D}_3$ also has a cyclic invariant subgroup $\mathbb{Z}_3$ with generator

(2.25) $$s_1 s_2 \sim \begin{pmatrix} 0 & 0 & -1 \\ 1 & 0 & 0 \\ 0 & -1 & 0 \end{pmatrix} \quad \text{or} \quad \begin{pmatrix} -1 & -1 \\ 1 & 0 \end{pmatrix} , \quad (s_1 s_2)^3 = 1 .$$

The inclusion $\mathcal{D}_3 \subset SO_3$ allows to write down the quadratic invariant

(2.26a) $$2J_2 = x_1^2 + x_2^2 + x_3^2 = 2(x_1^2 + x_2^2 + x_1 x_2)$$

or

(2.26b) $$J_2 = x_3^2\,(1 - x + x^2) \quad \text{where } x = x_1/x_3 .$$

The homogeneous polynomials of degree 3 contain another independent invariant

(2.27) $$J_3 = (x_1+x_3)(x_2+x_3)(x_1-x_2) = (2x_1+x_2)(x_1+2x_2)(x_1-x_2) ,$$

and a pseudoinvariant (that changes sign under reflection)

(2.28) $$Q_3 = x_1 x_2 x_3 = x_1 x_2 (x_1+x_2) = x_3^3\, x(1-x) .$$

Each invariant polynomial in $\vec{x}$ is a polynomial in the basic invariants J_2 and J_3 ; in particular,

(2.29) $$27\, Q_3^2 = 4\, J_2^3 - J_3^2 .$$

To exhibit the $\mathcal{D}_3$ properties of the operator **K** (2.19) we need the relations $\Omega_{ij} = \Omega_{ji}$,

(2.30) $$\Omega_{12} = \Omega_{34} , \quad \Omega_{14} = \Omega_{23} , \quad \Omega_{13} = \Omega_{24}$$

and

(2.31) $$\Omega_{12} + \Omega_{23} + \Omega_{24} + \frac{1}{2}\lambda(\lambda+2) = 0 \quad .$$

The derivatives $\partial/\partial x_i$ $(i=1,2)$ transform under the contragradient 2×2 representation of $\mathcal{D}_3$ so that the "homogeneity degree operator"

(2.32) $$x_i \frac{\partial}{\partial x_i} = x_1 \frac{\partial}{\partial x_1} + x_2 \frac{\partial}{\partial x_2}$$

remains invariant. If $g \in \mathcal{D}_3$ acts on (x_1, x_2) by

(2.33) $$g : x_j \to x_i g_{ij} \qquad \text{then} \qquad g : \frac{\partial}{\partial x_i} \to (g^{-1})_{ij} \frac{\partial}{\partial x_j} \, .$$

In particular, the reflections act on $\partial/\partial x_i$ by the matrices transposed to their 2×2 representatives in (2.24). The s-invariance of **K** is manifest from (2.24c) and the relations

(2.34) $$s : \Omega_{12} \to \Omega_{32} = \Omega_{23} \quad , \quad \Omega_{23} \to \Omega_{21} = \Omega_{12} \, .$$

On the other hand, s_2 acts on the two terms of (2.19) as follows:

(2.35a) $$s_2 : p\eta_3 (\eta_1 \frac{\partial}{\partial \eta_1} - \eta_2 \frac{\partial}{\partial \eta_2}) \to p\eta_3 (\eta_1 \frac{\partial}{\partial \eta_1} - \eta_2 \frac{\partial}{\partial \eta_2}) + \lambda(\lambda+2)\eta_2$$

(2.35b) $$s_2 : 2(\eta_2 \Omega_{12} - \eta_1 \Omega_{23}) \to 2(\eta_2 \Omega_{12} - \eta_1 \Omega_{23}) + \lambda(\lambda+2) \; ;$$

this again implies the invariance of the difference. Thus the KZ operator (2.19) is $\mathcal{D}_3$ invariant.

Expressing, finaly, H in terms of $f(\xi,\eta)$ (2.7) and using the relation

(2.36) $$p(\eta_1 \frac{\partial}{\partial \eta_1} - \eta_2 \frac{\partial}{\partial \eta_2})\eta_3^{\nu} f = \eta_3^{\nu}(2p\eta(1-\eta)d + (2\eta-1)\lambda(\lambda+2)\} f \, ,$$

($d=\partial/\partial\eta$), we recover Eq. (1.1) with

(2.37) $$C_{ij} = \Omega_{ij} + \frac{1}{2}\lambda(\lambda+2) \quad (= (\vec{I}_i + \vec{I}_j)^2 \quad) \; ;$$

if $\tilde{C}_{ij}$ are the Casimir operators in the homogeneous polynomial realization then $\tilde{C}_{ij}\ \xi_3^{\lambda} = \xi_3^{\lambda}\ C_{ij}$,

$$C_{12} = \lambda(\lambda+1) - \lambda^2\xi - [2\lambda\xi(1-\xi) + \xi^2]\frac{\partial}{\partial\xi} + \xi^2(1-\xi)\frac{\partial^2}{\partial\xi^2}\ , \tag{2.38a}$$

$$C_{23} = \lambda(\lambda+1) - \lambda^2(1-\xi) + (2\lambda\xi+1-\xi)(1-\xi)\frac{\partial}{\partial\xi} + \xi(1-\xi)^2\frac{\partial^2}{\partial\xi^2}\ . \tag{2.38b}$$

We note that while the homogeneous variables operator (2.19) is $\mathcal{D}_3$ invariant its (ξ,η) counterpart (1.8) obeys a more general covariance law, given by Eq. (1.9) with multipliers

$$\mu(s_2,\eta) = \eta\ ,\quad \mu(s,\eta) = -1\ . \tag{2.39}$$

3. Classification of the polynomial solutions of the KZ equation

From (1.8) and (2.38) (with $p = k+2$) we find the equation for the function $f(\xi,\eta)$ (defined in (2.5) (2.7)) to be :

$$\begin{aligned}(k+2)\eta(1-\eta)\frac{\partial}{\partial\eta} f(\xi,\eta) = \{\ &\lambda[\lambda + 1 - (\lambda+2)\eta - \lambda\xi] - \\ - [\xi^2 - 2\xi\eta + \eta + 2\lambda\xi(1-\xi)]\frac{\partial}{\partial\xi} &+ \xi(1-\xi)(\xi-\eta)\frac{\partial^2}{\partial\xi^2}\ \}\ f(\xi,\eta)\ .\end{aligned} \tag{3.1}$$

The standard normalization of the 2-point functions and the factorization property of the 4-point function W_4 for $\eta\to 0$ into a product of 2-point functions implies also the initial condition

$$f(\xi,0) = \xi^\lambda \tag{3.2a}$$

or equivalently

$$f(\xi,1) = (\xi-1)^\lambda\ . \tag{3.2b}$$

If the field V_λ with su_2 weight $\lambda=2I$ is observable then its 4-point function W_4 (1.7) should be single valued. This implies that the dimension of V_λ

$$\Delta_\lambda = \frac{\lambda(\lambda+2)}{4(k+2)} \tag{3.3}$$

is integer and $f(\xi,\eta)$ is a polynomial in η of degree 4Δ. We shall analyze also the case of halfinteger Δ, which (as follows from (2.5)) corresponds to anticomuting (Fermi) fields.

Permutation invariance of W_4 implies the following $\mathcal{D}_3$ invariance conditions for the polynomials $f(\xi,\eta)$:

(3.4a) $$s_1 \ : \ f(\xi,\eta) = (1-\xi)^{\lambda}(\eta-1)^{\nu} f(\frac{\xi}{\xi-1}, \frac{\eta}{\eta-1}) , \quad (\nu = 4\Delta_{\lambda})$$

(3.4b) $$s_2 \ : \ f(\xi,\eta) = \xi^{\lambda} (-\eta)^{\nu} f(\frac{1}{\xi}, \frac{1}{\eta}) ,$$

(3.4c) $$s \ : \ f(\xi,\eta) = (-1)^{\lambda} f(1-\xi,1-\eta) .$$

The polynomial solutions of Eq. (3.1) should be consistent with a restricted Gepner-Witten fusion rule

(3.5) $$V_{\lambda} \times V_{\lambda} = \sum_{I=0}^{\min(\lambda,k-\lambda)} N_{\lambda\lambda 2I} V_{2I} ,$$

where $N_{\lambda\lambda 2I}$ is nonvanishing only if the primary field V_{2I} is local, so that in particular, it also has an integer dimension

(3.6) $$\Delta_{2I} = \frac{I(I+1)}{k+2} \in \mathbb{Z}_+ .$$

If we expand $f(\xi,\eta)$ as

(3.7) $$f(\xi,\eta) = \sum_{j=0}^{\lambda} \binom{\lambda}{j} \xi^{\lambda-j} (1-\xi)^j f_j(\eta) ,$$

where all $f_j(\eta)$ are polynomials in η of degree less or equal to ν and insert the expression into (3.1) we find the following system of ordinary differential equations for f_j :

(3.8) $$\begin{aligned}(k+2)\eta(1-\eta)\frac{d}{d\eta} f_j &= \{ j(j+1)(1-\eta) - (\lambda-j)(\lambda-j+1)\eta \} f_j + \\ &+ (j+1)(\lambda-j)(1-\eta) f_{j+1} - j(\lambda-j+1)\eta f_{j-1}\end{aligned}$$

for $j=0,\ldots,\lambda$. The partial waves have simple transformation properties under s (3.4c) :

$$f_j(1-\eta) = (-1)^{\lambda} f_{\lambda-j}(\eta) .$$

The classification of all local 4-point functions is given by the following result [1]

Theorem. The polynomial solutions of Eq. (3.1) satisfying the boundary conditions (3.2) are given by

$$(3.9) \qquad f_k(\xi,\eta) = (\xi-\eta)^k \quad \text{for} \quad \lambda = \nu = k = 0,1,2,\ldots$$

or they correspond to one of the following exceptional cases :

$$(3.10a) \qquad k = 10 \quad (\lambda,\nu) = (4,2)\ ,\ (6,4)\ ,$$

$$(3.10b) \qquad k = 28 \quad (\lambda,\nu) = (10,4)\ ,\ (18,12)\ .$$

Remark. For integer Δ $(= \frac{1}{4}k)$ the solution (3.9) is associated with the $\mathbf{D}_{2\Delta+2}$ series, while (3.10a) and (3.10b) correspond to the simple Lie algebras $\mathbf{E}_6$ and $\mathbf{E}_8$ in the **ADE** classification [7].

The *proof* is based on the observation that only integer dimensions may contribute to the s-channel expansion of W_4 if $f(\xi,\eta)$ is a polynomial and proceeds in three steps.

Lemma **1.** Every polynomial solution of (3.8), (3.4), (3.2) satisfies

$$(3.11) \qquad f_0(\eta) = (1-\eta)^\nu \quad , \quad f_\lambda(\eta) = (-1)^\lambda \eta^\nu \quad .$$

Proof. The small η behaviour of the s-channel contribution of a primary field of isospin I to f_λ is

$$(3.12) \qquad f_{\lambda I} \sim \eta^{\lambda - I + \Delta_{2I}} ,$$

hence to prove the lemma it is enough to show that

$$(3.13) \qquad \lambda - I + \frac{I(I+1)}{k+2} \geq \frac{\lambda(\lambda+2)}{k+2} = \nu$$

since the degree of $f_\lambda(\eta)$ does not exceed ν . For integer Δ_{2I} (3.6) and ν Eq. (3.13) follows from the weaker inequality

$$(3.14) \qquad \lambda - I + \frac{I(I+1)}{k+2} \geq \frac{\lambda(\lambda+1)}{k+2} ,$$

noting that ν is the smallest integer larger than the right hand side. But (3.14) is equivalent to the quadratic inequality $I(k+1-I) \leq \lambda(k+1-\lambda)$ for $I \leq \min(k-\lambda,\lambda)$ which can easily be checked.

The normalization of $f_\lambda(\eta)$ follows from the equality $f_\ell(\eta) = (-1)^\lambda f_{\lambda-\ell}(1-\eta)$ implied by (3.14c) and the initial condition (3.2a).

Lemma **2.** If the system (3.8) admits a $\mathcal{D}_3$ invariant polynomial solution, then either $\lambda = k = \nu$ (and $f = f_k$ is given by (3.9)) or there exists a local primary field V_{2I} of dimension $\Delta_{2I} = 1$ -i.e. with isospin I satisfying $I(I+1) = k+2$, $I \geq 2$, which contributes to the operator product expansion

$$z_{12}^{2\Delta} V_\lambda(z_1,\zeta_1)V_\lambda(z_2,\zeta_2) = \zeta_{12}^\lambda - \frac{\lambda}{k}\zeta_{12}^{\lambda-1}(1+\tfrac{1}{2}\zeta_{12}\partial_{\zeta_2})\int_{z_2}^{z_1} J(z,\zeta_2)dz +$$

(3.15)

$$+ C_{\lambda\lambda 2I}\,\zeta_{12}^{\lambda-I}\sum_{n=0}^{I}\binom{I}{n}\frac{(2I-n)!}{(2I)!}(\zeta_{12}\partial_{\zeta_2})^n\int_{z_2}^{z_1} V_{2I}(z,\zeta_2)dz + O(z_{12}^2) ,$$

with a nonvanishing structure constant $C_{\lambda\lambda 2I}$.

Remark. The expansion (3.15) is a consequence of the Ward identities (2.12) and the current-current commutation relation (2.10).

Proof. We will prove that if $C_{\lambda\lambda 2I}$ is equal to zero, then a solution exists only for $\lambda=\nu$. If we insert $f_0(\eta)$ (3.11) into (3.8) for $\ell=0$ we find

(3.16) $$f_1(\eta) = -\eta(1-\eta)^{\nu-1} \quad (= f_{\lambda-1}(1-\eta)).$$

On the other hand if $C_{\lambda\lambda 2I}=0$, then it follows from (3.15) that the small η behaviour of f_ℓ is $f_\ell(\eta)=O(\eta^2)$ for $\ell\geq 2$, hence (using (3.7))

(3.17) $$f(\xi,\eta) = \xi^\lambda + [(\lambda - \nu)\xi - \lambda\,]\,\xi^{\lambda-1}\eta + O(\eta^2) .$$

Applying the symmetry property (3.4a) we find

(3.18) $$f(\xi,\eta) = \xi^\lambda - \lambda\xi^{\lambda-1}\eta + O(\eta^2) ,$$

which is compatible with (3.17) only for $\lambda=\nu$ ($=k$). In this case the solution is given by (3.9). It thus remains to list the models involving a local primary current with dimension $\Delta=1$ and $\lambda>4$ ($\lambda=4$ is the $\mathbf{D}_4$ model).

Lemma **3.** The system (3.8) has a solution for $\Delta=1$, $\lambda>4$ only in two cases (corresponding to the alternative (3.10))

(3.19) $$k = 10 \ , \ \lambda = 6 \quad ,$$

(3.20) $$k = 28 \ , \ \lambda = 10 \quad .$$

Proof. From the initial condition (3.2a) follows that

(3.21) $$f_{\ell}(0) = \delta_{\ell 0} .$$

On the other hand, from Lemma 1 and Eq. (3.16) we have $f_{\lambda}(\eta) = \eta^4$, $f_{\lambda-1}(\eta) = -\eta^3(1-\eta)$. Using again (3.8) we find

$$f_{\lambda-4}(0) = -\frac{(\lambda-6)(\lambda-10)\lambda(3\lambda-4)(\lambda^2-8\lambda+4)}{768\,(\lambda-1)(\lambda-2)(\lambda-3)} \quad ,$$

which is compatible with (3.21) only for $\lambda=6$ or $\lambda=10$.

To complete the proof of the Theorem we have to check that these possible $\Delta=1$ solutions (as well as the other $k=10$ and $k=28$ solutions listed in (3.10)) really exist. This has been done in [1], where the expressions for $f(\xi,\eta)$ in the exceptional cases are also written down.

Only the solutions with integer Δ refer to the original problem of classifying the local extensions of the chiral observable algebra. The case of odd 2Δ gives rise to a $\mathbb{Z}_2$ graded (extended) quasiobservable algebra whose odd elements (of half integer dimension) locally anticommute. The solutions of type (3.9) with odd k ($=4\Delta$) correspond to the generalized $\mathbb{Z}_4$ parafermions described in Sec.4A of [9].

4. Discussion

We end up with an addition to the locality argument of Sec.3 and with a couple comments.

The theorem of Sec.3 is not quite sufficient to classify the local extensions of $\mathcal{A}_k = \mathcal{A}_k(su_2)$. For k=28 (the $\mathbf{E}_8$ case) we have a possible extension of $\mathcal{A}_k$ by 3 primary fields (of dimensions 1,3 and 7). Our Theorem (including the construction of Sec.5 of [1]) establishes the local commutativity of each primary field with itself. In order to verify the relative locality of different primary fields we need to consider a more general 4-point function

$$W_4^{(\lambda_1,\lambda_2)} = \langle 0|V_{\lambda_1}(z_1,\zeta_1)V_{\lambda_2}(z_2,\zeta_2)V_{\lambda_1}(z_3,\zeta_3)V_{\lambda_2}(z_4,\zeta_4)|0\rangle = \tag{4.1}$$

$$= \frac{(z_{13})^{2\Delta_2}(z_{24})^{2\Delta_1}\zeta_{13}^{\lambda_1}\zeta_{24}^{\lambda_2}}{(z_{12}z_{23}z_{34}z_{14})^{\Delta_1+\Delta_2}}[\eta(1-\eta)]^{\Delta_0} f(\xi,\eta) .$$

Here Δ_i are the dimensions of the fields V_{λ_i} ,

$$4p\Delta_i = \lambda_i(\lambda_i+2), \quad i=1,2, \quad (p=k+2) , \tag{4.2}$$

Δ_0 is the minimal dimension of a primary field in the expansion of $V_{\lambda_1}V_{\lambda_2}|0\rangle$; in our case (if Δ_1 and Δ_2 are not both equal to 1/4 k) according to Lemma 2, Δ_0=1. The operator $\mathbf{K}_{\lambda_1\lambda_2}$ appearing in the KZ equation

$$\mathbf{K}_{\lambda_1\lambda_2} f_{\lambda_1\lambda_2}(\xi,\eta) = 0 \tag{4.3}$$

where $f_{\lambda_1\lambda_2}$ is a polynomial in ξ of degree $\frac{1}{2}(\lambda_1+\lambda_2)$ (which is an integer), reads :

$$\mathbf{K}_{\lambda_1\lambda_2} = p\eta(1-\eta)\frac{\partial}{\partial\eta} - [\, p(\Delta_1+\Delta_2-\Delta_0)(1-2\eta) + \lambda_1\lambda_2(\tfrac{1}{2}-\xi)\,] + \tag{4.4}$$

$$+ [\xi^2 - 2\xi\eta + \eta + (\lambda_1+\lambda_2)\xi(1-\xi)]\frac{\partial}{\partial\xi} - \xi(1-\xi)(\xi-\eta)\frac{\partial^2}{\partial\xi^2}$$

(for $\lambda_1=\lambda_2=\lambda$, $p(\Delta_1+\Delta_2-\Delta_0) = 2p\Delta = \frac{1}{2}\lambda(\lambda+2)$ it reproduces (3.1)). It is a straightforward exercise to verify that in the $\mathbf{E}_8$ case, for $p=30$, $\Delta_1, \Delta_2 \in \{1,3,7\}$, $\Delta_0=1$, there exists a solution of (4.3) that is a polynomial in η of degree

$$\nu = \nu(\lambda_1, \lambda_2) = 2(\Delta_1+\Delta_2-\Delta_0), \tag{4.5}$$

thus completing the proof of the announced classification of maximal local extensions of $\mathcal{A}_k$.

It is instructive to note that $\mathcal{D}_3$ invariance strongly restricts the form of a partial differential equation $\mathbf{L}f=0$ of the KZ type (3.1). More precisely, if the leading order derivative terms in $\mathbf{L}$ are given by

$$\mathbf{L}_0 = p\eta(1-\eta)\frac{\partial}{\partial\eta} - \xi(1-\xi)(\xi-\eta)\frac{\partial^2}{\partial\xi^2}, \tag{4.6}$$

then the $\mathcal{D}_3$ transformation law with homogeneity degrees ν and λ in η and ξ) determines the operator $\mathbf{L}$ to be

$$\begin{aligned}\mathbf{L} = \mathbf{L}_0 &+ (\lambda-1)[2\xi(1-\xi)\frac{\partial}{\partial\xi} + \lambda(\xi+\eta-1)] + \frac{1}{3}[\nu p - \lambda(\lambda-1)](2\eta-1) + \\ &+ a\,[\,(2\xi+\eta-\xi^2-2\xi\eta)\frac{\partial}{\partial\xi} + \lambda(\xi+\eta-1)\,],\end{aligned} \tag{4.7}$$

where a is a free parameter.

The KZ equation (3.1) is reproduced for $a=1$, $p=k+2$, $\nu p=\lambda(\lambda+2)$. Another interesting case is obtained for $a=1$, rational p and $\nu p = \lambda(\lambda+2) - 3\lambda p$. For such values of the parameters (following the reduction procedure described in [8]) one can derive from (4.1) the $\mathcal{D}_3$ - covariant ordinary differential equations implied by the "null vector condition" for 4-point functions of a primary field in a minimal conformal model.

One of the authors, I.T., takes the pleasure to thank Moshe Flato, James Lepowsky and Paul Sally for their hospitality in Mount Holyoke and to the American Mathematical Society for financial support.

REFERENCES

1. L.Michel,Ya.S.Stanev,I.T.Todorov, **D-E** Classification of the local extensions of su_2 current algebras, Bures-sur-Yvette preprint IHES/P/92/47; Theor.Math.Phys. (Moskow) **92** (1992) 507-521 (Volume dedicated to M.C.Polivanov).
2. V.G.Knizhnik, A.B.Zamolodchikov, Current algebra and Wess-Zumino model in two dimensions, Nucl.Phys. **B247** (1984) 83-103.
3. A.Tsuchiya,Y.Kanie, Vertex operators in the conformal field theory on $\mathbb{P}^1$ and monodromy representations of the braid group, in: *Conformal Field Theory and Solvable Lattice Models*, Advanced Studies in Pure Mathematics **16** (1988) 297-372; Lett.Math.Phys. **13** (1987) 303-312.
4. P.Furlan,G.M.Sotkov,I.T.Todorov, Two-dimensional conformal quantum field theory, Rivista Nuovo Cim. **12:6** (1989) 1-202.
5. I.B.Frenkel,J.Lepowsky,A.Meurman, *Vertex Operator Algebra and the Monster*, Pure and Appl.Math. **134**, Academic Press, Boston 1989; Yi-Zhi Huang, J.Lepowsky, Toward a theory of tensor products for representations of a vertex operator algebra, Rutgers University preprint, New Brunswick, 1992.
6. G.Moore,N.Seiberg, Naturality in conformal field theory, Nucl. Phys. **B313** (1989) 16-40.
7. A.Cappelli,C.Itzykson,J.B.Zuber, The **A-D-E** classification of minimal and $A_1^{(1)}$ conformal invariant theories, Commun.Math.Phys. **113** (1987) 1-26.
8. P.Furlan,A.Ganchev,R.Paunov,V.Petkova, Reduction of the rational spin sl(2,$\mathbb{C}$) WZNW conformal theory, Phys.Lett **B267** (1991) 63-70; B293 (1992) 56-66.
9. P.Furlan,R.R.Paunov,I.T.Todorov, Extended U(1) conformal field theories and $\mathbb{Z}_k$ -parafermions, Fortschr.d.Physik **40**, N.3 (1992) 211-271.

Institute for Nuclear Research and Nuclear Energy,
Bulgarian Academy of Sciences,
Sofia, Bulgaria, Tzarigradsko Chaussee 72, BG 1784

Recent Titles in This Series

(Continued from the front of this publication)

147 **Neil Robertson and Paul Seymour, Editors,** Graph structure theory, 1993

146 **Martin C. Tangora, Editor,** Algebraic topology, 1993

145 **Jeffrey Adams, Rebecca Herb, Stephen Kudla, Jian-Shu Li, Ron Lipsman, and Jonathan Rosenberg, Editors,** Representation theory of groups and algebras, 1993

144 **Bor-Luh Lin and William B. Johnson, Editors,** Banach spaces, 1993

143 **Marvin Knopp and Mark Sheingorn, Editors,** A tribute to Emil Grosswald: Number theory and related analysis, 1993

142 **Chung-Chun Yang and Sheng Gong, Editors,** Several complex variables in China, 1993

141 **A. Y. Cheer and C. P. van Dam, Editors,** Fluid dynamics in biology, 1993

140 **Eric L. Grinberg, Editor,** Geometric analysis, 1992

139 **Vinay Deodhar, Editor,** Kazhdan-Lusztig theory and related topics, 1992

138 **Donald St. P. Richards, Editor,** Hypergeometric functions on domains of positivity, Jack polynomials, and applications, 1992

137 **Alexander Nagel and Edgar Lee Stout, Editors,** The Madison symposium on complex analysis, 1992

136 **Ron Donagi, Editor,** Curves, Jacobians, and Abelian varieties, 1992

135 **Peter Walters, Editor,** Symbolic dynamics and its applications, 1992

134 **Murray Gerstenhaber and Jim Stasheff, Editors,** Deformation theory and quantum groups with applications to mathematical physics, 1992

133 **Alan Adolphson, Steven Sperber, and Marvin Tretkoff, Editors,** p-adic methods in number theory and algebraic geometry, 1992

132 **Mark Gotay, Jerrold Marsden, and Vincent Moncrief, Editors,** Mathematical aspects of classical field theory, 1992

131 **L. A. Bokut′, Yu. L. Ershov, and A. I. Kostrikin, Editors,** Proceedings of the International Conference on Algebra Dedicated to the Memory of A. I. Mal′cev, Parts 1, 2, and 3, 1992

130 **L. Fuchs, K. R. Goodearl, J. T. Stafford, and C. Vinsonhaler, Editors,** Abelian groups and noncommutative rings, 1992

129 **John R. Graef and Jack K. Hale, Editors,** Oscillation and dynamics in delay equations, 1992

128 **Ridgley Lange and Shengwang Wang,** New approaches in spectral decomposition, 1992

127 **Vladimir Oliker and Andrejs Treibergs, Editors,** Geometry and nonlinear partial differential equations, 1992

126 **R. Keith Dennis, Claudio Pedrini, and Michael R. Stein, Editors,** Algebraic K-theory, commutative algebra, and algebraic geometry, 1992

125 **F. Thomas Bruss, Thomas S. Ferguson, and Stephen M. Samuels, Editors,** Strategies for sequential search and selection in real time, 1992

124 **Darrell Haile and James Osterburg, Editors,** Azumaya algebras, actions, and modules, 1992

123 **Steven L. Kleiman and Anders Thorup, Editors,** Enumerative algebraic geometry, 1991

122 **D. H. Sattinger, C. A. Tracy, and S. Venakides, Editors,** Inverse scattering and applications, 1991

121 **Alex J. Feingold, Igor B. Frenkel, and John F. X. Ries,** Spinor construction of vertex operator algebras, triality, and $E_8^{(1)}$, 1991

120 **Robert S. Doran, Editor,** Selfadjoint and nonselfadjoint operator algebras and operator theory, 1991

119 **Robert A. Melter, Azriel Rosenfeld, and Prabir Bhattacharya, Editors,** Vision geometry, 1991

(See the AMS catalog for earlier titles)